Biodiversity Monitoring and Conservation

Conservation Science and Practice Series

Published in association with the Zoological Society of London

Wiley-Blackwell and the Zoological Society of London are proud to present our *Conservation Science and Practice* series. Each book in the series reviews a key issue in conservation today from a multidisciplinary viewpoint.

Books in the series can be single or multi-authored and proposals should be sent to:
Ward Cooper, Senior Commissioning Editor. Email: ward.cooper@wiley.com

Each book proposal will be assessed by independent academic referees, as well as our Series Editorial Panel. Members of the Panel include:
Richard Cowling, Nelson Mandela Metropolitan University, Port Elizabeth, South Africa
John Gittleman, Institute of Ecology, University of Georgia, USA
Andrew Knight, University of Stellenbosch, South Africa
Nigel Leader-Williams, University of Cambridge, UK
Georgina Mace, University College London, UK
Daniel Pauly, University of British Columbia, Canada
Stuart Pimm, Duke University, USA
Hugh Possingham, University of Queensland, Australia
Peter Raven, Missouri Botanical Gardens, USA
Helen Regan, University of California, Riverside, USA
Alex Rogers, University of Oxford, UK
Michael Samways, University of Stellenbosch, South Africa
Nigel Stork, Griffith University, Australia.

Previously published

Biodiversity Conservation and Poverty Alleviation: Exploring the Evidence for a Link
Edited by Dilys Roe, Joanna Elliott, Chris Sandbrook and Matt Walpole
ISBN: 978-0-470-67478-9 Paperback;
ISBN: 978-0-470-67479-6 Hardback; December 2012

Applied Population and Community Ecology: The Case of Feral Pigs in Australia
Edited by Jim Hone
ISBN: 978-0-470-65864-2 Hardcover; July 2012

Tropical Forest Conservation and Industry Partnership: An Experience from the Congo Basin
Edited by Connie J. Clark and John R. Poulsen
ISBN: 978-0-4706-7373-7 Hardcover; March 2012

Reintroduction Biology: Integrating Science and Management
Edited by John G. Ewen, Doug. P. Armstrong, Kevin A. Parker and Philip J. Seddon
ISBN: 978-1-4051-8674-2 Paperback;
ISBN: 978-1-4443-6156-8 Hardcover; January 2012

Trade-offs in Conservation: Deciding What to Save
Edited by Nigel Leader-Williams, William M. Adams and Robert J. Smith
ISBN: 978-1-4051-9383-2 Paperback;
ISBN: 978-1-4051-9384-9 Hardcover; September 2010

Urban Biodiversity and Design
Edited by Norbert Müller, Peter Werner and John G. Kelcey
ISBN: 978-1-4443-3267-4 Paperback;
ISBN: 978-1-4443-3266-7 Hardcover; April 2010

Wild Rangelands: Conserving Wildlife While Maintaining Livestock in Semi-Arid Ecosystems
Edited by Johan T. du Toit, Richard Kock and James C. Deutsch
ISBN: 978-1-4051-7785-6 Paperback;
ISBN: 978-1-4051-9488-4 Hardcover; January 2010

Reintroduction of Top-Order Predators
Edited by Matt W. Hayward and Michael J. Somers
ISBN: 978-1-4051-7680-4 Paperback;
ISBN: 978-1-4051-9273-6 Hardcover; April 2009

Recreational Hunting, Conservation and Rural Livelihoods: Science and Practice
Edited by Barney Dickson, Jonathan Hutton and Bill Adams
ISBN: 978-1-4051-6785-7 Paperback;
ISBN: 978-1-4051-9142-5 Hardcover; March 2009

Participatory Research in Conservation and Rural Livelihoods: Doing Science Together
Edited by Louise Fortmann
ISBN: 978-1-4051-7679-8 Paperback; October 2008

Bushmeat and Livelihoods: Wildlife Management and Poverty Reduction
Edited by Glyn Davies and David Brown
ISBN: 978-1-4051-6779-6 Paperback; December 2007

Managing and Designing Landscapes for Conservation: Moving from Perspectives to Principles
Edited by David Lindenmayer and Richard Hobbs
ISBN: 978-1-4051-5914-2 Paperback; December 2007

Conservation Science and Practice Series

Biodiversity Monitoring and Conservation: Bridging the Gap between Global Commitment and Local Action

Edited by

Ben Collen
Institute of Zoology, Zoological Society of London, London, UK

Nathalie Pettorelli
Institute of Zoology, Zoological Society of London, London, UK

Jonathan E.M. Baillie
Conservation Programmes, Zoological Society of London, London, UK

Sarah M. Durant
Institute of Zoology, Zoological Society of London, London, UK

WILEY-BLACKWELL
A John Wiley & Sons, Inc., Publication

This edition first published 2013 © 2013 by John Wiley & Sons, Ltd

Blackwell Publishing was acquired by John Wiley & Sons in February 2007. Blackwell's publishing program has been merged with Wiley's global Scientific, Technical and Medical business to form Wiley-Blackwell.

Registered office: John Wiley & Sons, Ltd, The Atrium, Southern Gate, Chichester, West Sussex, PO19 8SQ, UK

Editorial offices: 9600 Garsington Road, Oxford, OX4 2DQ, UK
The Atrium, Southern Gate, Chichester, West Sussex, PO19 8SQ, UK
111 River Street, Hoboken, NJ 07030-5774, USA

For details of our global editorial offices, for customer services and for information about how to apply for permission to reuse the copyright material in this book please see our website at www.wiley.com/wiley-blackwell.

The right of the author to be identified as the author of this work has been asserted in accordance with the UK Copyright, Designs and Patents Act 1988.

All rights reserved. No part of this publication may be reproduced, stored in a retrieval system, or transmitted, in any form or by any means, electronic, mechanical, photocopying, recording or otherwise, except as permitted by the UK Copyright, Designs and Patents Act 1988, without the prior permission of the publisher.

Designations used by companies to distinguish their products are often claimed as trademarks. All brand names and product names used in this book are trade names, service marks, trademarks or registered trademarks of their respective owners. The publisher is not associated with any product or vendor mentioned in this book.

Limit of Liability/Disclaimer of Warranty: While the publisher and author(s) have used their best efforts in preparing this book, they make no representations or warranties with respect to the accuracy or completeness of the contents of this book and specifically disclaim any implied warranties of merchantability or fitness for a particular purpose. It is sold on the understanding that the publisher is not engaged in rendering professional services and neither the publisher nor the author shall be liable for damages arising herefrom. If professional advice or other expert assistance is required, the services of a competent professional should be sought.

Library of Congress Cataloging-in-Publication Data

Biodiversity monitoring and conservation : bridging the gap between global commitment and local action / Ben Collen, Nathalie Pettorelli, Jonathan Baillie, Sarah Durant.
      pages cm
  Includes bibliographical references and index.
  ISBN 978-1-4443-3291-9 (cloth) – ISBN 978-1-4443-3292-6 (pbk.) 1. Biodiversity–Monitoring.
2. Biodiversity conservation. I. Collen, Ben.
  QH541.15.B56B5786 2013
  333.95′16 – dc23
                                                                                              2012033654

A catalogue record for this book is available from the British Library.

Wiley also publishes its books in a variety of electronic formats. Some content that appears in print may not be available in electronic books.

Cover Image:
Front Cover: Red slender loris, *Loris tardigradus tardigradus*, Sri Lanka © James T. Reardon / ZSL
Back Cover: Cheetah, *Acinonyx jubatus*, Tanzania © Sarah Durant / ZSL
Cover design by Design Deluxe

Set in 9.5/11.5 pt Minion by Laserwords Private Limited, Chennai, India
Printed and bound in Malaysia by Vivar Printing Sdn Bhd

1   2013

Contents

Contributors	xi
Acknowledgements	xv

1. Biodiversity Monitoring and Conservation: Bridging the Gaps Between Global Commitment and Local Action   1
   Ben Collen, Nathalie Pettorelli, Jonathan E.M. Baillie and Sarah M. Durant

Part I   Species-Based Indicators of Biodiversity Change   17

2. Tracking Change in National-Level Conservation Status: National Red Lists   19
   Ben Collen, Janine Griffiths, Yolan Friedmann, Jon Paul Rodriguez, Franklin Rojas-Suárez and Jonathan E.M. Baillie

3. The Wildlife Picture Index: A Biodiversity Indicator for Top Trophic Levels   45
   Timothy G. O'Brien and Margaret F. Kinnaird

4. Tracking Change in Abundance: The Living Planet Index   71
   Ben Collen, Louise McRae, Jonathan Loh, Stefanie Deinet, Adriana De Palma, Robyn Manley and Jonathan E.M. Baillie

Part II   Indicators of the Pressures on Biodiversity   95

5. Satellite Data-Based Indices to Monitor Land Use and Habitat Changes   97
   Nathalie Pettorelli

6. Indicators of Climate Change Impacts on Biodiversity   120
   Wendy B. Foden, Georgina M. Mace and Stuart H.M. Butchart

7. Monitoring Trends in Biological Invasion, its Impact and Policy Responses   138
   Piero Genovesi, Stuart H.M. Butchart, Melodie A. McGeoch and David B. Roy

8. Exploitation Indices: Developing Global and National Metrics of Wildlife Use and Trade   159
Rosamunde E.A. Almond, Stuart H.M. Butchart, Thomasina E.E. Oldfield, Louise McRae and Steven de Bie

9. Personalized Measures of Consumption and Development in the Context of Biodiversity Conservation: Connecting the Ecological Footprint Calculation with the Human Footprint Map   189
Eric W. Sanderson

Part III The Next Generation of Biodiversity Indicators   211

10. Indicator Bats Program: A System for the Global Acoustic Monitoring of Bats   213
Kate E. Jones, Jon A. Russ, Andriy-Taras Bashta, Zoltán Bilhari, Colin Catto, István Csősz, Alexander Gorbachev, Péter Győrfi, Alice Hughes, Igor Ivashkiv, Natalia Koryagina, Anikó Kurali, Steve Langton, Alanna Collen, Georgiana Margiean, Ivan Pandourski, Stuart Parsons, Igor Prokofev, Abigel Szodoray-Paradi, Farkas Szodoray-Paradi, Elena Tilova, Charlotte L. Walters, Aidan Weatherill and Oleg Zavarzin

11. Occupancy Methods for Conservation Management   248
Darryl I. MacKenzie and James T. Reardon

12. Monitoring and Evaluating the Socioeconomic Impacts of Conservation Projects on Local Communities   265
Katherine Homewood

13. Science to Policy Linkages for the Post-2010 Biodiversity Targets   291
Georgina M. Mace, Charles Perrings, Philippe Le Prestre, Wolfgang Cramer, Sandra Díaz, Anne Larigauderie, Robert J. Scholes and Harold A. Mooney

Part IV Biodiversity Monitoring in Practice   311

14. Building Sustainable National Monitoring Networks   313
Sarah M. Durant

15. Monitoring in the Real World   335
Julia P.G. Jones

16. Monitoring in UNDP-GEF Biodiversity Projects: Balancing Conservation Priorities, Financial Realities, and Scientific Rigour 348
    Sultana Bashir

17. Scaling Up or Down? Linking Global and National Biodiversity Indicators and Reporting 402
    Philip Bubb

18. Conserving Biodiversity in a Target-Driven World 421
    Simon N. Stuart and Ben Collen

Index 439

Contributors

Rosamunde Almond United Nations Environment Programme World Conservation Monitoring Centre (UNEP-WCMC), 219 Huntingdon Road, Cambridge, CB3 0DL, UK.

Jonathan E.M. Baillie Conservation Programmes, Zoological Society of London, Regent's Park, London, NW1 4RY, UK.

Sultana Bashir Institute of Zoology, Zoological Society of London, Regent's Park, London NW1 4RY, UK.

Andriy-Taras Bashta Animals Research and Protection Association "Fauna", Trylovsky st. 7/54, Lviv 79049, Ukraine; Institute of Ecology of the Carpathians, National Academy of Sciences of Ukraine, Kozelnytska st. 4, Lviv 79026, Ukraine.

Zoltán Bilhari The Nature Foundation, Toldi út 63, Nyíregyháza 4400, Hungary.

Philip Bubb United Nations Environment Programme World Conservation Monitoring Centre (UNEP-WCMC), 219 Huntingdon Road, Cambridge, CB3 0DL, UK.

Stuart Butchart BirdLife International, Wellbrook Court, Cambridge, CB3 0NA, UK.

Colin Catto The Bat Conservation Trust, 15 Cloisters Business Park, Battersea, London, UK.

Alanna Collen Institute of Zoology, Zoological Society of London, Regent's Park, London, NW1 4RY, UK.

Ben Collen Institute of Zoology, Zoological Society of London, Regent's Park, London, NW1 4RY, UK.

Wolfgang Cramer Mediterranean Institute of Biodiversity and Ecology (IMBE), Aix-en-Provence, France.

István Csősz Romanian Bat Protection Association, 440014 str. I. Budai Deleanu nr. 2, Satu Mare, Romania.

Steven de Bie Wageningen University, Droevendaalsesteeg 4, Wageningen, The Netherlands.

Stefanie Deinet Institute of Zoology, Zoological Society of London, Regent's Park, London, NW1 4RY, UK.

Adriana De Palma Institute of Zoology, Zoological Society of London, Regent's Park, London, NW1 4RY, UK.

Sandra Díaz Instituto Multidisciplinario de Biología Vegetal, (CONICET-UNC) and FCEFyN, Universidad Nacional de Córdoba, 5000 Cordoba, Argentina.

Sarah M. Durant Institute of Zoology, Zoological Society of London, Regent's Park, London, NW1 4RY, UK.

Wendy B. Foden IUCN, 219c Huntingdon Road, Cambridge, CB3 0DL, UK; Animal, Plant and Environmental Sciences, University of the Witwatersrand, Johannesburg, 2050, South Africa.

Yolan Friedmann Endangered Wildlife Trust, Private Bag X11, Parkview 2122, South Africa.

Piero Genovesi IUCN SSC Invasive Species Specialist Group; ISPRA, Via Curtatone 3, I-00185 Rome, Italy.

Alexander Gorbachev PERESVET, Olega Koshevogo St., 80, kv. 12, 241029, Bryansk, Russia; University of Bryansk, Bezhitskaya St., 14, 241036, Bryansk, Russia.

Janine Griffiths Institute of Zoology, Zoological Society of London, Regent's Park, London, NW1 4RY, UK; Conservation Programmes, Zoological Society of London, Regent's Park, London, NW1 4RY, UK.

Péter Győrfi The Nature Foundation, Toldi út 63, Nyíregyháza 4400, Hungary.

Katherine Homewood University College London, Gower Street, London, WC1E 6BT, UK.

Alice Hughes School of Biological Sciences, Woodland Road, University of Bristol, Bristol BS8 1UG, UK.

Igor Ivashkiv Animals Research and Protection Association "Fauna", Trylovsky st. 7/54, Lviv 79049, Ukraine; Institute of Ecology of the Carpathians, National Academy of Sciences of Ukraine, Kozelnytska st. 4, Lviv 79026, Ukraine.

Julia Jones School of the Environment, Natural Resources and Geography, Bangor University, Bangor, UK.

Kate Jones Institute of Zoology, Zoological Society of London, Regent's Park, London, NW1 4RY, UK; University College London, Gower Street, London, WC1E 6BT, UK.

Margaret Kinnaird Mpala Research Centre, PO Box 555, Nanyuki, Kenya 10400.

Natalia Koryagina PERESVET, Olega Koshevogo St., 80, kv. 12, 241029, Bryansk, Russia.

Anikó Kurali The Nature Foundation, Toldi út 63, Nyíregyháza 4400, Hungary.

Steve Langton The Bat Conservation Trust, 15 Cloisters Business Park, Battersea, London, UK.

Anne Larigauderie DIVERSITAS, Muséum National d'Histoire Naturelle (MNHN), 57, rue Cuvier, CP 41, 75231 Paris Cedex 05, France.

Philippe Le Prestre Institut Hydro-Québec en environnement, développement et société, Université Laval, Pavillon Des-Services 3800, Québec (Qc) G1V 0A6, Canada.

Jonathan Loh WWF International, Avenue du Mont-Blanc CH-1196, Gland, Switzerland.

Georgina Mace Imperial College London, Centre for Population Biology, Silwood Park, Ascot, SL5 7PY, UK; University College London, Gower Street, London, WC1E 6BT, UK.

Darryl MacKenzie Proteus Wildlife Research Consultants, PO Box 5193, Dunedin 9058, New Zealand.

Robyn Manley Institute of Zoology, Zoological Society of London, Regent's Park, London, NW1 4RY, UK.

Georgiana Margiean Romanian Bat Protection Association, 440014 str. I. Budai Deleanu nr. 2, Satu Mare, Romania.

Melodie McGeoch Centre for Invasion Biology and Cape Research Centre, South African National Parks, P.O. Box 216, Steenberg 7947, South Africa.

Louise McRae Institute of Zoology, Zoological Society of London, Regent's Park, London, NW1 4RY, UK.

Harold Mooney Department of Biology, Stanford University, Stanford, California, 94306, USA.

Timothy O'Brien Wildlife Conservation Society, Global Conservation Programs, 2300 Southern Blvd., Bronx, New York 10460, USA.

Thomasina Oldfield TRAFFIC International, 219a Huntingdon Rd, Cambridge, CB3 ODL, UK.

Ivan Pandourski Institute of Zoology, Bulgaria Academy of Sciences, 1 Tsar Osvoboditel Blvd., 1000 Sofia, Bulgaria.

Stuart Parsons School of Biological Sciences, University of Auckland, Private Bag 92019, Auckland, New Zealand.

Charles Perrings ecoSERVICES Group, School of Life Sciences, PO Box 874501, Arizona State University, Tempe, AZ 85287, USA.

Nathalie Pettorelli Institute of Zoology, Zoological Society of London, Regent's Park, London, NW1 4RY, UK.

Igor Prokofev PERESVET, Olega Koshevogo St., 80, kv. 12, 241029, Bryansk, Russia; University of Bryansk, Bezhitskaya St., 14, 241036, Bryansk, Russia.

James Reardon Conservations Programmes, Zoological Society of London, Regent's Park, London, NW1 4RY, England, UK; Southland Conservancy, Department of Conservation, PO Box 743, Invercargill 9840, New Zealand.

Jon Paul Rodriguez Centro de Ecología, Instituto Venezolano de Investigaciones Científicas, Apdo. 20632, Caracas 1020-A, Venezuela.

Franklin Rojas-Suárez Provita, Apdo. 47552, Caracas 1041-A, Venezuela.

David Roy Centre for Ecology and Hydrology, Maclean Building, Benson Lane, Crowmarsh Gifford, Wallingford, Oxfordshire, OX10 8BB, UK.

Jon Russ The Bat Conservation Trust, 15 Cloisters Business Park, Battersea, London, UK.

Eric Sanderson Global Conservation Programs, Wildlife Conservation Society, 2300 Southern Blvd., Bronx, New York 10460, USA.

Robert Scholes CSIR Natural Resources and Environment, PO Box 395, Pretoria 0001, South Africa.

Simon Stuart IUCN Species Survival Commission, Rue Mauverney 28, 1196 Gland, Switzerland; UNEP World Conservation Monitoring Centre, 219 Huntingdon Road, Cambridge, CB3 0DL, UK; Conservation International, 2011 Crystal Drive, Arlington, VA 22202, USA; Department of Biology and Biochemistry, University of Bath, Bath BA2 7AY, UK; Al Ain Zoo, P.O. Box 45553, Abu Dhabi, United Arab Emirates.

Abigel Szodoray-Paradi Romanian Bat Protection Association, 440014 str. I. Budai Deleanu nr. 2, Satu Mare, Romania.

Farkas Szodoray-Paradi Romanian Bat Protection Association, 440014 str. I. Budai Deleanu nr. 2, Satu Mare, Romania.

Elena Tilova The Green Balkans – Stara Zagora, 9 Stara Planina Str., Stara Zagora 6000, Bulgaria.

Charlotte L. Walters The Institute of Zoology, Zoological Society of London, Regent's Park, London, NW1 4RY, UK; The Bat Conservation Trust, 15 Cloisters Business Park, Battersea, London, UK.

Aidan Weatherill The Institute of Zoology, Zoological Society of London, Regent's Park, London, NW1 4RY, UK.

Oleg Zavarzin PERESVET, Olega Koshevogo St., 80, kv. 12, 241029, Bryansk, Russia.

Acknowledgements

This book is the result of a symposium held at the Zoological Society of London, and Ben Collen, Nathalie Pettorelli, Jonathan Baillie, and Sarah Durant are extremely grateful to the many individuals and institutions that contributed to this symposium. The editors are also grateful to Joy Hayward and Linda DaVolls from the Zoological Society of London for their help in organizing the symposium. Matthew Hatchwell and Linda Krueger from the Wildlife Conservation Society helped shape our original ideas on the symposium and chair the meeting. The international participation at the symposium would not have been possible without funding received from the Zoological Society of London and the Wildlife Conservation Society. B.C. is supported by the Rufford Foundation and is an Honorary Fellow of the United Nations Environment Programme – World Conservation Monitoring Centre. Three anonymous reviewers gave us insightful comments on our first proposal, which helped shape this book. Each chapter of this book was evaluated by multiple referees, and we thank them for their time, input, and advice. The editors appreciate the guidance throughout the writing of this publication of Wiley-Blackwell, particularly Kelvin Matthews, Ward Cooper, and Delia Sandford.

1

Biodiversity Monitoring and Conservation: Bridging the Gaps Between Global Commitment and Local Action

Ben Collen[1], Nathalie Pettorelli[1], Jonathan E.M. Baillie[2] and Sarah M. Durant[1]

[1]Institute of Zoology, Zoological Society of London, London, UK
[2]Conservation Programmes, Zoological Society of London, London, UK

Why a book on biodiversity monitoring and conservation?

As the impacts of anthropogenic activities increase in both magnitude and extent, biodiversity is under increasing pressure. Habitats available to wildlife have undergone dramatic modifications, and significant biodiversity has already been lost over modern times, while we are yet to experience the full impacts of anthropogenic climate change (Mace *et al.*, 2005; Dawson *et al.*, 2011; Pereira *et al.*, 2010b). Over the past few hundred years, humans have increased species extinction rates by as much as 1000 times compared with background rates that were typical over Earth's history (Regan *et al.*, 2001; Millennium Ecosystem Assessment, 2005), and accelerating increases in anthropogenic pressures on biodiversity may further increase species extinction rates (Balmford and Bond, 2005). In developing means to address these challenges, scientists are hampered by a lack of information on biological systems, particularly information relating to long-term trends, which is crucial to developing an understanding of how these systems may respond to global environmental change. Such serious knowledge gaps make it very difficult to develop effective policies and legislation to reduce and reverse biodiversity loss.

A further impetus for conservation action has been gained through an increasing realization that declines in biodiversity have detrimental impacts on ecosystem structures and functions as well as human well-being, particularly for the world's most

---

Biodiversity Monitoring and Conservation: Bridging the Gap between Global Commitment and Local Action, First Edition. Edited by Ben Collen, Nathalie Pettorelli, Jonathan E.M. Baillie and Sarah M. Durant.
© 2013 John Wiley & Sons, Ltd. Published 2013 by John Wiley & Sons, Ltd.

marginalized and impoverished communities (Millennium Ecosystem Assessment, 2005). Biodiversity provides many products – often plants, animals, and fungi – that directly contribute to incomes and human livelihoods. Biodiversity also provides genetic resources for the pharmaceutical industry, which can be key in maintaining human health, while the growth of nature tourism has meant that biodiversity conservation has become a major contributor to many national economies, including those of some of the world's poorest countries. As well as delivering these ecosystem services, biodiversity underpins the functioning of ecosystems, and hence the delivery of services such as access to fresh water or climate regulation. Biodiversity is therefore key to security, resilience, social relations, and human health and hence affects people not only by way of material livelihoods and macroeconomics.

In order to counter global biodiversity loss and consequent impacts on human well-being, there have been several recent high-profile international political commitments to improve biodiversity conservation. These have mainly consisted of goal setting, in the form of conservation targets to which governments, decision-makers, and the international community are committed; the most notable example of which are the targets set by the Convention on Biological Diversity (CBD; Convention on Biological Diversity, 2011; UNEP, 2002). However, because of the complexity of biological systems, and a lack of long-term biodiversity data, nations are hampered not only in assessing progress towards such targets, but also in developing appropriate policy and legislative responses to reverse biodiversity declines.

Global commitments to stemming biodiversity loss have contributed to the development of methods to track changes in many metrics of biodiversity, and addressing biodiversity information requirements has become one of the fastest growing areas of research in the field of conservation biology. This information is critical for increasing our understanding of the manner in which biodiversity is changing, and how changes can be influenced and reversed. It is also required for setting priorities for biodiversity conservation, such as protected area placement (e.g., Araújo 1999; Possingham *et al.*, 1993; Rodrigues *et al.*, 2004), species and ecosystem priority setting among the many deserving causes of conservation attention (e.g., Isaac *et al.*, 2007; Myers *et al.*, 2000), and for the biodiversity assessments required to provide the data for such activities (Baillie *et al.*, 2008; Collen *et al.*, 2012; Mace *et al.*, 2008; Pereira *et al.*, 2010a).

The process of reversing decline in biodiversity, at the outset, might appear straightforward. We should simply measure what is happening to the components of biodiversity that we wish to conserve; put in place conservation actions to counteract declines in the taxa and places that are changing most rapidly, or which we are least willing to lose; monitor and evaluate the impacts of these actions; and continue to manage adaptively. Yet our first collective attempt to measure and slow biodiversity change (the Convention on Biological Diversity 2010 Target) met with almost universal agreement that we had failed (Butchart *et al.*, 2010; Convention on Biological Diversity, 2010). That there were only eight years between the agreement of that target ('to achieve, by 2010, a slowing in the rate of biodiversity loss') by parties to the CBD, and the deadline by which a change should have taken place, must at least partly explain why we failed to meet this target. Even with the strongest political will, a substantial slowing in biodiversity declines would not have been possible in the timeframe, unless the many and complex underlying drivers of decline were effectively tackled.

It has become clear though, in the myriad of post-2010 papers, reports, and evaluations, that there are some problems in the overall approach. First of all, the target set was not action orientated, nor tied to appropriate activities from which the impact of changing pressures on biodiversity could be measured. This has to some degree been addressed in the newly agreed Aichi Target and Strategic Plan for 2020. Secondly, there appeared to be a disconnection between these laudable global commitments to improving the status of biodiversity, and the local-scale action required to ultimately ensure their achievement. From a research perspective, there has been a focus on identifying the most effective means to generate the metrics of biodiversity required to measure significant change (Dobson, 2005; Mace and Baillie, 2007), and how best to fill the many gaps in biodiversity data (Collen *et al.*, 2008; Pereira and Cooper, 2006). However, from a policy perspective it remains unclear how global targets should be harmonized with the many national responsibilities to biodiversity conservation and vice versa (Jones *et al.*, 2011; Nicholson *et al.*, 2012). Moreover, from a practical perspective there is a need to better coordinate biodiversity monitoring and conservation, at all scales, for increased efficiency and greater impact.

As the Aichi Target becomes agreed and implemented, it is extremely timely to reflect not only on lessons learned from the 2010 targets, but also on how we might better integrate national and global biodiversity monitoring and indicators over the coming decade. Such complex policy objectives present many challenges to conservation scientists and policy-makers alike. A key issue is how best to monitor progress towards such global-scale targets. There is also growing recognition of a need for biodiversity monitoring at a national, as well as a global scale, and better coordination between different monitoring approaches so as to make optimal use of all forms of biodiversity data. Although several indicators have been developed for use at the global scale, the data on which these indicators are based frequently come from monitoring schemes carried out with quite different objectives than monitoring global biodiversity change. While a dedicated global monitoring system may be ideal, would it be prohibitively expensive? Might a more cost-effective solution be to implement monitoring at a national scale, according to national priorities, and aggregate national measures to a global indicator? The scale at which monitoring takes place may need to be taken into account when assessing progress towards both global and national targets. At the local level, the theory of optimal monitoring is advancing fast; focusing on how best to allocate limited resources in the face of the inevitable trade-offs between monitoring and intervention, and explicitly considering uncertainty. This approach could potentially be applicable to promote more cost-effective monitoring across larger regional or national scales.

In this book, which results from a symposium held at the Zoological Society of London in summer 2009, we have addressed two key themes in biodiversity conservation and monitoring, bringing together insights from science and policy spheres: evaluating a variety of approaches to biodiversity monitoring that could help to provide indicators at national to global scales, and the steps needed to reduce the barriers for successful implementation of such approaches. Specifically we have focused on addressing challenges faced by countries in meeting their obligations under the biodiversity conventions, particularly CBD, and to help bridge the gap between international commitments and local action. We have structured this book around four areas: first, we examine the use of species-based indicators, and what they can tell us about the status and trends of several important metrics directly related to the overall health of biodiversity (Balmford *et al.*, 2005; Green *et al.*, 2005). These chapters describe how each measure of change in status might be appropriately used at the national level. Second, we evaluate indices of the extent and magnitude of threatening processes and the drivers of biodiversity loss, and how they might provide knowledge of how and where to prioritize conservation action (Mace and Baillie, 2007; McGeoch *et al.*, 2010). These indicators are in general far less well developed than their biodiversity counterparts, and these chapters identify opportunities for their further development and implementation over the coming decade. Third, we examine indices of important components of biodiversity that are amenable to monitoring but are not yet being widely measured, and how they can contribute to future understanding of biodiversity change. Finally we explore how best to ensure that global commitment to biodiversity conservation and monitoring is aligned with local and national action. We focus on terrestrial biodiversity for this book, although many chapters are also relevant to the marine environment.

Biodiversity and human well-being

Societies value many different aspects of biodiversity. We concentrate here on two roles. Firstly, there is a view that biodiversity should be valued for its own sake. The simple existence of species, populations, and habitats is deemed a sufficient enough justification for their continued protection. Secondly, there is a growing more utilitarian view of our natural world. Biodiversity underpins the functioning of the ecosystems on which humans depend for a variety of services including food and fresh water, health and recreation, and protection from natural disasters. Current trends of biodiversity loss are thought to be endangering these services, such that continued loss may bring us to a point where the capacity of ecosystems to provide these essential services is catastrophically reduced (Diaz *et al.*, 2006; Mace *et al.*, 2012). Marine biodiversity loss, for example, is increasingly impairing the ocean's capacity to provide humans with food, maintain water quality, and recover from anthropogenic perturbations (Worm *et al.*, 2006). As well as being detrimental to human well-being, biodiversity loss is costly for society as a whole, particularly for economic actors in sectors that depend directly on ecosystem services. For example, insect pollination in the European Union (EU) has an estimated economic value of €15 billion per year. The continued decline in bees and other pollinators could have serious consequences for Europe's farmers and agribusiness sector (Gallai *et al.*, 2009; TEEB, 2010), and, ultimately, for our ability to feed ourselves. The conservation of biodiversity also makes a critical contribution to moderating the scale of climate change and reducing its negative impacts on the functioning of ecosystems. This makes biodiversity loss the most critical global environmental threat alongside climate change; and these two threats are inextricably linked.

It has been suggested that effective conservation requires addressing three fundamental questions (Salafsky *et al.*, 2002), namely:

- what should our goals be and how do we measure progress in reaching them?
- how can we most effectively take action to achieve conservation?
- how can we learn to do conservation better?

The effectiveness of biodiversity conservation therefore depends on our ability to define, measure, and monitor biodiversity change, and on adaptive responses to biodiversity loss of a wide group of stakeholders and actors, including governments, local communities, and international society. Yet ensuring that appropriate monitoring systems are in place and translating monitoring results into effective conservation on the ground remains a major global challenge for a number of reasons, including financial and technical capacity constraints and policy and legal barriers.

It should also be noted that biodiversity conservation is ultimately implemented by humans for human society, be it for economic, cultural, or other reasons. Biodiversity conservation is thus inextricably linked to human behaviour. Its effective implementation therefore also depends on a sound understanding of the influence that social factors (such as markets, cultural beliefs and values, laws and policies, or demographic change) exert on human interactions with the environment and choices to exploit or conserve biodiversity (Mascia *et al.*, 2003).

Species-based indicators of biodiversity change

Biodiversity comprises a range of features that are important for evolution and the effective functioning of ecosystems. These features include species richness, ecological diversity, genetic diversity, phylogenetic diversity, and functional diversity. It can be argued, however, that the natural units of biodiversity conservation are species (Agapow *et al.*, 2004), and the severity of the extinction crisis is frequently expressed in terms of number of species lost (e.g., Collen *et al.*, 2010; Regan *et al.*, 2001), threatened (Hilton-Taylor *et al.*, 2009), or depleted (Loh *et al.*, 2005). There are three key metrics that are generally thought to reflect the conservation health of a given species, or set of species. These are geographical range size, population abundance, and variation in genetic diversity. During biodiversity indicator development, all have been evaluated to some degree or another. However, the lack of broad-scale data on genetic diversity has meant that indicators of genetic health have been largely limited to a handful of domestic species (Walpole *et al.*, 2009), while indicators of change in geographical range size while feasible, have been fraught with the vagaries of primary occurrence data, and moreover have limited sensitivity to population change (Boakes *et al.*, 2010). Abundance data, be they one-off measures of population size, or measures of change in abundance over time, are particularly appropriate for use as relatively sensitive measures of change (Collen *et al.*, 2009; Pereira and Cooper, 2006). While the compilation and development of indicators of such metrics has been mixed, many are a component of aggregated measures of extinction risk of species. Measuring the relative risk of extinction of a species or set of species is a measure that has been tackled by many, but pioneered by the International Union for the Conservation of Nature (IUCN) for its Red List (IUCN, 2001; Mace *et al.*, 2008; Mace and Lande, 1991). By evaluating the symptoms of risk, a composite measure of relative risk of extinction can be evaluated, regardless of the taxonomic affiliation of the species.

The three chapters in Part I of this book consider three different aspects of indicators of change currently used to assess the status of species. Chapter 2 evaluates how an index of relative extinction risk of species can be used as a biodiversity indicator at the national level. Using the widely applied IUCN Red List as a metric for extinction risk, and evaluating change in extinction risk over time (Butchart *et al.*, 2004, 2007), Collen *et al.* in Chapter 2 evaluate how the growing number of national-level Red List Assessments (Miller *et al.*, 2007; Zamin *et al.*, 2010) could be harnessed to develop national indices, which can be used to track species of national and global conservation concern. Chapter 2 concludes with two national case studies, Venezuela and South Africa, and provides lessons learned from the two decades of species extinction risk assessment carried out in these countries.

Risk assessment relies on relatively coarse measures of population change in order to place species in categories of risk, as changes between categories of risk reflect large changes in population size. A more direct measure of absolute or relative abundance provides a potentially more sensitive indicator. In Chapters 3 and 4, two different indicators that track indices of abundance over time are evaluated as scaleable indicators of biodiversity change. Drawing on an increasingly widely used technology, O'Brien and Kinnaird in Chapter 3 evaluate how remote-triggered camera traps are being used to understand trends in wildlife, particularly in tropical forest and savanna habitats. The authors argue that the technology can integrate local-scale vertebrate monitoring in a cost-effective standardized manner, to meet national responsibilities to biodiversity reporting, which can in turn feed into measuring progress to global-scale targets. While the method monitors only one component of biodiversity (ground-dwelling vertebrates), this component is often of economic and cultural significance to local communities and national governments.

Finally in Part I, Collen *et al.* in Chapter 4 evaluate the approaches of the Living Planet Index to develop national-level indicators of biodiversity change. The technique aggregates a variety of measures of population change into a single index, and is already widely used as an index of global and regional change in biodiversity. The chapter evaluates the use of the indicator of change in wildlife abundance at a national level. Using examples from Uganda, the authors explore the different levels at which meaningful measures of change in abundance can inform global, national, and sub-national biodiversity targets. The authors make the case that the loss of populations is a prelude to species extinction, and tends to reduce taxonomic, genetic, and functional diversity, and therefore is integrally related to many different elements of biodiversity. A national Living Planet Index has an additional key attribute in that it is generally likely to be well aligned to national and local biodiversity conservation priorities.

Indicators of the pressures on biodiversity

Indicators of pressure can encompass both positive and negative alterations in the direct and indirect causes of biodiversity loss (Mace and Baillie, 2007) and provide useful measures for assessing changes in ongoing threats to biodiversity. Changes in pressure may be detectable before change can be perceived in metrics of the state of biodiversity, and hence indicators of pressure can be particularly useful for instigating timely and effective conservation interventions (Failing and Gregory, 2003; McGeoch *et al.*, 2010). However, impacts of threats on biodiversity are often complex, and the removal of a threat does not necessarily lead to an immediate increase or recovery of biodiversity, hence there needs to be a clear understanding of the mechanistic links between biodiversity state and pressure (Jones *et al.*, 2011). For these reasons, it is often not sufficient to monitor pressures alone, and such monitoring should ideally be supplemented either by fully independent measures of state or studies to confirm causal linkages between pressure and state (Mace and Baillie, 2007). The development of national and global indicators of the pressures of biodiversity has lagged behind the growth of those of status, though there are good sub-national-level examples (e.g., measures of snaring). Nevertheless, several indicators are becoming available, and further development is likely to be catalysed by the focus of six of the 20 Aichi targets under goal B, which is to 'reduce the direct pressures on biodiversity and promote sustainable use' (Convention on Biological Diversity, 2011). In Part II of this book we therefore consider the development of indicators of pressures on biodiversity.

The greatest driver of threats to species has consistently been shown to be habitat loss (Hoffmann *et al.*, 2010; Millennium Ecosystem Assessment, 2005). In Chapter 5, Pettorelli examines how satellite technology and remote sensing information might be used to track changes in ecosystem distribution and functioning, and provide measures of habitat degradation and loss. In particular, Pettorelli demonstrates how satellite-derived vegetation indices, such as the Normalized Difference Vegetation Index (NDVI), can not only provide indicators of pressure, but also help predict spatio-temporal trends in species distribution and abundance, and provide important information for decision-makers. The author argues that the accessibility of satellite data means that satellite-based indicators can be both cheap and sustainable.

Climate change is one of the most rapidly growing threats to species persistence, and will likely be one of the greatest drivers of biodiversity change over the coming decades (Thomas *et al.*, 2004). Our collective understanding of the nature and extent of the susceptibility of species to climate change lags far behind that of many of the other indicators (Dawson *et al.*, 2011; Foden *et al.*, 2009; Gregory *et al.*, 2009). In Chapter 6, Foden *et al.* assess the possible ways of attributing and measuring climate change impacts to species. The authors argue that given the limited empirical information on species' responses to climate change, it is necessary to consider alternative ways for measuring how species are being or will be affected, and crucially, the degree to which these are positive or negative effects. They show how, to date, four different kinds of methods are in common use, and develop a theoretical framework for climate change indicators. They conclude that evaluating the veracity of model-based studies is central to further development, though attributing cause and effect remains problematic due to complex and poorly understood interactions between multiple threats to species.

Genovesi *et al.* in Chapter 7 address the issue of measuring the impact of invasive alien species (IAS). The problem is enormous; IAS are cited as a factor in over 50% of animal extinctions where the cause is known, and are the second most important threat to birds (BirdLife International, 2010; Hilton-Taylor *et al.*, 2009). With the number of documented impacts of IAS rising, the need to combat IAS grows. Among their recommendations, the authors contend that, alongside more direct measures of management effectiveness, detailing the economic consequences of IAS, and their impacts on ecosystem services and human health and well-being, will aid arguments for better control.

Direct unsustainable exploitation of natural resources is a key pressure on ecosystems. Wild animals and plants are essential for human livelihoods and well-being. According to the United Nations Food and Agriculture Organization (FAO), 40% of the world's economy is based directly and indirectly on the use of biological resources. However, as the world's human population increases, our use of species is having a greater and greater impact on both the species being targeted and the ecosystems in which they live, and in many cases this impact is becoming unsustainable. Almond *et al.* in Chapter 8 show how indicators of sustainably managed forests and marine fisheries can be calculated. They argue that to be nationally relevant, these indicators should provide a measure not only of the sustainable use of a natural resource, but also of the value of the resource to lives and livelihoods. They propose a pressure-state-response framework for evaluating exploitation of species, developing and adapting a number of existing indicators, to provide more complete insights into exploitation and sustainable use.

Finally, in Chapter 9, Sanderson evaluates how combinations of different measures of aggregated anthropogenic pressures can be represented in conceptual 'footprints' of consumption. Sanderson combines two key measures of human impact on the biosphere: the human footprint (Sanderson *et al.*, 2002), and the ecological footprint (Wackernagel and Rees, 1996). The former is calculated by combining spatial measures of human impact (e.g., human population density, land use, access from roads) and is presented in a mapped form. The latter is calculated at the level of the individual, through combining consumption patterns derived from where a person lives. By demonstrating how these two footprints might be combined, Sanderson provides a framework for a powerful tool that can act as a surrogate for human consumption patterns at national and global scales.

The next generation of biodiversity indicators

Parts I and II cover many of the indicators of biodiversity state and pressures currently in use. While it is increasingly widely accepted that biodiversity loss and ecosystem degradation jeopardize human well-being, not all of the relevant parts of biodiversity that matter are currently systematically measured, and the short time period between the date that the CBD 2010 target was set, and the breadth of the target, has meant that many indicators were not developed in time (Walpole *et al.*, 2009). In order to better understand and address negative human impact on biodiversity, it is important to continue to develop the field of biodiversity monitoring; taking advantage of new technologies, methods and approaches that are likely to enable the development of new indicators. As the CBD and its technical bodies, scientists, and policy-makers navigate the path towards the Aichi Target and Strategic Plan for 2020, in Part III of the book we ask what the next generation of biodiversity indicators might be. Although it is generally accepted that it is impossible to monitor all aspects of biodiversity state and pressures, we identify three key themes that are likely to be important for achieving the new target: new technologies and metrics for biodiversity, socioeconomic impacts, and how best to establish effective linkages between biodiversity information and policy.

Chapters 10 (Jones *et al.*) and 11 (McKenzie and Reardon) assess how two new metrics that might be used to monitor aspects of biodiversity to fill gaps in the current set of national and global biodiversity indicators, have so far been neglected. Jones *et al.* explore how acoustic monitoring might be used to develop measures of biodiversity change. Using the echolocation calls of bats, they show how this metric could provide a catch-all for a number of important aspects of biodiversity (abundance, geographical range, habitat and foraging niches). Given that bats comprise one-fifth of all mammal species (Simmons, 2005), their relative scarcity in monitoring schemes to date is cause for concern. The authors' argument that the need for bat monitoring due to their provision of a range of ecosystem services (e.g. pollination, seed dispersal, and insect regulation) is convincing. Yet even more powerful is the framework they set out for implementing a sustainable citizen science-based acoustic national monitoring scheme. They use examples from Romania, Bulgaria, and Hungary to illustrate their approach.

McKenzie and Reardon (Chapter 11) sustain the theme of new metrics in their evaluation of the use of occupancy, a measure of the proportion of an area occupied by a species, to develop an index of abundance. The authors carefully designed a methodology for the use of occupancy methods to monitor the slender loris, a species of primate from Sri Lanka. This animal is small and nocturnal and, like most small nocturnal mammals, has previously been very difficult to monitor. They make a compelling argument for how less labour-intensive metrics of abundance, such as occupancy, might be useful in obtaining reliable estimates of population change over time. Crucially, they establish that in order to account for imperfect detection of animals, a key requirement of these methods is that the monitoring programme is designed such that there are repeated opportunities of detecting the species at each location within a relatively short timeframe.

Strategic goal D of the Aichi targets emphasizes the need to enhance benefits to all, from biodiversity and ecosystem services (Convention on Biological Diversity, 2011). While access and benefits sharing has been widely discussed both in political forums and the scientific literature, little progress has been made as yet towards its integration into biodiversity target setting. The social and economic impacts of conservation projects on local communities in the developing world have been the subject of substantial debate. In Chapter 12, Homewood focuses on evaluating why and how socioeconomic monitoring for conservation initiatives should be carried out, appraising both rapid light-touch, and more in-depth quantitative analytical options. The conclusions reached are, in essence, that no single tool will effectively monitor socioeconomic impact, and that, like other areas of biodiversity research, a range of metrics are required, and these need to be tailored to each situation. Moreover, unlike plant or animal subjects, humans (including the researcher) are prone to bias and influence, and hence particular care must be taken to ensure monitoring is truly independent.

Finally in Part III, Chapter 13 (Mace *et al.*) looks at how creating targets that are likely to have more impact, by enhancing the links between science and policy, are central to biodiversity conservation. The authors argue that scientists tend to evaluate targets objectively in relation to their scientific relevance, how measurable they are, and how those measures meet the policy goal. This approach might not always align with that of governments, managers, and policy-makers, who are understandably sensitive to what it might mean if targets are not met. The authors make out a clear case for a need to adopt a small set of focused, relevant, efficient, and achievable targets, using an iterative process, in which targets are adapted as underlying conditions change. This process needs to exploit the existing science base while involving key decision-makers. The authors provide a colour-coded categorization system for different types of targets to aid this process.

Biodiversity monitoring in practice

Biodiversity targets, such as those set by the Convention on Biological Diversity in Aichi, can only be met through effective conservation action that is monitored appropriately and evaluated to track progress towards targets. However, there remain a number of major challenges to effective biodiversity conservation, monitoring, and evaluation. These include financial constraints, lack of technical capacity, policy and legal barriers, and perverse incentives. The final part of the book explores these practical constraints and challenges confronting biodiversity conservation and monitoring. In this the authors bring together several of the themes explored in this book to evaluate how some of the recommendations made in previous chapters might play out in the real world.

Durant argues in Chapter 14 that any global target is likely to be difficult to achieve without national-level buy-in. This chapter demonstrates that, in order to be sustainable and long-term, biodiversity monitoring needs to be firmly embedded within a national context. This is best achieved when monitoring is aligned with local and national priorities, and engages managers and policy-makers from the beginning, to ensure ownership of a biodiversity monitoring plan. The chapter outlines the different stages that should be considered in an effective monitoring plan that is grounded in good science but focused on clear conservation goals. Durant presents a framework where local capacity lies at the centre of biodiversity monitoring, and argues that local ownership and effective institutionalization of biodiversity monitoring is key to long-term sustainability.

Conservation management involves a number of difficult decisions, and in Chapter 15 Jones presents a further reality check. The resources available for conservation are insufficient to prevent the loss of much of the world's threatened biodiversity during this crisis (Vane-Wright *et al.*, 1991). Conservation planners are therefore forced to prioritize their activities under this reality, and this also in the context of great uncertainty about future change – this has become known as 'the agony of choice'. Practitioners are unlikely ever to have sufficient funds to monitor and protect everything, and Jones argues that monitoring programmes must be designed with sufficient power to detect the desired level of change in the metric of interest; at times the best decision might be not to monitor but rather to put the money into more direct conservation action.

The problems caused by financial constraints to all scales of biodiversity monitoring is a thread that runs through many chapters in this book. Bashir examines this issue in Chapter 16, drawing on the example of the key financier of biodiversity conservation projects, the Global Environment Facility (GEF), for insight. Since 1991, it has provided US$3.1 billion in grants, and $8.3 billion in leveraged co-financing to support the implementation of around 1000 biodiversity projects in more than 155 developing countries and countries with economies in transition (GEF, 2010). Although biodiversity monitoring is not a requirement of GEF, many GEF-funded projects include some form of biodiversity monitoring. One of the main conclusions of the chapter is that there is substantial scope to increase the contribution of GEF projects to broader national or international biodiversity monitoring efforts. The author recommends that GEF build on this potential, together with established monitoring capacity within the GEF, to enable GEF projects to contribute to a standardized biodiversity monitoring database. This would enable better monitoring of the biodiversity conservation contribution played by GEF as well as better overall biodiversity monitoring. The author argues that this could be achieved at relatively low cost, by improving coordination and collaboration with the scientific community and establishing mentoring programmes.

National-level biodiversity monitoring and reporting will be central to our collective efforts to stem global biodiversity loss. Yet limited attention has been paid to integrating the two processes. Bubb in Chapter 17 evaluates how linkages between these two apparently rather separate entities might be exploited. While there are conceptually two directions in which indicators could be built (disaggregating global indicators to the national level, and aggregating national-level indicators to create a global index), Bubb argues strongly for the latter. Only with national integration of biodiversity target setting into a set of streamlined biodiversity indicators, is significant progress likely to be made.

We must accept that we live in a target-driven world. Perhaps the greatest challenge for biodiversity conservation is integrating care for wildlife, the environment, and ecosystems and the benefits that humans derive from them, into the mindset of other sectors. In the final chapter, Stuart and Collen (Chapter 18) draw on some of the lessons learned from both the process that was undertaken, and wording of the CBD 2010 target, and set out a vision for how the pitfalls identified in that process can be avoided in the run-up to 2020. These include a better understanding of the complex relationship between biodiversity and ecosystem services (Mace *et al.*, 2012), quantifying the economic costs of inaction, and reconnecting the public with wildlife. Only by linking targets to actions can the rapid decline of biodiversity be halted.

Conclusions

We set out with the aims of tackling the issue of measuring and assessing biodiversity decline at multiple levels, demonstrating the importance of biodiversity monitoring from the national to global scale, and highlighting approaches for biodiversity monitoring that will provide information needed to plan, prioritize, adapt, and respond to our rapidly changing world. Demonstrating the importance of monitoring our natural world, providing options for national and global biodiversity monitoring, and exploring ways in which the results can most effectively influence decision-making are thus major themes of this book. Developing an understanding of how countries can assess national biodiversity change, and bridge the gap between international commitments and local action will, we hope, help galvanize progress towards a more proactive approach to biodiversity conservation.

References

Agapow, P.-M., Bininda-Emonds, O.R.P., Crandall, K.A., et al. (2004) The impact of species concept on biodiversity studies. *Quarterly Review of Biology*, 79, 161–179.

Araújo, M.B. (1999) Distribution patterns of biodiversity and the design of a representative reserve network in Portugal. *Diversity and Distributions*, 5, 151–163.

Baillie, J.E.M., Collen, B., Amin, R., et al. (2008) Towards monitoring global biodiversity. *Conservation Letters*, 1, 18–26.

Balmford, A. and Bond, W. (2005) Trends in the state of nature and their implications for human well-being. *Ecology Letters*, 8, 1218–1234.

Balmford, A., Bennun, L.A., ten Brink, B., et al. (2005) The Convention on Biological Diversity's 2010 target. *Science*, 307, 212–213.

BirdLife International (2010) State of the world's birds. Available at: http://www.birdlife.org/sowb. BirdLife International, Cambridge, UK.

Boakes, E., McGowan, P.J.K., Fuller, R.A., et al. (2010) Distorted views of biodiversity: spatial and temporal bias in species occurrence data. *PLoS Biology*, 8, 1–11.

Butchart, S.H.M., Stattersfield, A.J., Bennun, L.A., et al. (2004) Measuring global trends in the status of biodiversity: Red List Indices for birds. *PLoS Biology*, 2, 2294–2304.

Butchart, S.H.M., Akcakaya, H.R., Chanson, J.S., et al. (2007) Improvements to the Red List Index. *PLoS ONE*, 2, e140. doi:110.1371/journal.pone.0000140.

Butchart, S.H.M., Walpole, M., Collen, B., et al. (2010) Global biodiversity decline continues. *Science*, 328, 1164–1168.

Collen, B., Ram, M., Zamin, T., and McRae, L. (2008) The tropical biodiversity data gap: addressing disparity in global monitoring. *Tropical Conservation Science*, 1, 75–88 Available online at: tropicalconservationscience.org.

Collen, B., Loh, J., Whitmee, S., McRae, L., Amin, R., and Baillie, J.E.M. (2009) Monitoring change in vertebrate abundance: the Living Planet Index. *Conservation Biology*, 23, 317–327.

Collen, B., Purvis, A., and Mace, G.M. (2010) When is a species really extinct? Testing extinction inference from a sightings record to inform conservation assessment. *Diversity and Distributions*, 16, 755–764.

Collen, B., Böhm, M., Kemp, R. and Baillie, J.E.M. (2012) Spineless: status and trends of the world's invertebrates. Zoological Society of London, United Kingdom, pp. 86.

Convention on Biological Diversity (2010) Global Biodiversity Outlook 3. UNEP. Available at: http://www.cbd.int/gbo3/.

Convention on Biological Diversity (2011) Strategic Plan for Biodiversity 2011–2020, including Aichi Biodiversity Targets. Available at: http://www.cbd.int/sp/.

Dawson, T.P., Jackson, S.T., House, J.I., Prentice, I.C., and Mace, G.M. (2011) Beyond predictions: biodiversity conservation in a changing climate. *Science*, 332, 53–58.

Diaz, S., Fargione, J., Chapin III, F.S., and Tilman, D. (2006) Biodiversity loss threatens human well-being. *PLoS Biology*, 4, 1300–1305.

Dobson, A. (2005) Monitoring global rates of biodiversity change: challenges that arise in meeting the Convention on Biological Diversity (CBD) 2010 goals. *Philosophical Transactions of the Royal Society of London B*, 360, 229–241.

Failing, L. and Gregory, R. (2003) Ten common mistakes in designing biodiversity indicators for forest policy. *Journal of Environmental Management*, 68, 121–132.

Foden, W., Mace, G.M., Vié, J.-C., et al. (2009) Species susceptibility to climate change impacts. In: Vié, J.-C., Hilton-Taylor, C., and Stuart, S.N. (eds), *Wildlife in a Changing World – an Analysis of the 2008 IUCN Red List of Threatened Species*. IUCN, Gland, pp. 77–87.

Gallai, N., Salles, J.M., Settele, J., and Vaissière, B.E. (2009) Economic valuation of the vulnerability of world agriculture confronted with pollinator decline. *Ecological Economics*, 68, 810–821.

GEF (2010) Behind the numbers: a closer look at GEF achievements. Global Environment Facility, Washington, DC.

Green, R.E., Balmford, A., Crane, P.R., Mace, G.M., Reynolds, J.D., and Turner, R.K. (2005) A framework for improved monitoring of biodiversity: responses to the World Summit on Sustainable Development. *Conservation Biology*, 19, 56–65.

Gregory, R.D., Willis, S.G., Jiguet, F., et al. (2009) An indicator of the impact of climate change on European bird populations. *PLoS ONE*, 4, e4678.

Hilton-Taylor, C., Pollock, C.M., Chanson, J.S., Butchart, S.H.M., Oldfield, T.E.E., and Katariya, V. (2009) State of the world's species. In: Vié, J.-C., Hilton-Taylor, C., and Stuart, S.N. (eds), *Wildlife in a Changing World: an Analysis of the 2008 IUCN Red List of Threatened Species*. IUCN, Gland, pp. 15–42.

Hoffmann, M., Hilton-Taylor, C., Angulo, A., et al. (2010) The impact and shortfall of conservation on the status of the world's vertebrates. *Science*, 330, 1503–1509.

Isaac, N.J.B., Turvey, S.T., Collen, B., Waterman, C., and Baillie, J.E.M. (2007) Mammals on the EDGE: Conservation priorities based on threat and phylogeny. *PLoS ONE*, 2, e296. doi:210.1371/journal.pone.0000296.

IUCN (2001) IUCN Red List Categories and Criteria version 3.1. IUCN, Gland, Switzerland. Available at: http://www.iucnredlist.org/technical-documents/categories-and-criteria/2001-categories-criteria.

Jones, J.P.G., Collen, B., Atkinson, G., et al. (2011) The why, what and how of global biodiversity indicators beyond the 2010 target. *Conservation Biology*, 25, 450–457.

Loh, J., Green, R.E., Ricketts, T., *et al.* (2005) The Living Planet Index: using species population time series to track trends in biodiversity. *Philosophical Transactions of the Royal Society of London B*, 360, 289–295.

Mace, G.M. and Baillie, J.E.M. (2007) The 2010 Biodiversity Indicators: challenges for science and policy. *Conservation Biology*, 21, 1406–1413.

Mace, G.M. and Lande, R. (1991) Assessing extinction threats – toward a reevaluation of IUCN Threatened Species categories. *Conservation Biology*, 5, 148–157.

Mace, G.M., Masundire, H., and Baillie, J.E.M. (2005) Biodiversity. In: Scholes, B. and Hassan, R. (eds), *Ecosystems and Human Well-Being: Current State and Trends*. Island Press, Washington, DC, pp. 77–122.

Mace, G.M., Collar, N.J., Gaston, K.J., *et al.* (2008) Quantification of extinction risk: IUCN's system for classifying threatened species. *Conservation Biology*, 22, 1424–1442.

Mace, G.M., Norris, K., and Fitter, A.H. (2012) Biodiversity and ecosystem services: a multi-layered relationship. *Trends in Ecology and Evolution*, 27, 19–26.

Mascia, M.B., Brosius, J.P., Dobson, T.A., *et al.* (2003) Conservation and the social science. *Conservation Biology*, 17, 649–650.

McGeoch, M.A., Butchart, S.H.M., Spear, D., *et al.* (2010) Global indicators of biological invasion: species numbers, biodiversity impact and policy responses. *Diversity and Distributions*, 16, 95–108.

Millennium Ecosystem Assessment (2005) *Ecosystems and Human Well-Being: Biodiversity Synthesis*. World Resources Institute, Washington, DC.

Miller, R.M., Rodríguez, J.P., Aniskowicz-Fowler, T., *et al.* (2007) National threatened species listing based on IUCN criteria and regional guidelines: current status and future perspectives. *Conservation Biology*, 21, 684–696.

Myers, N., Mittermeier, R.A., Mittermeier, C.G., de Fonseca, G.A.B., and Kent, J. (2000) Biodiverity hotspots for conservation priorities. *Nature*, 403, 853–858.

Nicholson, E., Collen, B., Barausse, A., *et al.* (2012) Making robust policy decisions using global biodiversity indicators. *PLoS ONE*, 7, e41128. doi:10.1371/journal.pone.0041128.

Pereira, H.M. and Cooper, H.D. (2006) Towards the global monitoring of biodiversity change. *Trends in Ecology and Evolution*, 21, 123–129.

Pereira, H.M., Proença, V., Belnapp, J., *et al.* (2010a) Global biodiversity monitoring: filling the gap where it counts the most. *Frontiers in Ecology and the Environment*, 8, 459–460.

Pereira, H.M., Leadley, P.W., Proenca, V., *et al.* (2010b) Scenarios for global biodiversity in the 21st century. *Science*, 330, 1496–1501.

Possingham, H., Day, J., Goldfinch, M., and Salzborn, F. (1993) The mathematics of designing a network of protected areas for conservation. In: Sulton, D., Cousins, E., and Pierce, C. (eds), *Proceedings of the 12th National ASOR Conference*. Australian Operations Research Conference, Adelaide University, pp. 536–545.

Regan, H.M., Lupia, R., Drinnan, A.N., and Burgman, M.A. (2001) The currency and tempo of extinction. *American Naturalist*, 157, 1–10.

Rodrigues, A.S.L., Akçakaya, H.R., Andelman, S.J., *et al.* (2004) Global gap analysis – priority regions for expanding the global protected area network. *BioScience*, 54, 1092–1100.

Salafsky, N., Margoluis, R., Redford, K.H., and Robinson, J.G. (2002) Improving the practice of conservation: a conceptual framework and research agenda for conservation science. *Conservation Biology*, 16, 1469–1479.

Sanderson, E.W., Jaiteh, M., Levy, M.A., Redford, K.H., Wannebo, A.V., and Woolmer, G. (2002) The human footprint and the last of the wild. *BioScience*, 52, 891–904.

Simmons, N.B. (2005) Order Chiroptera. In: Wilson, D.E. and Reeder, D.M., (eds), *Mammal Species of the World: a Taxonomic and Geographic Reference*. Smithsonian Institution Press, Washington, DC, pp. 312–529

TEEB (2010) *The Economics of Ecosystems and Biodiversity: Mainstreaming the Economics of Nature: A Synthesis of the Approach, Conclusions and Recommendations of the TEEB*. Progress Press, Malta.

Thomas, C.D., Cameron, A., Green, R.E., *et al.* (2004) Extinction risk from climate change. *Nature*, 427, 145–148.

UNEP (2002) Report on the sixth meeting of the Conference of the Parties to the Convention on Biological Diversity (UNEP/CBD/COP/20/Part 2) Strategic Plan Decision VI/26. Convention on Biological Diversity. Available at: https://www.cbd.int/doc/?meeting=cop-06.

Vane-Wright, R.I., Humphries, C.J., and Williams, P.H. (1991) What to protect – systematics and the agony of choice. *Biological Conservation*, 55, 235–254.

Wackernagel, M. and Rees, W. (1996) *Our Ecological Footprint: Reducing Human Impact on the Earth*. New Society Publishers, Gabriola Island, Canada.

Walpole, M., Almond, R., Besançon, C., *et al.* (2009) Tracking progress towards the 2010 biodiversity target and beyond. *Science*, 325, 1503–1504.

Worm, B., Barbier, E.B., Beaumont, N., *et al.* (2006) Impacts of biodiversity loss on ocean ecosystem services. *Science*, 314, 787–790.

Zamin, T., Baillie, J.E.M., Miller, R., Rodriguez, J.P., Ardid, A., and Collen, B. (2010) National Red Listing beyond the 2010 target. *Conservation Biology*, 24, 1012–1020.

Part I
Species-Based Indicators of Biodiversity Change

2

Tracking Change in National-Level Conservation Status: National Red Lists

Ben Collen[1], Janine Griffiths[1,2], Yolan Friedmann[3], Jon Paul Rodriguez[4,5], Franklin Rojas-Suárez[5] and Jonathan E.M. Baillie[2]

[1]Institute of Zoology, Zoological Society of London, London, UK
[2]Conservation Programmes, Zoological Society of London, London, UK
[3]Endangered Wildlife Trust, Parkview, South Africa
[4]Centro de Ecología, Instituto Venezolano de Investigaciones Científicas, Caracas, Venezuela
[5]Provita, Caracas, Venezuela

Introduction

Tracking change in species extinction risk provides an effective metric with which to follow trends in biodiversity. One of the most widely applied classifications of the relative risk of extinction of species is the International Union for the Conservation of Nature (IUCN) Red List (IUCN, 2010; Mace *et al.*, 2008). Species are assigned to Red List categories through detailed assessment of biological information against a set of objective, standard, quantitative criteria (IUCN, 2001). Multiple assessments of the same species or species set enable the calculation of an index of biodiversity change, termed the Red List Index (Figure 2.1) (Butchart *et al.*, 2004).

The concept of a Red List of threatened species was first defined by IUCN in the 1950s. Though used widely in scientific, political, and popular contexts as a means of highlighting the world's most threatened species, these early lists lacked objective criteria and guidelines, instead relying on assessors to use their own judgement in correctly categorizing species (Mace *et al.*, 2008). The subjectivity of this early method

---

Biodiversity Monitoring and Conservation: Bridging the Gap between Global Commitment and Local Action, First Edition. Edited by Ben Collen, Nathalie Pettorelli, Jonathan E.M. Baillie and Sarah M. Durant.
© 2013 John Wiley & Sons, Ltd. Published 2013 by John Wiley & Sons, Ltd.

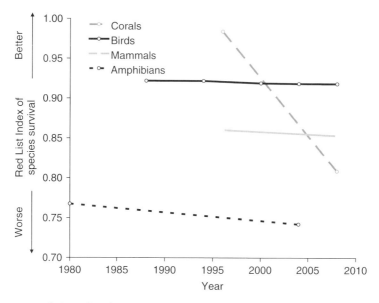

Figure 2.1 **Red List Index of species survival, PLoS Biology.**

was recognized (Fitter and Fitter, 1987; Mace and Lande, 1991), and subsequently developed into a set of objective criteria, concluding with the publication of the 1994 IUCN Red List of Threatened Species (IUCN, 1994), which was later updated in 2001 (IUCN, 2001). Today the IUCN Red List of Threatened Species is the foremost comprehensive inventory for classifying species' extinction risk at the global scale and has also been exceptionally successful at raising awareness of biodiversity loss (e.g. Baillie et al., 2004; BirdLife International, 2010; Convention on Biological Diversity, 2010b).

Due to the distribution and movement of species across political borders, the global Red List is unsuitable for assessing extinction risk at the sub-global level. If the global Red List Criteria were directly applied to species at a national level in a small nation or region, many species would be classed as threatened due to the apparent small ranges (Gärdenfors, 2001). To overcome this problem, guidelines were designed for use in national and regional Red Lists, adapted from the global Criteria (IUCN, 2003). National Red Lists provide a practical means of assessing species status, with a range of different taxonomic groups and geographical units being eligible for assessment (Miller et al., 2006). As with the global Red List, National Red Lists are based on categories that use the same threshold values for population reduction over time, reduced and fragmented geographical range, and small, restricted, and declining populations. As a first stage species are assessed according to the global IUCN Red List Categories and

Criteria (IUCN, 2001). A second stage then requires the evaluation of whether a rescue effect might be apparent from populations of the same species over the national or regional border. The initial category assignment can then be upgraded or downgraded depending on immigration and emigration between populations (IUCN, 2003). It follows that if a population is isolated or endemic, the global Red List is still applicable and as certain regions are important as migratory stopovers or feeding sites, even non-breeding taxa should be included in National Red Lists (Gärdenfors, 2001).

Increased awareness and intensified concern over the loss of global biodiversity (Butchart *et al.*, 2010b; Collen *et al.*, 2009a) has led to an advance in international policy agreements highlighting the need to measure trends of species at the national level. The United Nations eight Millennium Development Goals (MDG) aim to achieve significant and measurable improvements in the lives of the world's poor by 2015; MDG 7 – 'ensuring environmental stability' – requires signatories to monitor the status and trends of native and migratory species in their respective countries. Likewise, the Convention on Biodiversity's (CBD) 2020 Strategic Plan for Biodiversity was adopted by 193 signatory nations, all committed to periodically monitoring the status of their biodiversity (Convention on Biological Diversity, 2010a). The need for national responsibility for reporting has come to the fore as nations are invited to set their own targets within this flexible framework, taking into account national needs and priorities, while also bearing in mind national contributions to the achievement of the global targets.

Despite this rapidly growing requirement for countries to accurately gauge status and trends in national level biodiversity, there are few tools available to achieve this, but National Red Lists can provide this information (Zamin *et al.*, 2010). By using quantitative information from a Red Listing, two consecutive Red Lists on a given set of species can illustrate the change in extinction risk using the Red List Index (Butchart *et al.*, 2004). The National Red List Index can be calculated based on the same process (e.g. Szabo *et al.*, 2012), and compiled at regular intervals National Red Lists are valuable biodiversity indicators, reporting biodiversity trends at the national level and providing critical information for reporting to international conventions such as the CBD (Bubb *et al.*, 2009; Zamin *et al.*, 2010).

Unfortunately, countries vary in their capacity to monitor biodiversity (Collen *et al.*, 2009b; see also Chapter 17) and those countries with the least capacity tend to be custodians of the greatest quantity of wildlife. Least is known about the status of species where diversity is greatest. A recent study by Zamin *et al.* (2010) evaluated the taxonomic and geographical gaps in national threatened species list coverage. Geographical gaps were most apparent in western and central Africa, Oceania and the Caribbean. Although historically invertebrates and plants are poorly studied, plants are relatively well documented in threatened species list coverage such as National Red Lists and certain invertebrate groups such as butterflies, dragonflies, and molluscs are also well represented.

In this chapter we explore the prospect of taking the Red List Index, a global indicator of change in species extinction risk, and applying it at the regional and national level. We highlight some examples where this has been done, and discuss how data coverage and data integrity are important factors in extending the utility of this index. We look in detail at the National Red List process in two countries, Venezuela and South Africa, and look at how lessons may be learned from their experiences, and applied to other countries. We set out the potential strengths and limitations to using extinction risk indices to monitor change in biodiversity.

Methods

The IUCN Red List Index (RLI) measures trends in the overall extinction risk of a set of species. The RLI is based on movement of species status through the IUCN Red List Categories, and so requires a good knowledge of these Categories and Criteria for assessment of extinction risk. Mathematically the calculation of the RLI can be expressed as:

$$RLI_t = 1 - \frac{\sum_s W_{c(t,s)}}{W_{EX} \cdot N}$$

where $W_{c(t,s)}$ is the weight of category c for species s at time t, W_{EX} is the weight for Extinct, and N is the number of assessed species excluding those considered Data Deficient in the current time period and those considered to be Extinct in the year the set of species was first assessed (Butchart et al., 2004, 2007). The RLI is calculated from the number of species in each Red List Category (Least Concern, LC; Near Threatened, NT; Vulnerable, VU; Endangered, EN; Critically Endangered, CR), and the number changing categories between assessments as a result of genuine improvement or deterioration in status (category changes owing to improved knowledge or revised taxonomy are excluded; Butchart et al., 2004).

Put simply, for a given group, the number of species in each Red List Category is multiplied by the Category weight (0 = Least Concern, 1 = Near Threatened, 2 = Vulnerable, 3 = Endangered, 4 = Critically Endangered, and 5 = Extinct in the Wild and Extinct). These products are summed, divided by the maximum possible product (the number of species multiplied by the maximum weight), and subtracted from one. This produces an index that ranges from 0 to 1 (see Figure 2.1). In order to calculate an RLI, a number of criteria must be met, requiring that:

1. the same set of species is included in all time steps; and
2. only category changes resulting from genuine improvement or deterioration in status (i.e. excluding changes resulting from improved knowledge or taxonomic revisions) are included.

In practice, species lists can be relatively fluid and will often change slightly from one assessment to the next owing to splitting and lumping through taxonomic revisions (Collen *et al.*, 2011; Isaac *et al.*, 2004). Consequently many species change category between assessments owing to improved knowledge of their population size, trends, distribution, and threats (Collen *et al.*, 2011). The conditions can therefore be met by retrospectively correcting earlier Red List categorizations using current information and taxonomy (e.g. Hoffmann *et al.*, 2010). This is achieved through assuming that the current Red List Categories for the taxa have applied since the set of species was first assessed, unless there is information to the contrary that genuine status changes have occurred. Such information is often contextual, for example relating to the known history of habitat loss within the range of the species (see Butchart *et al.*, 2007).

Interpreting the index in the context of biodiversity change?

The RLI measures the rate change in extinction risk of a set of species, rather than the status of species, as interpreted through changing aggregated risk of extinction. Although some of the Red List Criteria are based on absolute population size or range size, others are based on rates of decline in these values or combinations of absolute size and rates of decline (Mace *et al.*, 2008). A downward trend in the RLI over time means that the expected rate of future species extinctions is worsening (i.e. the rate of biodiversity loss is increasing). An upward trend means that the expected rate of species extinctions is abating (i.e. the rate of biodiversity loss is decreasing), and a horizontal line means that the expected rate of species extinctions is remaining the same, although in both cases it does not mean that biodiversity loss has stopped.

Global Red List and RLI results

On the global IUCN Red List, 55,926 species have been assessed (IUCN, 2010). The results (Figure 2.2) show that 18,351 are listed as threatened with extinction, with approximately 19% of vertebrates threatened (Baillie *et al.*, 2010), ranging from 13% of birds to 41% of amphibians (Hoffmann *et al.*, 2010). These proportions are broadly comparable with the range observed in the few invertebrate and plant taxa completely or representatively assessed to date (Brummitt *et al.*, 2008; Carpenter *et al.*, 2008; Clausnitzer *et al.*, 2009; Collen *et al.*, 2009b; Cumberlidge *et al.*, 2009). However, the proportion of Data Deficient (DD) species can have some influence on these conclusions. Evidence suggests that the probable status of DD species in many taxonomic groups might be more likely to be threatened (Collen *et al.*, 2009b); at the very least, some unknown proportion of DD species are threatened.

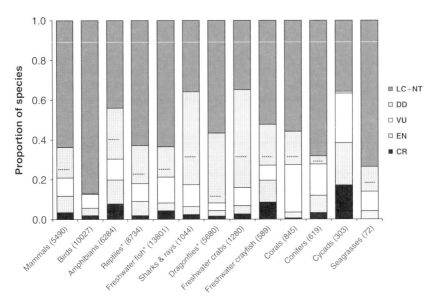

Figure 2.2 Proportion of species in different Red List categories on the 2010 IUCN Red List. Taxa are ordered by estimated proportion of extant species considered Threatened if Data Deficient (DD) species are threatened in the same proportion as non-DD species (horizontal black dotted lines). Numbers beneath the bars are number of extant species assessed in the group; *Denotes estimates are derived from a randomized sampling approach. EW, Extinct in the Wild; CR, Critically Endangered; EN, Endangered; VU, Vulnerable; NT, Near Threatened; LC, Least Concern; DD, Data Deficient. Extinct species are excluded. Data from Baillie *et al.* 2010; Hoffmann *et al.* 2010; IUCN 2010.

The RLI (Figure 2.1) shows that the proportion of warm-water coral, bird, mammal, and amphibian species expected to survive into the near future without additional conservation actions has declined over time. Coral species appear to be moving most rapidly towards greater extinction risk (Carpenter *et al.*, 2008), while amphibians are, on average, the group most threatened (Stuart *et al.*, 2004). A Red List Index value of 1.0 indicates that all species in a group are classified as Least Concern. At the other extreme, a value of 0 indicates that all species in a group have gone extinct.

The RLI can be disaggregated for subsets of data to evaluate sub-global trends in extinction risk at several different scales. Examples include trends in extinction risk for particular taxonomic groups (e.g. Figure 2.1), trends in extinction risk for species relevant to particular environmental policy, such as the Convention on Migratory Species (Convention on Migratory Species, 2008), identifying continents, regions, or biogeographical realms where the extinction risk of species is changing most rapidly

(BirdLife International, 2008; Butchart *et al.*, 2004), and to explore trends in the importance and impacts of specific threats (McGeoch *et al.*, 2010; see also Chapter 7).

Despite all this new and developing species information, the conservation status of less than 2.5% of the world's described biodiversity is currently known. Clearly this limits understanding of the impact of humans on biodiversity, and with it the ability to make informed decisions on conservation planning and action. One of the major challenges for the IUCN Red List is assessing the larger groups that represent the majority of the world's biodiversity (Baillie *et al.*, 2008; Collen and Baillie 2010). With these larger groups of less well-known organisms, a comprehensive survey of extinction risk for the whole group is often not feasible. A new approach has been developed that takes a large random sample of species groups (Baillie *et al.*, 2008). Using a random sample of 1500 species from a group, this approach allows for the identification of the general level of threat to each group, the mapping of areas likely to contain the most threatened species, and the identification of the main drivers of threat, and helps pinpoint what key actions are required to address declines in the group. While this approach is yet to be tested at a national scale, simulations show that there are no obvious barriers to its implementation at a sub-global level (Baillie *et al.*, 2008), if species groups are too large to be efficiently or cost-effectively assessed.

Implementation of the index: national case studies

It is at regional and local scales that human actions collide most obviously with biodiversity conservation (Pimm *et al.*, 2001; see also Chapter 14). Threatened species lists, based on extinction risk, are becoming increasingly influential. While it has long been clear that priority setting is a societal process, and that risk assessment (as a scientific endeavour) should not be confounded (Mace and Lande, 1991), there is a clear need for biodiversity information at regional, national, and local levels as part of the conservation planning tool kit. Consequently, interest in producing regional and national threatened species lists has soared (Collen *et al.*, 2008; Miller *et al.*, 2007; Zamin *et al.*, 2010).

The importance of national-level monitoring of biodiversity trends is now enshrined more than ever in global biodiversity policy, with the overarching biodiversity protection framework of the Convention on Biological Diversity incorporating national monitoring and action plans into its 2020 Strategic Plan (Convention on Biological Diversity, 2011). Even so, a gap exists, not only in the coverage of national-level biodiversity data (Collen *et al.*, 2008; Zamin *et al.*, 2010), but also in the implementation of national schemes to redress this issue. To the best of our knowledge, only two countries have produced a national Red List Index to date (Sweden: Gärdenfors 2010; and China: Xu *et al.*, 2009 – though see Butchart *et al.*, 2010a). We examine two case studies of contrasting countries that have well-developed National Red Listing programmes, reporting the processes that they underwent and drawing conclusions and recommendations for countries that may wish to follow suit.

Two decades of extinction risk assessment for the Venezuelan fauna

The publication of a Red Data Book for Venezuela was identified as a priority in the 1988–1992 National Species Conservation Plan (Fudena, 1989). In 1989, the Venezuelan non-governmental organization Provita, with financial support from the Wildlife Conservation Society, launched Project EVE (for its name in Spanish *Especies Venezolanas en Extinción*), with the primary objective of identifying the animal taxa threatened with extinction in Venezuela (vertebrate and invertebrate), and assessing their conservation status. During the following few years, information was compiled, a database was developed, and the first edition of the *Libro Rojo de la Fauna Venezolana* (*Red Book of Venezuelan Fauna*) was published in collaboration with Fundación Polar (Rodríguez and Rojas-Suárez, 1995).

In 1999, a second edition was published, including an appendix updated with new information since the first edition (Rodríguez and Rojas-Suárez, 1999). Despite a second printing in 2003, the *Red Book of Venezuelan Fauna* sold out again, creating the expectation of a new edition. Since the time of publication of the first edition, however, the quantity and quality of information on Venezuelan threatened animals significantly increased due to new research carried out primarily by the scientific community and NGOs. Additionally, in 2001 IUCN implemented new Red List categories and criteria (IUCN, 2001), and in 2003 it published a set of guidelines for application of the criteria at regional and national levels (IUCN, 2003). It became apparent that it was necessary to renew the methodological approach and completely update the contents of the Red Book (Rodríguez and Rojas-Suárez, 2008). Below, we describe the process, results, and impacts of the *Red Book of Venezuelan Fauna*, and outline the plans for the future. Although our focus is on animals, it is important to mention that a similar process was followed for the production of the *Libro Rojo de la Flora Venezolana* (*Red Book of Venezuelan Flora*; Llamozas *et al.*, 2003).

The first two editions: 1995 and 1999

Prior to initiating the process that would lead to the first and second editions of the *Red Book of Venezuelan Fauna*, an Advisory Committee was established, chaired by Professor Juhani Ojasti (Universidad Central de Venezuela) and Dr Stuart D. Strahl (Wildlife Conservation Society), and including 29 other members chosen for their knowledge and experience on various animal taxonomic groups (Rodríguez and Rojas-Suárez, 1995).

Next, a preliminary Red List of potentially threatened Venezuelan animals was compiled, by integrating information from numerous published sources, such as international Red Lists and books (e.g. Collar *et al.*, 1992; Groombridge, 1982; IUCN, 1986, 1988; Thornback and Jenkins, 1982), and articles in national journals, books, and magazines (e.g. Gondelles *et al.*, 1981; Romero, 1985; Trebbau, 1985), regardless of their level of detail or reliability. This list was examined and further expanded by the Advisory Committee.

An extensive directory of experts was developed in conjunction with the Advisory Committee, in order to identify anyone able to provide information on a particular species or taxonomic group. Over 2000 questionnaires were distributed among 130 experts in Venezuela and abroad, requesting unpublished observations and bibliographic references on 367 taxa. Experts were also asked to recommend a category for each species, following one of the early versions of the quantitative criteria (IUCN, 1994; Mace *et al.*, 2008). Although most of the taxa assessed were species, in some cases experts proposed the inclusion of subspecies, either island forms or species with disjunct populations exposed to very different levels of risk.

Data processing and analysis consisted in summarizing the information available for each species (mainly classification, distribution, population size, trends, threats, and conservation actions implemented or recommended), comparing and contrasting the opinion of different experts regarding the categories assigned to each species, and validating the assessments. Reaching consensus regarding the correct category was often challenging. In some cases, five experts recommended five different categories for a single species, including instances where one expert believed that the species should be considered Endangered while another thought that Least Concern was appropriate. Fortunately, most assessments were similar, and in the majority of cases they complemented each other. Each species or subspecies assessed was written up by one of the authors (JPR or FRS) in a standard information sheet, using both the compiled literature and the questionnaires completed by experts.

Increased participation in the 2008 edition

The editorial team of the 2008 *Red Book of Venezuelan Fauna* was composed of two primary editors, eight associate editors, 81 authors of species summary data sheets, 20 illustrators, and 11 photographers. All species with summary data sheets were illustrated, every data sheet was authored by one or more person, and all chapters had an associate editor responsible for coordinating the work within a taxonomic group (Rodríguez and Rojas-Suárez, 2008). The process was decentralized, and once

Table 2.1 **Number of assessed and threatened animals in Venezuela: 1999–2008. From Rodríguez and Rojas-Suárez, 2008**

Taxon	No. assessed 1999	No. threatened 1999	No. assessed 2008	No. threatened 2008
Mammals	127	34 (27%)	351	44 (12%)
Birds	122	33 (27%)	1418	35 (2%)
Reptiles	23	13 (56%)	341	22 (6%)
Amphibians	11	7 (64%)	315	26 (8%)
Fishes	16	10 (62%)	1200*	37
Invertebrates	68	8 (12%)	Not systematic	34
Total	367	105 (29%)	3625	198 (5%)

*Includes only freshwater fishes.

a taxonomic group had been assessed and the data sheets written, all the information was sent to the primary editors for final processing.

In contrast to the first two editions, where a relatively small number of taxa were assessed (Table 2.1), for the 2008 edition all species of mammals, birds, amphibians, reptiles, and freshwater fishes were evaluated (available in pdf at www.nationalredlist.org). Marine fishes, particularly sharks, rays, chimaeras, and related species (class Chondrichthyes), could not be assessed due to lack of information, so the Venezuelan species included in the global IUCN Red List (IUCN, 2007) were identified, and the global category used as their national category. Given that several Venezuelan marine fishes in the IUCN Red List were cosmopolitan or at least widely distributed in the Caribbean, species where data for Venezuela were virtually non-existent were classified as Data Deficient (DD). Only a small sample of invertebrates was assessed in the third edition. This included those species in the first two editions, those included in international lists or other publications (such as IUCN, 2007) that reported Venezuela as part of their range, and any species recommended by an associate editor.

The two main products were the Red List of Venezuelan Fauna (RLVF) and a collection of species summary data sheets (Rodríguez and Rojas-Suárez, 2008). The RLVF includes all species listed in any category, with the exception of Least Concern (LC) and Not Evaluated (NE). All Venezuelan mammals, birds, reptiles, amphibians, and freshwater fishes described at the time of the assessments that do not appear on the Red List are listed as LC, because these groups were comprehensively assessed. In contrast, as the assessments of marine fishes and invertebrates were not comprehensive, species in these groups not included in the Red List were considered NE. The only exceptions were 11 insects, one crustacean and one bivalve, that after being assessed were judged to be LC (Rodríguez and Rojas-Suárez, 2008).

Assessments were carried out only at the level of species, using current categories, criteria, and regional guidelines (IUCN, 2001, 2003). For species that were abundant in the wild, widely distributed, or not believed to be threatened (CR, EN, or VU) or NT in the foreseeable future, LC listings were often assigned without collecting the full range of data required for a complete assessment. However, assessors were required to fully document and justify the criteria used in assigning risk categories. Afterwards, the primary associate editors reviewed all the assessments and made the necessary adjustments to ensure consistency among assessors, which meant that in some cases the final categories did not correspond to those proposed in the original assessments. The primary taxonomic reference used was the IUCN Red List of Threatened Species (IUCN, 2007). Our objective was to stay as close as possible to this publication, in order to facilitate transfer of information between national and global databases (Miller et al., 2007; Rodríguez, 2008; Rodríguez et al., 2000).

Temporal changes in the number of species listed

The 2008 RLVF includes 748 species: 4 regionally or globally extinct (EX, ER), 198 threatened (CR, EN, VU), 138 NT, and 408 DD (Rodríguez and Rojas-Suárez, 2008). This is more than double the number in the 1999 RLVF, which included 367 species: 3 possibly extinct (EX?), 102 threatened, 69 NT, 101 DD and 92 in other categories (Rodríguez and Rojas-Suárez, 1999). At least three mechanisms may have contributed to the increase in the number of species in the RLVF (Table 2.1): (i) generalized decline in the conservation status of species; (ii) increase in the number of species assessed; and (iii) improvement in knowledge about species in general or threatened species in particular. Although it is likely that the three mechanisms act jointly, in the case of amphibians it is clear that they have undergone a worldwide decline during the last few decades (Stuart et al., 2004). When the first and second editions of the *Red Book of Venezuelan Fauna* were published, no extinct amphibians had been reported, and seven were listed as threatened. Currently, 26 Venezuelan amphibians are threatened and one is extinct. The general change in status of amphibians in Venezuela is likely to be part of this global amphibian decline, a phenomenon that is still not fully understood (Bielby et al., 2008; Blaustein and Wake, 1990).

While it is possible that groups other than amphibians may have also declined in conservation status, the growth of the RLVF is probably due to a significant expansion of the data available. On the one hand, all species in five major taxonomic groups – mammals, birds, reptiles, amphibians, and freshwater fishes – were assessed, contrasting with the selective assessments of the past. On the other hand, one of the most important effects of the previous editions has been the growing interest of the scientific community, governmental agencies, and NGOs in research and conservation on threatened species. Combined with innovative and reliable research funding mechanisms, such as the *Iniciativa Especies Amenazadas* led by Provita, there has been a notable growth in the availability of information for species risk assessments (Giraldo *et al.*, 2009).

Societal impact of the Red Book of Venezuelan Fauna

Soon after the publication of the first edition, a major Venezuelan book foundation, Fundalibro, selected the *Red Book of Venezuelan Fauna* as the best educational book for a general audience of 1996. In the same year, the Ministry of the Environment and Natural Resources published two official decrees, listing all species banned for hunting and those in danger of extinction, respectively (Venezuela, 1996a, 1996b). A quick examination of the decrees showed that the official lists had been influenced by the contents of the book, demonstrating a rapid translation of conservation science results into public policy.

A few years later, in 2006, one of the most important Venezuelan newspapers, *El Nacional*, in commemoration of its 63rd anniversary, published a list of the 63 most influential books in Venezuela in the previous 63 years. Only two scientific books were chosen, the *Guide to the Birds of Venezuela* (Phelps and Meyer de Schauensee 1994) and the *Red Book of Venezuelan Fauna*. But perhaps the greatest sign of success came in 2008, when the Central Bank put in circulation new bills, all of them depicting illustrations of threatened species taken directly from the *Red Book of Venezuelan Fauna* (Figure 2.3). Newspapers and other media regularly cite this publication, schools use it throughout the country, the National Library has placed copies in all its branches, and the public frequently requests it in book stores. Whether any of this has improved the status of threatened species is hard to determine, but what is clear is that there is a firm demand for information, and that it appeals to a general audience.

Future assessments and analyses

The next versions of the *Red Book of Venezuelan Fauna* will probably follow a similar process to that of the IUCN Red List of Threatened Species, moving to an online format where data can be updated regularly, and summaries and analyses published every 4–5 years. In the meantime, other analyses continue, such as the calculation of a Red Index for the Venezuelan Fauna, one of the indicators that must be reported by all parties to the Convention on Biological Diversity (Butchart *et al.*, 2007, 2010b). Additionally, in collaboration with Shell Venezuela and Lenovo (Venezuela), we recently published the *Libro Rojo de los Ecosistemas Terrestres de Venezuela* (*Red Book of Venezuelan Terrestrial Ecosystems*; Rodríguez *et al.*, 2010), while the second edition of the *Red Book of Venezuelan Flora* is also in the works, with its publication expected in 2013.

Figure 2.3 Venezuelan currency depicting threatened species. From top left to bottom right: Amazon River dolphin (*Inia geoffrensis*), giant armadillo (*Priodontes maximus*), harpy eagle (*Harpia harpyja*), hawksbill turtle (*Eretmochelys imbricata*), spectacled bear (*Tremarctos ornatus*), and red siskin (*Carduelis cucullata*). From Rodríguez and Rojas-Suárez, 2008.

The Red Data Book for the mammals of South Africa

South Africa is ranked the third most biologically diverse country on Earth based on an index of species diversity and endemism, and is one of 12 megadiverse countries that collectively contain more than two-thirds of global biodiversity (Groombridge and Jenkins, 2002). South Africa occupies only 2% of the world's land area yet contains a significant share of global biodiversity, being home to nearly 10% of the planet's plants and 7% of its reptiles, birds, and mammals. As a means of tracking trends in the status of species and their populations, as well as overall biodiversity loss, the use of threatened species lists has become popular in recent years in South Africa. IUCN Red List assessments are one of the most widely recognized forms of threatened species lists and document and highlight biodiversity losses at the species level, thereby becoming important tools for guiding the conservation activities of governments and conservation organizations.

The previous South African Red Data Book for Mammals was published in 1986 (Smithers, 1986) and covered terrestrial mammals only. In the years between 1986 and 2004, South Africa experienced changes to its provincial borders, amendments to the taxonomic classification of many species were made, knowledge and understanding of species and trends improved, the tools for data accumulation and management vastly improved, and the IUCN Red List categories and criteria applied in the process of assessing and assigning threat status to species changed (IUCN, 2001). Therefore there was a need for a comprehensively updated publication assessing all terrestrial and marine mammals in South Africa. A comprehensive, scientifically sound publication providing updated information on the status of South African mammals was called for by the local conservation community and identified as a critical step towards setting conservation and management priorities, identifying threatened species and their habitats, and ensuring more effective conservation and management of species and their habitats.

In order to update the South African National Red List, a Conservation Assessment and Management Plan (CAMP) framework tool was used, developed by the Conservation Breeding Specialist Group (CBSG) of the IUCN for strategic conservation planning and the management of species and their habitat. A CAMP is a rapid, broad, comprehensive, and scientifically informed assessment of the taxonomic groups of a region or country using the IUCN Red List criteria to categorize the level of extinction risk facing species. In order to update the South African Red Data Book for Mammals, the CBSG's CAMP process was identified as being the most comprehensive and suitable means of assessing the mammals. CBSG Southern Africa, a regional CBSG network, and the Endangered Wildlife Trust (EWT) coordinated and managed the project, which included running the CAMP workshop and producing the resulting publication.

Importantly, the resulting Red Data Book of the Mammals of South Africa does not simply cover the threatened mammals but also includes a comprehensive conservation assessment for all species of terrestrial and marine mammals in South Africa. This decision to cover all South African mammals was taken for several important reasons. Firstly, while determining which threatened species should be assessed may be relatively easy, there are a number of species falling into 'grey' areas where it was not so easy to determine where the line between threatened and not threatened lies. Secondly, and perhaps more importantly, in order for this publication to serve as a useful indicator of biodiversity trends, including allowing trends in species populations and habitat to be effectively tracked, obtaining baseline data for every mammal species was essential. This would enable future updates of the publication to provide a means of identifying trends, measuring conservation success or failure across the entire range of mammals, and highlighting areas of concern indicated by trends in both the common as well as the threatened mammals. A total of 295 terrestrial and marine species and subspecies of mammals was reviewed, and species were evaluated within South African borders only, and excluded Swaziland and Lesotho.

In summary, the goals of the project were to:

- Update the current scientific knowledge on, and level of threat facing the mammals of South Africa.
- Assess all terrestrial and marine mammals, providing a baseline dataset for each.

- Publish an updated Red Data Book for South African mammals.
- Formulate recommendations for strategic conservation and management of threatened mammal species and their habitats.
- Improve the effectiveness and synergy of existing conservation efforts.

The assessment process

In January 2002, almost 90 South African mammal conservation practitioners, biologists, and taxonomists were invited to participate in the South African Mammal CAMP. Participants were identified by a small project steering committee based on their knowledge of and available data on the species being assessed. The project committee comprised experts in relevant fields including zoology, Red Listing and/or project management. The committee was initially constituted by five coordinating editors, eight thematic editors, and two advisors, but over time, some people fell away due to the length and complexity of the project.

Invitees were asked to collect relevant data on species within their areas of expertise and these data were submitted to the project manager in CBSG-developed CAMP taxon datasheets (TDS). These data pertained to the species' distribution, habitat, population status and trends, breeding and feeding characteristics, threats, and other criteria relevant to the Red Listing process, and they also included all available references and research findings. The completed TDS were submitted to CBSG Southern Africa in advance of the CAMP workshop in the form of hard copies of the TDS, which were then collated and categorized according to species and subspecies in preparation for the workshop.

The six-day CAMP workshop was held in Johannesburg in March 2002 and was attended by 33 participants with data having been submitted by an additional 27 contributors. A total of 35 organizations participated in the entire process, including South African National Parks, various Provincial Parks Authorities, research organizations, non-governmental organizations, national and provincial museums, academic institutions, private organizations, and governmental departments. The workshop was preceded by a three-day training workshop in the application of the IUCN Red List criteria and categories of threat. Participants were then asked to convene working groups based on taxonomic groupings, in which the data submitted to the project manager were evaluated, debated, and considered as part of the process of assigning a category of threat to each species.

A final, all-inclusive taxon datasheet for each species, and some subspecies, was completed by the groups, which were presented and discussed by the group as a whole in plenary sessions. A final assessment of the Red List category for each species or subspecies was then entered into the electronic database. Distribution maps compiled using Geographical Information System (GIS) were collated for all extant terrestrial species during and after the workshop by a dedicated GIS working group (which informed the assessments). Distributional records were collated using all available data sources contained in the SA-ISIS (South African Integrated Spatial Information System) mammal database. The distributions were generalized to quarter-degree square grid, which in turn were used to produce presence maps for individual species and subspecies within South Africa. The extent of occurrence (EOO) for each species and subspecies was derived from specialists' recent information on the most likely distribution of the species over approximately the last 10 years (1990–2000). Quarter-degree grid squares outside the EOO were validated and updated at appropriate resolution. Post-workshop review and editing was then carried out, particularly for species where consensus had not been achieved, in a process very similar to that undertaken during global Red Listing (http://www.iucnredlist.org/technical-documents/assessment-process).

Status of South African mammals

Of the 295 species and subspecies evaluated, 57 (~18%) were classified in threatened categories (CR, EN, VU: Table 2.2): 3% Critically Endangered, 6% Endangered and 10% Vulnerable. Fifty-one (17%) mammal species were assessed as Data Deficient.

Table 2.2 **Number of assessed and threatened mammals in South Africa. With permission from Endangered Wildlife Trust and Conservation Breeding Specialist Group**

Order	CR	EN	VU	NT	LC	DD	No. assessed	Proportion threatened
Artiodactyla	0	2	5	1	25	0	33	0.21
Carnivora	0	2	2	7	25	2	38	0.11
Cetacea	0	2	4	1	12	23	42	0.14
Chiroptera	2	2	6	19	19	2	50	0.20
Hyracoidea	0	0	1	0	2	0	3	0.33
Insectivora	5	4	5	4	1	14	33	0.42
Lagomorpha	1	0	0	0	6	0	7	0.14
Macroscelidae	0	1	0	0	4	2	7	0.14
Perissodactyla	1	1	2	0	2	0	6	0.67
Pholidota	0	0	1	0	0	0	1	1.00
Primates	0	1	2	0	4	0	7	0.43
Proboscidea	0	0	0	0	1	0	1	0.00
Rodentia	1	3	1	7	46	8	66	0.08
Tubilidentata	0	0	0	0	1	0	1	0.00
Total	10	18	29	39	148	51	295	0.19

CR, Critically Endangered; EN, Endangered; VU, Vulnerable; NT, Near Threatened; LC, Least Concern; DD, Data Deficient.

Table 2.3 **Management recommendations stemming from the national Red Data Book evaluation of South African mammals (Friedmann and Daly, 2004). With permission from Endangered Wildlife Trust and Conservation Breeding Specialist Group**

Management recommendation	No. of species
Population and habitat viability assessments	27
Captive breeding	8
Wild population management	79
Habitat management	136
Research	248
Monitoring	182

Thirty-eight (13%) species were assessed as Near Threatened, and the remaining 147 (50%) as Least Concern.

A range of management recommendations were devised through the Red Listing process. Of the recommendations, the vast majority concerned establishing greater monitoring networks to track populations, and conducting further research into various aspects of species biology, threat status, trends, and taxonomy (for smaller mammals). Habitat management was also considered a critical aspect of future management (Table 2.3). The primary threats causing decline in many mammals include habitat loss and land transformation through deforestation, agriculture, timber planting, and urban and industrial development. Poisoning, pollution, and hunting have also been listed as having a negative impact on a number of mammals (Friedmann and Daly, 2004).

Conservation actions

As a complementary process to the data collection for assessments, participants collated a range of concerns and priorities based on the work and discussions of the workshop, as well as long-standing concerns of the South African mammal community. Issues were collated into the following four thematic working groups:

1. Public education and awareness.
2. Information management and database initiatives.
3. Conservation management.
4. Research and capacity building.

Participants then identified conservation goals for addressing each issue, with particular reference to what is needed to strengthen preparation of the next Red List update, and what might decrease the risk of extinction of mammals and loss of biodiversity in South Africa. This resulted in around 20 key recommendations, ranging in issues from developing species-specific national management plans, creating monitoring initiatives and a national mammal survey, capacity building, interdisciplinary research efforts, and habitat restoration (Friedmann and Daly, 2004).

Future assessments and analyses

The South Africa Mammal CAMP process, from beginning to end, was successful in collating invaluable data and generating many ideas and resolutions. The final publication (Friedmann and Daly, 2004) was the result of almost 18 months of intensive internal and external data review and editing. A commitment was made during the project to review and update the publication within 5 years in order to keep it relevant and meaningful in a rapidly changing world. Many improvements have since been made to the database housing the species information and listings, which will result in a vastly different workshop and listing process being followed in the next iteration of the publication. The fundamental changes include the recent development of an online species listing database by the Endangered Wildlife Trust, which allows not only data contributors to enter their data directly into the database from any location, but also multiple contributors with password-controlled access to input data, and comment on, add to, or change other data contributions in such a way that all additions and alterations are tracked and monitored. This will no doubt speed up the process of participants arriving at complete and consolidated datasheets for all species and allow any workshop time to focus only on the Red Listing process. The online system will also facilitate periodic reviews of various species using the expertise and input of experts at intervals between workshops and will facilitate transparency and consensus building in the review and listing process.

Critique of strengths and weaknesses

Strengths

The IUCN Red List has a reputation as an objective and widely applicable method of assessing the relative risk of extinction of species, and is certainly the most high profile and widely used. Accurate extinction risk assessment is vital to determining conservation priorities and establishing biodiversity management plans at all governance levels, and as Red List conservation assessments are made at the global scale, it is not always possible to integrate the information they provide into conservation planning and priority-setting at the national scale. National Red Lists are based on the same objective IUCN categories and criteria, with modifications to make them suitable for the sub-global level. Trends at global and regional level can also be used as a proxy indicator of the state of the ecosystems that the species inhabit, and can be used as biodiversity indicators in a broader sense, rather than the state of species only (Butchart *et al.*, 2010b).

Using composite measures of the symptoms of extinction risk to generate standard classifications of risk enables broad species coverage of Red List assessments. Species have been assessed from groups as different as marine corals to terrestrial mammals, alpine plants to freshwater molluscs. With sufficient funding, comprehensive assessments of entire groups are possible in a reasonable timeframe (Jones *et al.*, 2011), with a sampled approach to Red Listing providing meaningful output for the more diverse groups (Clausnitzer *et al.*, 2009; Collen *et al.*, 2009b).

The outputs of a National Red Listing process do not stop at the production of a list of species and corresponding extinction risks classification. Additional outputs include agreed national taxonomic lists and a biodiversity database (often based on the IUCN Species Information Service), which provides a baseline for future assessments and conservation action plans, thereby building local capacity. When producing a National Red List, a wide range of sources are utilized, including published reports, unpublished data, field studies, and input from a range of experts and global networks. This provides detailed supporting information, which was previously widely dispersed and of limited availability, creating an invaluable data resource on national biodiversity. As National Red Lists are supplemented with such species-specific information including threats and impacts, these can be combined with the National Red List Index and factors driving change can be identified (Butchart *et al.*, 2005). In Mongolia, interactive geographical species search tools have been developed alongside the national Red List, which allows species distribution searches of the region using GIS information gathered during the National Red List workshops (see www.nationalredlist.org/mongolia). National Red Lists also contribute to more informed and evidence-based national policies through advocacy and lobbying with local governments, enhanced capacity through training of local scientists, and strengthened collaborations between institutions in the region.

National Red Lists have been utilized in many varied applications. They have been used to monitor biodiversity change through calculating the National Red List Index (Gärdenfors, 2010; Xu *et al.*, 2009), and they also measure the success of conservation projects that have already been implemented. National Red Lists have been used to influence national and international conventions and polices, including the Convention on International Trade in Endangered Species of Wild Fauna and Flora (CITES), the Convention on Migratory Species (CMS), and the Convention on Biological Diversity (CBD), and also in aspects of trade, protection, and development projects (Collar, 1996). By feeding knowledge on the conservation status of endemic species into the global Red List, National Red Lists are enhancing global knowledge about species and contributing to the growth of the global IUCN Red List (Gärdenfors, 2001; Gärdenfors *et al.*, 2001). Even in the case of non-endemic species, supporting information can be included in a global species account, providing vital information about a species throughout its range (Zamin *et al.*, 2010). National Red Lists have contributed to national and international conservation priority setting and resource management by the identification of key ecosystems, specific issues, threats, and key habitats and have supported protected area planning by indicating vulnerable, sensitive, and rare habitats (Rodrigues *et al.*, 2006). National Red Lists also have the potential to be used in conservation planning adaptations under different climate change scenarios (Zamin *et al.*, 2010). Listing all species, including those that are Data Deficient, has enabled specific research topics to be identified and gaps highlighted (Collen *et al.*, 2008). Foremost, regional extinctions are prevented by identification of species under threat and the recruitment of conservation attention.

Weaknesses and solutions

One source of limitation is discrepancies between the global IUCN Red List and National Red Lists, with debate over which are the more accurate (Hilton-Taylor *et al.*, 2000; Rodríguez *et al.*, 2000). Unfortunately, direct comparisons are few and far between (though see Milner-Gulland *et al.*, 2006), but it has also been commented that National Red Lists incorporate global Red List information more frequently than vice versa; an improvement of the flow of this information is considered essential (Rodríguez *et al.*, 2000). While there are many legitimate reasons for national and global listings to differ, in a worst case scenario, such inconsistency between global and National Red Lists could lead to an increased extinction risk if a species were overlooked. It would also be detrimental to the credibility of Red Listing and could also have repercussions if international funders consider one or other of the Red Listing processes to be more accurate (Rodríguez *et al.*, 2000).

Lack of understanding and training regarding the categories and criteria has been identified as one of the primary obstacles for National Red Listing (Miller *et al.*, 2007). Interpretation of the National Red List Categories and Criteria can differ between organizations and countries (Zamin *et al.*, 2010). For example, Gärdenfors (2001) found different interpretations of the Data Deficient category when comparing the National Red Lists of Finland and Sweden. As a result DD classified species were not included on the Finnish Red List (Gärdenfors, 2001). Other issues concerning interpretation and understanding of the criteria included Criterion A thresholds being met at a national level by apparently globally non-threatened species, vague interpretations of area of occupancy in Criterion B, and difficulties in defining 'individuals' when considering plants and fungi (Gärdenfors, 2001). Obtaining detailed information about the emigration and immigration of a population is also difficult, but an essential step to the National Red Listing process. Keller *et al.* (2005) therefore suggested a more formalized approach to downgrading and upgrading using a set of questions. Certainly, as assessors differ in interpretation of criteria and use different methods when preparing a National Red List, comparing National Red Lists can be problematic (Miller *et al.*, 2007; Possingham *et al.*, 2002). The advantage of the modern IUCN system is its transparency, so when discrepancies occur, the causes can be determined (Regan *et al.*, 2004). With the previous system based largely on expert subjective judgement, this was not necessarily the case.

Although there are many factors that influence the production of a National Red List, the availability of financial resources has been identified as the biggest limiting factor, reinforcing the assumption that the regions with the highest biodiversity are the very regions with limited financial resources (Zamin *et al.*, 2010). National Red Lists by their very nature need constant revision to be applicable and accurate, and to produce a National Red List multiple times is a labour-intensive and costly task, and is a project that few nations are able to undertake. A possible solution is a sampled approach, conducting Red List assessments on randomly selected species that will be representative of the entire taxonomic group (Baillie *et al.*, 2008). Thus far this has not been attempted for a sub-global group.

Invertebrates and plants are generally underrepresented in the IUCN global Red List (Baillie *et al.*, 2008; Vie *et al.*, 2009) and little is known about the status and trends of the majority of species comprising ecosystems globally, with high numbers of Data Deficient species in speciose and elusive groups. However, coverage in National Red Lists appears more even, with plants particularly well covered (Zamin *et al.*, 2010). Regardless, in general, conservation priorities often focus on large, charismatic species, and not necessarily those that require the most attention. Such sampling bias could result in disproportionate attention being paid to certain species and not those that are essential to the functioning of a healthy ecosystem.

Proactive conservation action requires data to be widely available. Although many countries throughout the world have developed national threatened species lists (either using IUCN criteria, or other similar frameworks), this information is often inaccessible (Collen *et al.*, 2008). As conservation efforts are hindered by the limited availability of such species information, a National Red List website and database has been recommended for over a decade to improve communication between Red List assessors, allowing National Red List information to be more easily accessible and facilitate online discussions (Miller *et al.*, 2007; Rodríguez, 2008; Rodríguez *et al.*, 2000). Launched in 2009, the National Red List website and database (www.nationalredlist.org) seeks to document all existing national species' assessments and conservation action plans, thereby increasing the effectiveness of conservation planning. This capacity-building website is the first central source of national-level biodiversity data, featuring an online forum, library, educational tools, and over 69 000 species accounts from 45 countries and regions to date.

Acknowledgements

We are grateful to the vast network of scientists producing National Red Lists. BC and JG are supported by the Rufford Foundation. For production of the Mammals of South Africa we would like to thank the Endangered Wildlife Trust, the Conservation Breeding Specialist Group (Species Survival Commission/IUCN), Vodacom, the National Research Foundation, the Lomas Wildlife Trust, the Red List Office of the IUCN/SSC, all CAMP participants and contributors, the participating organizations, and the project committee and editorial contributors. Provita is a local conservation partner of the EcoHealth Alliance, formerly known as Wildlife Trust.

References

Baillie, J.E.M., Hilton-Taylor, C., and Stuart, S.N. (2004) *2004 IUCN Red List of Threatened Species: a Global Species Assessment*. IUCN, Gland, Switzerland, and Cambridge, UK.

Baillie, J.E.M., Collen, B., Amin, R., *et al.* (2008) Towards monitoring global biodiversity. *Conservation Letters*, 1, 18–26.

Baillie, J.E.M., Griffiths, J., Turvey, S.T., Loh, J., and Collen, B. (2010) *Evolution Lost: Status and Trends of the World's Vertebrates*. Zoological Society of London, 72 pp.

Bielby, J., Cooper, N., Cunningham, A.A., Garner, T.W.J., and Purvis, A. (2008) Predicting declines in the world's frogs. *Conservation Letters*, 1, 82–90.

BirdLife International (2008) *State of the World's Birds: Indicators for Our Changing World*. BirdLife International, Cambridge, UK.

BirdLife International (2010) State of the world's birds. BirdLife International, Cambridge, UK. Available at: http://www.birdlife.org/sowb.

Blaustein, A.R. and Wake, D.B. (1990) Declining amphibian populations – a global phenomenon. *Trends in Ecology and Evolution*, 5, 203–204.

Brummitt, N., Bachman, S.P., and Moat, J. (2008) Applications of the IUCN Red List: towards a global barometer for plant diversity. *Endangered Species Research*, 6, 127–135.

Bubb, P.J., Butchart, S.H.M., Collen, B., *et al.* (2009) *IUCN Red List Index – Guidance for National and Regional Use*. IUCN, Gland, Switzerland.

Butchart, S.H.M., Stattersfield, A.J., Bennun, L.A., *et al.* (2004) Measuring global trends in the status of biodiversity: Red List Indices for birds. *PLoS Biology*, 2, 2294–2304.

Butchart, S.H.M., Stattersfield, A.J., Baillie, J.E.M., *et al.* (2005) Using Red List Indices to measure progress towards the 2010 target and beyond. *Philosophical Transactions of the Royal Society of London B*, 360, 255–268.

Butchart, S.H.M., Akcakaya, H.R., Chanson, J.S., *et al.* (2007) Improvements to the Red List Index. *PLoS ONE*, 2:e140. doi:110.1371/journal.pone.0000140.

Butchart, S.H.M., Baillie, J.E.M., Chenery, A., *et al.* (2010a) Response to Xu et al. *Science*, 329, 900–901.

Butchart, S.H.M., Walpole, M., Collen, B., *et al.* (2010b) Global biodiversity: indicators of recent declines. *Science*, 328, 1164–1168.

Carpenter, K.E., Abrar, M., Aeby, A., *et al.* (2008) One-third of reef-building corals face elevated risk from climate change and local impacts. *Science*, 321, 560–563.

Clausnitzer, V., Kalkman, V.J., Ram, M., *et al.* (2009) Odonata enter the biodiversity crisis debate: the first global assessment of an insect group. *Biological Conservaion*, 142, 1864–1869.

Collar, N.J. (1996) The reasons for Red Data Books. *Oryx*, 30, 121–130.

Collar, N.J., Gonzaga, L.P., Krabbe, K., *et al.* (1992) *Threatened Birds of the Americas. The ICBP/IUCN Red Data Book*. International Council for Bird Preservation, Cambridge, UK, 1150 pp.

Collen, B. and Baillie, J.E.M. (2010) The barometer of life: sampling. *Science*, 329, 140.

Collen, B., Ram, M., Zamin, T., and McRae, L. (2008) The tropical biodiversity data gap: addressing disparity in global monitoring. *Tropical Conservation Science*, 1, 75–88. Available at: tropicalconservationscience.org.

Collen, B., Loh, J., Holbrook, S., McRae, L., Amin, R., and Baillie, J.E.M. (2009a) Monitoring change in vertebrate abundance: the Living Planet Index. *Conservation Biology*, 23, 317–327.

Collen, B., Ram, M., Dewhurst, N., *et al.* (2009b) Broadening the coverage of biodiversity assessments. In: Vié, J.-C., Hilton-Taylor, C., and Stuart, S.N. (eds), *Wildlife in a Changing World – An Analysis of the 2008 IUCN Red List of Threatened Species*. IUCN, Gland, pp. 67–76.

Collen, B., Turvey, S.T., Waterman, C., Meredith, H.M.R., Baillie, J.E.M., and Isaac, N.J.B. (2011) Investing in evolutionary history: implementing a phylogenetic approach for mammal conservation. *Philosophical Transactions of the Royal Society of London B*, 366, 2611–2622.

Convention on Biological Diversity (2010a) COP 10 Decision X/2 – Strategic Plan for Biodiversity 2011–2020. Convention on Biological Diversity, Montreal, Canada. Available at: http://www.cbd.int/decision/cop/?id=12268.

Convention on Biological Diversity (2010b) *Global Biodiversity Outlook 3*. UNEP.

Convention on Biological Diversity (2011) Strategic plan for biodiversity 2011–2020, including Aichi biodiversity targets. Available at: http://www.cbd.int/sp/.

Convention on Migratory Species (2008) Report of the 15th meeting of the scientific council of the convention on the conservation of migratory species of wild animals. United Nations Environment Programme, Rome, Italy. Available at: http://www.cms.int/bodies/ScC/Reports/Eng/ScC_report_15_E.pdf.

Cumberlidge, N., Ng, P.K.L., Yeo, D.C.J., *et al.* (2009) Freshwater crabs and the biodiversity crisis: importance, threats, status, and conservation challenges. *Biological Conservation*, 142, 1665–1673.

Fitter, R. and Fitter, M. (1987) *The Road to Extinction*. International Union for the Conservation of Nature, Gland, Switzerland.

Friedmann, Y. and Daly, B. (eds) (2004) Red Data Book of the Mammals of South Africa: A Conservation Assessment. *CBSG Southern Africa, Conservation Breeding Specialist Group (SSC/IUCN)*, Endangered Wildlife Trust, South Africa, pp. 722.

Fudena (1989) *Hacia una Estrategia Nacional de Conservación: Plan de Acción para la Conservación de Especies*. Fundación para la Defensa de la Naturaleza (Fudena), Caracas, Venezuela, 82 pp.

Gärdenfors, U. (2001) Classifying threatened species at national versus global levels. *Trends in Ecology and Evolution*, 16, 511–516.

Gärdenfors, U. (ed.) (2010) Rödlistade arter i Sverige 2010 /The 2010 Red List of Swedish Species/. ArtDatabanken, SLU, Uppsala, 533 pp.

Gärdenfors, U., Hilton-Taylor, C., Mace, G.M., and Rodriguez, J.P. (2001) The application of IUCN Red List criteria at regional levels. *Conservation Biology*, 15, 1206–1212.

Giraldo, D., Rojas-Suárez, F., and RomeroV. (2009) *Una Mano a la Naturaleza: Conservando las Especies Amenazadas Venezolanas*. Provita y Shell Venezuela, Caracas, Venezuela, 220 pp.

Gondelles, R., Medina, G., Méndez-Arocha, J.L., and Rivero-Blanco, C. (1981) *Nuestros Animales de Caza, Guía para su Conservación*. Fundación de Educación Ambiental (MARNR), 119 pp.

Groombridge, B. (1982) *The IUCN Amphibia-Reptilia Red Data Book. Part 1: Testudines, Crocodylia, Rhynchocephalia*. IUCN, Gland, Switzerland, 426 pp.

Groombridge, B. and Jenkins, M.D. (2002) *World Atlas of Biodiversity*. University of California Press, Berkeley.

Hilton-Taylor, C., Mace, G.M., Capper, D.R., *et al.* (2000) Assessment mismatches must be sorted out – they leave species at risk. *Nature*, 404, 541.

Hoffmann, M., Hilton-Taylor, C., Angulo, A., *et al.* (2010) The impact and shortfall of conservation on the status of the world's vertebrates. *Science*, 330, 1503–1509.

Isaac, N.J.B., Mallet, J., and Mace, G.M. (2004) Taxonomic inflation: its influence on macroecology and conservation. *Trends in Ecology and Evolution*, 19, 464–469.

IUCN (1986) *1986 IUCN Red List of Threatened Animals*. IUCN, Gland, Switzerland, 105 pp.

IUCN (1988) *1988 IUCN Red List of Threatened Animals*. IUCN, Gland, Switzerland, 172 pp.

IUCN (1994) *IUCN Red List Categories*. IUCN, Gland, Switzerland.

IUCN (2001) *IUCN Red List Categories and Criteria version 3.1*. IUCN, Gland, Switzerland.

IUCN (2003) Guidelines for application of IUCN Red list criteria at regional levels version 3.0. IUCN, Gland, Switzerland.

IUCN (2007) 2007 IUCN Red List of Threatened Species. Available at: www.iucnredlist.org.

IUCN (2010) IUCN Red List of Threatened Species. Version 2010.3. Available at: www.iucnredlist.org.

Jones, J.P.G., Collen, B., Atkinson, G., *et al.* (2011) The why, what and how of global biodiversity indicators beyond the 2010 target. *Conservation Biology*, 25, 450–457.

Keller, V., Zbinden, N., Schmid, H., and Volet, B. (2005) A case study in applying the IUCN Regional Guidelines for National Red Lists and justifications for their modification. *Conservation Biology*, 19, 1827–1834.

Llamozas, S., Duno de Stefano, R., Meier, W., *et al.* (2003) *Libro Rojo de la Flora Venezolana*. PROVITA, Fundación Polar y Fundación Instituto Botánico de Venezuela, Caracas, Venezuela, 555 pp.

Mace, G.M. and Lande, R. (1991) Assessing extinction threats – toward a reevaluation of IUCN Threatened Species categories. *Conservation Biology*, 5, 148–157.

Mace, G.M., Collar, N.J., Gaston, K.J., *et al.* (2008) Quantification of extinction risk: IUCN's system for classifying threatened species. *Conservation Biology*, 22, 1424–1442.

McGeoch, M.A., Butchart, S.H.M., Spear, D., *et al.* (2010) Global indicators of biological invasion: species numbers, biodiversity impact and policy responses. *Diversity and Distributions*, 16, 95–108.

Miller, R.M., Rodríguez, J.P., Aniskowicz-Fowler, T., *et al.* (2006) Extinction risk and conservation priorities. *Science*, 313, 441.

Miller, R.M., Rodríguez, J.P., Aniskowicz-Fowler, T., *et al.* (2007) National Threatened Species listing based on IUCN criteria and regional guidelines: current status and future perspectives. *Conservation Biology*, 21, 684–696.

Milner-Gulland, E.J., Kreuzberg-Mukhina, E., Grebot, B., *et al.* (2006) Application of IUCN red listing criteria at the regional and national levels: A case study from Central Asia. *Biodiversity and Conservation*, 15, 1873–1886.

Phelps, W.H.J. and Meyer de Schauensee, R. (1994) *Una Guía de las Aves de Venezuela*. Editorial Ex Libris, Caracas, Venezuela.

Pimm, S.L., Ayres, M., Balmford, A., *et al.* (2001). Can we defy nature's end? *Science*, 293, 2207–2208.

Possingham, H.P., Andelman, S.J., Burgman, M.A., Medellin, R.A., Master, L.L., and Keith, D.A. (2002) Limits to the use of threatened species lists. *Trends in Ecology and Evolution*, 17, 503–506.

Regan, T.J., Master, L.L., and Hammerson, G.A. (2004) Capturing expert knowledge for threatened species assessments: a case study using NatureServe conservation status ranks. *Acta Oecologia*, 26, 95–107.

Rodrigues, A.S.L., Pilgrim, J.D., Lamoreux, J.F., Hoffmann, M., and Brooks, T.M. (2006) The value of the IUCN Red List for conservation. *Trends in Ecology and Evolution*, 21, 71–76.

Rodríguez, J.P. (2008) National Red Lists: the largest global market for IUCN Red List Categories and Criteria. *Endangered Species Research*, 6, 193–198.

Rodríguez, J.P. and Rojas-Suárez, F. (1995) *Libro Rojo de la Fauna Venezolana*, 1st edn. PROVITA, Fundación Polar, Caracas, Venezuela, 444 pp.

Rodríguez, J.P. and Rojas-Suárez. F. (1999) *Libro Rojo de la Fauna Venezolana*, 2nd edn. PROVITA, Fundación Polar, Caracas, Venezuela, 444 pp.

Rodríguez, J.P. and Rojas-Suárez. F. (2008) *Libro Rojo de la Fauna Venezolana*, 3rd edn. Provita, Shell Venezuela, Caracas, Venezuela, 364 pp.

Rodríguez, J.P., Ashenfelter, F.R., Fernández, J.J.G., Suárez, L., and Dobson, A.P. (2000) Local data are vital to worldwide conservation. *Nature*, 403, 241.

Rodríguez, J.P., Rojas-Suárez, F., and Giraldo Hernández, D. (eds) (2010) *Libro Rojo de los Ecosistemas Terrestres de Venezuela*. Provita, Shell Venezuela and Lenovo (Venezuela), Caracas, Venezuela, 324 pp.

Romero, A. (1985) Especies en Peligro de Extinción. *Carta Ecológica*, 24, 7–10.

Smithers, R.H.N. (1986) South African red data book – terrestrial mammals. South African National Scientific Programmes Report No.125, Pretoria, 216 pp.

Stuart, S.N., Chanson, J.S., Cox, N.A., *et al.* (2004) Status and trends of amphibian declines and extinctions worldwide. *Science*, 306, 1783–1786.

Szabo, J.K., Butchart, S.H.M., Possingham, H.P., and Garnett, S.T. (2012) Adapting global biodiversity indicators to the national scale: a Red List Index for Australian birds. *Biological Conservation*, 148, 61–68.

Thornback, J. and Jenkins, M. (1982) *The IUCN Mammal Red Data Book, Part 1: Threatened Mammalian Taxa of the Americas and the Australasian Zoogeographic Region (Excluding Cetacea)*. IUCN, Gland, Switzerland, 516 pp.

Trebbau, P. (1985) *10 Especies Venezolanas en Vías de Extinción*. Corpoven, Caracas, Venezuela.

Venezuela (1996a) Decreto 1485: Animales Vedados para la Caza. Gaceta Oficial No. 36.059 – 7 de octubre de 1996, Caracas, Venezuela.

Venezuela . (1996b) Decreto 1486: Especies en Peligro de Extinción. Gaceta Oficial No. 36.062 – 10 de octubre de 1996, Caracas, Venezuela.

Vie, J.-C., Hilton-Taylor, C., and Stuart, S.N. (2009) *Wildlife in a Changing World: an Analysis of the 2008 Red List of Threatened Species*. IUCN, Gland, 180 pp.

Xu, H., Tang, X., Liu, J., *et al*. (2009) China's progress toward the significant reduction of the rate of biodiversity loss. *BioScience*, 59, 843–852.

Zamin, T., Baillie, J.E.M., Miller, R., Rodriguez, J.P., Ardid, A., and Collen, B. (2010) National Red Listing beyond the 2010 target. *Conservation Biology*, 24, 1012–1020.

3

The Wildlife Picture Index: A Biodiversity Indicator for Top Trophic Levels

Timothy G. O'Brien[1,2] *and Margaret F. Kinnaird*[1,2]

[1]Wildlife Conservation Society, New York, USA
[2]Mpala Research Centre, Nanyuki, Kenya

Introduction

Most conservationists agree that tropical regions harbor most of the Earth's biodiversity. Two of the most important biomes in the terrestrial tropics are the rainforests and the tropical savannas. Tropical rainforests occur in five major tropical regions, covering 6.25 million km^2 (~4.1% of the Earth's land surface) and contain >50% of total biodiversity, including the majority of birds and mammals. Tropical savannas occur in six major tropical regions covering 27.6 million km^2 (~18.5% of the Earth's land surface) and, while less diverse than rainforests, they contain significant numbers of mammals and birds, especially the large iconic species. Forests and savannas face severe threats from the usual agents of biodiversity loss including human population growth, biomass consumption for fuel, natural resource extraction, habitat loss, agricultural development, and climate change (Millennium Ecosystem Assessment, 2005).

Recognizing the threats to the Earth's biodiversity, the Convention on Biological Diversity committed 190 countries to achieve a significant reduction in the rate of loss of biodiversity by 2010 (Decision VI/26; CBD Strategic Plan). Along with this commitment came the realization that monitoring change in biodiversity was an extremely complex issue, and that a toolkit of monitoring techniques was needed to track all the different components that make up our notion of biodiversity (Dobson, 2005). As a first step to developing this toolbox, the CBD encouraged the compilation of existing datasets into composite indicators to establish the baseline of knowledge about biodiversity change and to track that change into the future. Problems with this approach have been recognized and discussed (Balmford *et al.*, 2005; Dobson, 2005; Mace and Baillie, 2007) along with the recognition that new indicators may be needed that directly address the needs of biodiversity monitoring.

The CBD call for action presents an opportunity and a dilemma for conservation. The chance to collate the body of acquired knowledge should not be passed up, nor should methods for analysing this wealth of data. But, as Balmford *et al.* (2005) and Dobson (2005) point out, most species and habitats have not been evaluated once, and we require three time steps to detect a change in the rate of change. While the evaluation of existing data is a necessary short-term solution, we need new methods for rapidly collecting new data on species not currently under the microscope of conservation. It is unlikely that we will find datasets that are long-running, unbiased, sensitive to change, and relevant to biodiversity monitoring sitting unused on the shelves in a backroom. We therefore need new data, collected using methods consistent with the goals and objectives of biodiversity monitoring, to address these issues (Collen *et al.*, 2008b). We also need new indicators that are designed to be informative, sensitive, and robust to variation in the underlying data.

A number of authors have tackled the question of ideal datasets and ideal indicators. Yoccoz *et al.* (2001) and Balmford *et al.* (2005), for example, stress the need for indicators based on a representative, randomly selected subset of taxa, collected at a stratified random sample of habitats or sites of interest and corrected for detection bias. Buckland *et al.* (2005) outline desirable statistical properties of indicators. For a group of species, appropriately sampled, the index should not show a trend when abundance of individual species varies but number of species, overall abundance, and species evenness stay constant; the index should decrease if overall abundance, or species evenness, or number of species decreases; and the index should be insensitive to sample size and should have good and measurable precision.

It is unlikely that such an ideal indicator can be developed using available data (Dobson, 2005). The spatial coverage of current datasets is uneven and unrepresentative (Collen *et al.*, 2008b). While most biodiversity resides in the tropics, most data come from the temperate region. We have no Brazilian Wild Bird Index to compare to the UK Wild Bird Index (Gregory *et al.*, 2003). The rate of data publication is slow and uneven, and many relevant datasets are not publicly available. Often, the ability to even search for datasets published in the grey literature and for unpublished datasets is not available. Finally, there is no control over the data collection of published datasets, requiring untestable assumptions about the underlying data quality in order to justify comparisons (O'Brien *et al.*, 2010).

Looking beyond the near-term needs of the CBD to develop 2010 indicators from available data, many have called for the development of new indicators that are designed to meet the specific goals and objectives of biodiversity monitoring. One such proposed composite indicator is the Wildlife Picture Index (WPI: O'Brien *et al.*, 2010; O'Brien, 2010a), based on detection-nondetection data for mammals and birds provided by camera trap surveys. The WPI targets communities of terrestrial forest and savanna birds and mammals that occupy top trophic levels in their ecosystems, and are often the target of exploitation and management. The WPI is based on the collection of primary data in a statistically rigorous manner designed to meet the needs of a composite biodiversity indicator. Here we describe some of the properties of the WPI.

Component of biodiversity: medium to large terrestrial birds and mammals

Of the currently recognized 9702 species of extant birds (Sibley and Monroe, 1990), ~600 species are found in tropical regions and spend a considerable amount of time resting, foraging, and moving on the ground. The average bird (excluding rhea, cassowary, and ostrich) weighs approximately 0.04 kg, so a 0.1 kg bird (30% of species) might be considered a large bird (Dunning, 2008). Camera traps have photographed birds as small as 0.013 kg *Muscicapa* flycatchers and as large as ostriches (O'Brien and Kinnaird, 2008). If we examine the size distribution of tropical terrestrial birds and tropical birds that appear in camera traps (Figure 3.1), we see that, although the 0.5–0.999 kg birds are underrepresented, the overall size distributions of large tropical birds and tropical birds in camera traps do not differ significantly ($\chi^2 = 5.44$, $P = 0.49$). Importantly, it appears that birds in the 0.1–0.499 kg class may be captured reliably by camera traps.

Of 5488 recognized mammals, one-quarter may be considered terrestrial or semi-terrestrial tropical, and weigh more than 0.1 kg. Within this subset, the average body mass is 32 kg (median 0.9 kg, Figure 3.2), with 59% of species weighing 0.5 kg or more (MOM vers. 3.1: Smith *et al.*, 2003). Although not all of these species can be monitored adequately using camera traps, many can. O'Brien *et al.* (2010) have previously defined the Wildlife Picture Index as appropriate for medium- to large-bodied terrestrial birds and mammals. Based on observations from many camera trap datasets (compiled primarily from large felid studies), we believe that the appropriate range of body size for mammals is 0.5 kg upwards, and may be as low as 0.1 kg.

Many of the bird and mammal species in the larger size classes are considered strong interactors (Power *et al.*, 1996) and occupy top trophic levels in their respective food webs. They perform important ecosystem services such as predation, grazing, browsing, seed dispersal, and ecosystem engineering (Dobson *et al.*, 2006). Many of these species also are important to humans as food, pests, and objects of tourism, and therefore are vulnerable to exploitation, consumptive use, and habitat loss. Finally, many of these species are the object of management interventions and regulations, an indication of their importance to humans.

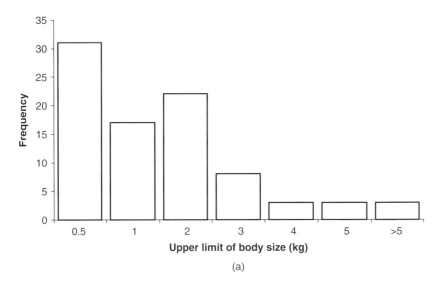

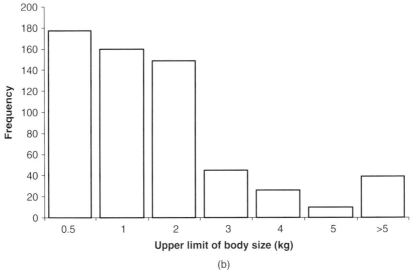

Figure 3.1 Body size distribution of tropical birds (a) in camera traps and (b) of terrestrial tropical birds >0.1 kg. From O'Brien & Kinnaird 2008, unpublished data.

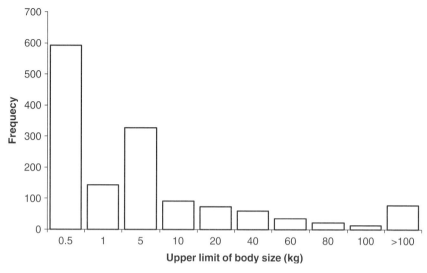

Figure 3.2 **Body size distribution for mammals.**

Terrestrial birds and mammals can be difficult to study in rainforest and savanna ecosystems. Many species are nocturnal, which presents special challenges for monitoring. Many species are rare and/or cryptic, requiring specialized sampling techniques relative to more common species in order to collect comparable data (Thompson, 2004). Many species react negatively to humans, either fleeing or hiding. To adequately sample this diverse group of mammalian and avian species might require a combination of live-trapping and mist-netting, cue counts and sign counts, line transects, and point counts. Low encounter rates can result in imprecise abundance estimates and in some cases might preclude abundance or density estimation. Remote and rugged terrain can compromise coverage, and wariness of humans may reduce detections. Use of indices based on sign and counts violates many principles of monitoring (Yoccoz et al., 2001; Pollock et al., 2002; Buckland et al., 2005) because detection probability is not properly considered.

One solution to the problem of multiple sampling methods in multispecies surveys is to use camera traps as the sampling method (Karanth et al., 2004). Camera traps have a number of features that make them useful as a data collection tool. They are passive data collection units that sit quietly in the study area. Most digital cameras use a silent electronic shutter and an infrared flash that do not scare animals. Camera traps are increasingly designed for deployment in extreme habitats under wet and hot conditions. New generation models employ digital camera technology and the latest advances in battery and power consumption technology, enabling them to be deployed continuously for months before batteries and memory cards need to be downloaded (www.trailcampro.com). Prices of camera traps have declined as more models have become available, and high-quality digital camera traps are now priced at US$200 or less. Using camera traps in conjunction with a randomized or systematic sampling design at an appropriate spatial scale should result in the generation of representative data on terrestrial bird and mammal communities that can be incorporated into a composite index of state variables of interest.

The Wildlife Picture Index

The Wildlife Picture Index (WPI) has been described in detail in O'Brien *et al.* (2010) and O'Brien (2010a). The WPI is a composite index similar to the UK Wild Bird Index (Gregory *et al.*, 2003), or the Living Planet Index (Loh *et al.*, 2005; see also Chapter 4). It is based on the state variable of occupancy, rather than abundance (or density) because of the difficulty of developing unbiased estimates of abundance for a set of species when animals are not individually recognizable. Typically, a community of terrestrial birds and mammals will contain a few species that can be recognized as individuals and tracked though space and time. For these species, it is possible to develop unbiased estimates of abundance or density using capture-recapture methods and spatially explicit capture-recapture models. Many other species, however, are difficult to identify reliably as individuals, and some relative abundance index must be used in place of density or abundance estimates. The result is a set of abundances composed of unbiased estimates and many relative abundance indices with unknown, species-specific biases. A composite index based on such a mix of biased and unbiased estimates will be biased to an unknown degree, and trends developed from such indices will be open to several interpretations, depending on the strength of the underlying assumptions (O'Brien, 2011).

Occupancy data are easily collected using camera traps, and the area occupied by a species, or a community of species, is a useful metric for monitoring biodiversity (Yoccoz *et al.*, 2001; Buckland *et al.*, 2005). We can develop unbiased estimates of occupancy for most species possessing detection probabilities greater than 0.02 and true habitat occupancy greater than 10%, given sufficient sampling (O'Brien, 2010a). Occupancy can be sensitive to changes in abundance and is sometimes used as a surrogate of abundance (MacKenzie *et al.*, 2006; Royle and Nichols, 2003; see also Chapter 11). This makes sense when we consider that the more individuals using a point, the more likely we are to detect at least one of them at least once during a sample. Expanding populations tend to occupy more space over time, whereas contracting populations tend to occupy smaller areas. Occupancy modeling also allows the incorporation of covariates that may help to explain some of the underlying variation in the proportion of habitat occupied and the detection probabilities. For species that are especially cryptic and/or rare on the landscape, O'Brien (2010a) discusses methods for improving these species' occupancy estimates by borrowing information from similar species, post-stratification, and, in extreme situations, substituting observed occupancy.

The choice of species to include in a WPI is subjective and a function of the characteristics of the sampling and the assumptions of occupancy analysis. Camera trap model, setup, and trap sensitivity may all affect the size range of species detected. More importantly, a species' preferred substrate may influence its suitability as a target species. While we may be able to photograph a siamang (*Symphalangus syndactylus*) any time it crosses the camera, this strictly arboreal species is unlikely to be moving terrestrially and is not regularly available for sampling. Alternatively, many macaque species (*Macaca* spp.) spend a significant amount of time on the ground and do not violate closure assumptions regarding movement in and out of the sampling area. We refer to the target community as terrestrial birds and mammals of medium to large size, to emphasize that this is the group of species that we are most likely to include in a sample. We do not, however, exclude semi-terrestrial species that spend considerable time on the ground and are regularly exposed to the possibility of detection, and we rely on local knowledge of the species in question to decide whether to include it.

The index is constructed following the methods of Fewster *et al.* (2000) and Buckland *et al.* (2005). Buckland *et al.* (2005) found that a modified Shannon Index and the geometric mean of relative abundance (defined as abundance at time t divided by abundance at time $t = 1$) were most satisfactory for composite indices of biodiversity. We chose the geometric mean because the modified Shannon Index had no theoretical justification other than fitting Buckland *et al.* (2005) performance criteria for well-behaved composite indices. We substitute the state variable of occupancy for abundance and use a relative occupancy index as the basic measure. Note that the form of the relative occupancy index anchors the index to the initial value of the occupancy estimate at time $t = 1$.

To develop a WPI, we begin with a set of n species-specific occupancy estimates (each denoted by ψ_{ijk} for the species i at location j in year k) calculated using closed season models over k years. A species that is present but not detected has an occupancy estimate of zero for the season. The geometric mean is restricted to values greater than 0, however, so the occupancy estimates must be adjusted to eliminate 0-values. Adjustment terms are arbitrary, and O'Brien (2010a) recommends that all estimates of $\psi = 0$ be adjusted by:

$$\psi^* = \psi + \frac{1}{2x} \tag{3.1}$$

for an occupancy estimate based on x camera trap points. This ensures a distribution of ψ values that is strictly non-zero, non-negative and has minimal effect on the variance of the distribution. The next step is to develop an index of relative occupancy for each species i at survey site j in year k. We do this by dividing occupancy in year k by the estimated occupancy at the initial season:

$$o_{ijk} = \frac{\hat{\psi}_{ijk}}{\hat{\psi}_{ij1}} \quad (3.2)$$

This creates a relative occupancy index that measures the change in occupancy from initial conditions for each of n species. The estimate for $k = 1$ is always 1. The WPI for year k and site j and n species is the geometric mean of scaled occupancy statistics for n species:

$$WPI_{jk} = \sqrt[n]{\prod_{i=1}^{n} o_{ijk}} \quad (3.3)$$

Equation 3.3 can also be expressed as:

$$WPI_{jk} = \exp\left(\frac{1}{n}\sum_{i=1}^{n} \log\left(o_{ijk}\right)\right) \quad (3.4)$$

This formulation has several advantages. First, it possesses most of the favorable characteristics of a biodiversity index outlined by Buckland et al. (2005). Second, it is intuitively understandable, since each point is expressed as a proportional shift from initial conditions for the n species. Third, the index can be disassembled and restructured to develop associated indices that track subsets of the community. For instance, it would be relatively straightforward to develop a tiger prey index by restricting the analysis to those species at a site that are considered tiger prey. Fourth, the index is insensitive to species-specific variation in abundance and occupancy, because each species is scaled before entering the site index. Finally, by scaling to the initial year, the ratio is robust to missing years of data. Most ratio estimators use a process called chaining that requires evenly spaced observations because ratios are calculated sequentially. Missing data in one year affect the ratio estimates for two years. The WPI does not depend on chaining as all estimates are calculated based on the temporal distance from the initial condition (Fewster et al., 2000).

Community dynamics and sampling error can create situations in which a species may disappear (local extinction) or appear (local colonization) or simply be missed (sampling error) during some sampling period. An example is the Sumatran short-eared rabbit (*Nesolagus netscheri*) that was photographed at two locations during a camera trap survey in Bukit Barisan National Park in 2000, missed during three subsequent surveys, and photographed again in 2007 (T.G. O'Brien, unpublished data). For species that are first recorded at time $t > 1$ as a result of colonization or sampling error, we follow the recommendation of the LPI (Collen et al., 2009) and recalculate the WPI with the new species' pre-detection occupancy estimates set to ψ^*. For species that disappear from the index due to extinction or sampling error, we face the subjective problem of deciding when a species is extinct. IUCN (2001) defines a species as presumed extinct when 'exhaustive surveys in known and/or expected habitat, at appropriate times . . . have failed to record an individual. Surveys should be over a time frame appropriate to the taxon's life cycle and life form.' Local extinction, however, is more closely related to the loss of populations (Ceballos and Erhlich, 2002) and should be easier to detect than global extinction. For the purposes of the WPI, we consider a species to be locally extinct if it has not been detected in a survey for 5 years. Until that time, a species in the index that becomes a non-detected species should be assigned a value of ψ^*. For any species believed to be locally extinct or present but undetectable using camera trap survey methods, more intensive surveys using methods appropriate for the species should be undertaken to confirm its status. If intensive targeted surveys indicate local extinction, then the species should be removed from the index.

Measuring the WPI

To develop a WPI, we begin with a camera trap survey designed to be (i) spatially representative of the area of interest, (ii) of sufficient sampling intensity to detect a representative sample of the species in the target community, and (iii) of sufficient duration to develop unbiased occupancy estimates for the sampled species. Spatial coverage and representativeness is, to some degree, a function of the species in the target community. A community of Malagasy birds and mammals, for example, is restricted in body size to less than 5 kg, and typical home ranges of lemurs, the dominant mammals, rarely exceed 1 km^2. Such a community might be adequately sampled at a scale of 100 km^2, whereas a community of birds and mammals that includes African elephants (*Loxodonta africana*: home range 10–8700 km^2) and cheetahs (*Acinonyx jubatus*: home range 34–1500 km^2) may require sampling at a landscape scale of several hundred km^2. O'Brien (2010a) recommends a sampling area (or site) of 200 km^2 as a compromise between adequate coverage, logistical feasibility of sampling, and maintaining population closure. Within this site, occupancy surveys are relatively easy to carry out and to interpret. We start with the objective of estimating the proportion of area occupied at a site by each target mammal and bird. The sampling units are camera trap points within the site that are arranged in a random, stratified random, or systematic fashion at a density of 1 point/2 km^2. Stratified random or systematic designs are preferred since they have desirable statistical properties and ensure uniform coverage of the site. Cameras are deployed by navigating to a predetermined sampling point using a GPS unit, and then setting the camera at an optimal location within 50 m of the predetermined point. The new point is georeferenced and this becomes the permanent sampling point (TEAM Network, 2008; O'Brien, 2010a). The site may have to be subdivided into sampling blocks and surveyed sequentially using repeated deployments of camera traps if the number of camera traps is less than the number of points.

The camera traps remain in the field for S survey days and populations of the target species in the community are assumed to be closed to changes in state of occupancy during that period. The period of population closure is considered a 'season' and for most species of medium- to large-sized mammals and birds, population closure may be between 1 and 5 months. The definition of season as a period of population closure requires that we be familiar with the behaviour of all species within the community. Some species are territorial, some residential, some nomadic, and some migratory. It is therefore likely that not all species using an area of interest are present at any given point or period in time. Careful consideration is required to ensure that the 'season' of closure coincides with the time that the maximum number of species occupies the area of interest, and avoids transition periods when species may be moving in and out of the area in an unpredictable, and possibly non-random, manner. For the purposes of the WPI, we suggest that S be set to 30 days. The S repeated surveys within a season are used to establish a capture history for each of the species at each sample point.

Once the cameras have been deployed and picked up, digital photographs need to be identified to species, and data prepared for occupancy estimates for each species. Species status can take three states: present, absent, and present but not detected. We assume that no species is detected at a site when it is absent (no false positives). Occupancy estimates are calculated using the free software PRESENCE. A desirable feature of occupancy analysis is that it is robust to missing data (MacKenzie *et al.*, 2006). If a camera trap fails during the survey, it affects precision of the occupancy estimates but not the accuracy. After processing we have occupancy estimates for n species over k years at j sites. A simple example for $j = 1$ site is of the form:

$$\begin{matrix} \hat{\psi}_{1,1} & \hat{\psi}_{2,1} & \hat{\psi}_{3,1} & \cdots & \hat{\psi}_{n,1} \\ \hat{\psi}_{1,2} & \hat{\psi}_{2,2} & \hat{\psi}_{3,2} & \cdots & \hat{\psi}_{n,2} \\ \hat{\psi}_{1,3} & \hat{\psi}_{2,3} & \hat{\psi}_{3,3} & \cdots & \hat{\psi}_{n,3} \\ \\ \hat{\psi}_{1,k} & \hat{\psi}_{2,k} & \hat{\psi}_{3,k} & \cdots & \hat{\psi}_{n,k} \end{matrix}$$

These are transformed to species-specific relative occupancies using Equation 3.2:

$$\begin{matrix} o_{1,1} & o_{2,1} & o_{3,1} & \cdots & o_{n,1} \\ o_{1,2} & o_{2,2} & o_{3,2} & \cdots & o_{n,2} \\ o_{1,3} & o_{2,3} & o_{3,3} & \cdots & o_{n,3} \\ \\ o_{1,k} & o_{2,k} & o_{3,k} & \cdots & o_{n,k} \end{matrix}$$

And finally, we take the geometric mean of these relative occupancies across species to develop k WPI estimates, denoted as I here:

$$I_1 \ I_2 \ I_3 \ \cdots \ I_k$$

Trend analysis using the WPI

To determine trends in the WPI, we follow Fewster *et al.* (2000) and Buckland *et al.* (2005). They recommend generalized additive models (GAMs) to model trends as a smooth non-linear function of time. GAMs are similar to regressions but they do not require that the data be normally distributed and they assume that the relationship between the index and time is smooth but not necessarily linear. GAMs incorporate smoothing procedures into the model fitting process, allow a range of curves to be considered, and allow for direct incorporation of covariates to test hypotheses of factors influencing trends. GAMs also allow for a statistical test of changes in direction of the index trajectory, thus satisfying the criteria of a CBD 2010 indicator.

A simple regression model has the structure $y_i = \alpha + \beta x_i + \varepsilon_i$ with the assumption that the error terms ε_i are normally distributed. Generalized Linear Models (GLMs) are structured in this manner but allow the distribution of the error term to vary (McCullagh and Nelder, 1989). Braak *et al.* (1994) used log-linear Poisson regression models, a type of GLM, to fit count data of birds. They assumed that an observation y_{it} at site i and time t comes from a Poisson distribution with mean μ_{it}. Their model resembles a linear regression:

$$\log(\mu_{it}) = \alpha_i + \beta_t \quad (3.5)$$

where α_i is called the site effect for site i, and β_t refers to the year effect for year t. Both the normal linear regression and the log-linear Poisson regression model can be considered as types of general additive models. In a generalized additive model:

$$y = \alpha_i + f(t) + \varepsilon \quad (3.6)$$

the error terms are not assumed to be normally distributed, and the f(t) is some non-linear smoothing function of time. The predictor function f(t) is the principal difference between a GAM and a GLM. The GAM is fitted by estimating the parameters α_i and the smooth function f in the same way that a linear regression is fitted by estimating the parameters α and β. For a linear trend over time, the regression f(t) = βt has a single parameter β to be estimated. For an annual model, the regression f(t) = β_t has many β parameters. In this case, the function is jagged and represented by joining βs with straight lines. Between these limits (linear and maximally jagged) are functions that are non-linear, smoother than the annual model, and of greater utility for detecting long-term, non-linear trends.

Before the function f can be estimated, the level of smoothing must be specified. This is a subjective process that has few agreed guidelines to assist decision-making. The degree of smoothing is flexible and controlled by the degrees of freedom (df) in the time series dataset, ranging from a linear trend (df = 1) to an unsmoothed trend representing the annual change during a time series of k years (df = $k - 1$). Between these two extremes, the function f is determined non-parametrically from the data. GAMs thus allow us to explore linear trends in short time series and develop more complicated non-linear trends as the time series increases. The choice of df value is an important part of the modeling process and depends on the objectives of a particular analysis as well as the length of the time series. GAMs are used to separate underlying trends from short-term fluctuations (noise in the data), but the point at which this separation occurs is subjective, relating to the desired level of detail that is to be retained in the trend. For long-term trends, a smooth index curve is desirable and df should be set low. If information about annual fluctuation is required the index should be set high at $k - 1$ to produce a curve of maximum fluctuations. The length of the time series is also important; it will be hard to detect non-linear trends in short time series. Fewster *et al.* (2000) have suggested that a df of $0.3k$ be used for long time series, but caution that this is based on the needs of their analyses. They avoid setting rules for model selection and advise plotting indices from GAMs with a range of df values before settling on a final value.

The 95% confidence limits for the GAM trend are determined by a non-parametric bootstrap process. To develop a bootstrap confidence interval, we first select a random sample with replacement from the species that make up the sample for a specific time point. We repeat this process 999 times. We then analyse each sample as if it had been our real data. The variation in estimates of the index among bootstrap samples should give a good guide to the variation we would expect if we could take new samples of the community. The standard deviation of the bootstrap samples is used to estimate the standard error of our index at each time step. If we take the 999 bootstrap estimates for each year in the time series, and order each bootstrap sample from smallest to largest, the 25th smallest and 25th largest estimates represent the lower and upper 2.5% quantiles and are approximate 95% confidence limits for the index at each point in the time series.

The rate of change in the diversity index is measured by the slope of the smoothed trend. The rate of change of a non-linear trend is measured by the slope of a line tangent to a point in time along the curve. Non-linear trends allow for changes in the rate of change over time. Changes in the rate of change (a benchmark of CBD 2010 indicators) are measured by deriving a numerical estimate of the second derivative of the trend. A crude approximation of the second derivative of the slope at time t can be obtained using three points and the equation:

$$D_t = I_{t-1} - 2(I_t) + I_{t+1} \tag{3.7}$$

where D_t is the second derivative evaluated at time t, and I is the smoothed index value at $t-1$, t, and $t+1$. If the time series is lengthy, a more precise second derivative can be estimated using the index value at $t-2$, $t-1$, t, $t+1$, and $t+2$ (R. Fewster, personal communication):

$$D_t = 2\left(I_{t-2}\right) - 1\left(I_{t-1}\right) - 2\left(I_t\right) - 1\left(I_{t+1}\right) + 2\left(I_{t+2}\right) \tag{3.8}$$

A negative D_t indicates the rate of decline is accelerating, and a positive D_t indicates the rate of decline is slowing. To test the significance of the D_t value, we develop a bootstrapped 95% confidence interval around the second derivative. If, in a given year t, the confidence interval does not include 0, then we have evidence that the rate of change is changing. The sign (+/−) of the confidence interval indicates the direction of the change. If the interval includes only negative values, the change is for the worse; if the interval includes only positive values, the change is for the better.

All procedures for implementing an occupancy analysis are available in the free software package PRESENCE (http://www.mbr-pwrc.usgs.gov/software/presence.html). GAM modelling software is available in the mcgv software package (Wood, 2006) in R (R Development Core Team, 2012; check the R website for the latest version: www.R-project.org). Rachel Fewster provides GAM modeling software for monitoring of wildlife populations on her website (http://www.stat.auckland.ac.nz/~fewster/gams/R/). Jorge Ahumada, technical director of the Tropical Ecology, Assessment and Monitoring Network (TEAM), has written a program in R to calculate the WPI, the bootstrap confidence intervals, and the significance of changes in slopes (O'Brien, 2010a).

Implementation of the WPI

Unlike most CBD 2010 indicators, implementation of the WPI is still in its infancy. Much of the development of the WPI (O'Brien *et al.*, 2010; O'Brien, 2010a) arose from the challenges of monitoring tigers and prey (O'Brien *et al.*, 2003) and species richness (O'Brien *et al.*, 2011) across large landscapes in a representative manner using camera traps (Karanth *et al.*, 2008). Our demonstration project was based on the sampling methods outlined in O'Brien *et al.*, (2003). In this study, we set up a monitoring programme for Sumatran tigers (*Panthera tigris sumatrae*) and their prey. The sampling took place in the Bukit Barisan Selatan National Park (BBSNP), a 3568 km^2 park in southwest Sumatra. BBSNP is one of the largest remaining blocks of lowland rainforest on Sumatra and home to a wide variety of terrestrial mammals and birds, including species of global priority such as Asian elephant (*Elephas maximus*), Sumatran rhinoceros (*Dicerorhinus sumatrensis*) and Sumatran ground cuckoo (*Carpococcyx viridis*). We set up 10 sampling blocks at 10–15 km intervals oriented from the edge of the park to the interior. Within each block, we randomized 20 camera trap point coordinates within 1 km^2 subunits in each block. This allowed for a uniform coverage of the park along an elevation gradient south to north, and an edge gradient west to east. The project eventually conducted five park-wide surveys between 1998 and 2006, along with five additional camera trapping surveys at the Way Canguk Research Station. We photographed 12 bird species and more than 40 mammal species (not all identifiable to species). It was the surprising amount of data generated by the camera traps that inspired us to evaluate the potential further and develop the WPI. O'Brien et al. (2010) details an application of the WPI to BBSNP. The results indicated a continuous decline in wildlife diversity in BBSNP that has accelerated over time. The linear rate of decline has exceeded the rate of forest loss during the same period suggesting that other factors, most likely hunting for commercial purposes as well as subsistence and pest control, contributed significantly to the decline (Kinnaird et al., 2003; O'Brien et al., 2010)

The TEAM programme is a consortium consisting of Conservation International, the Missouri Botanical Garden, the Smithsonian Tropical Research Institute, and the Wildlife Conservation Society, along with many partner institutions around the world (www.teamnetwork.org). The mission of TEAM is 'to generate real time data for monitoring long-term trends in tropical biodiversity through a global network of field stations, providing an early warning system on the status of biodiversity to effectively guide conservation action.' TEAM's Terrestrial Vertebrate Monitoring Protocol (TEAM Network, 2008) is designed to be compatible with the WPI sampling design and analysis. This protocol currently is being implemented annually at 16 sites in Central and South America, Africa, and Asia, with plans for implementation at 50 sites by 2015.

The Zoological Society of London (ZSL) has initiated a number of projects using the WPI sampling protocol. Collen et al. (2009) have developed a biodiversity monitoring programme built around an Evolutionarily Distinctive Globally Endangered (EDGE) priority species (Isaac et al., 2007), the pygmy hippopotamus (*Choeropsis liberiensis*) in Sapo National Park, Liberia (Collen et al., 2011). The Steppe Forward Program is a ZSL project to measure and monitor biodiversity in the Mongolian grasslands (Townsend and Baillie, 2009). Both projects use the WPI sampling protocol and are introducing camera trap monitoring to national protected area staff. Wacher et al. (2008) have implemented WPI monitoring in Niger in conjunction with protected area development and an addax (*Addax nasomaculatus*) conservation programme.

Finally, the Mpala Research Centre and Wildlife Conservation Society are using the WPI sampling and analytical protocols to monitor carnivore, grazer, and browser communities in the rangelands of Laikipia District Kenya, and to evaluate the impact of livestock management practices on wildlife (Kinnaird and O'Brien, in press). Results to date have shown that the WPI methods are equally useful in rainforest and savannas. While it is too early to evaluate the implementation of the

WPI globally, we expect that such an evaluation will be possible within the next few years.

Preliminary simulations

We conducted a preliminary bootstrapping analysis of the WPI to examine the precision and robustness of the index. We approached the problem in two ways:

1. evaluating the usefulness of the geometric mean as a way to track changes in community composition; and
2. evaluating the ability of the WPI to track a constant decline of occupancy across species similar to a scenario of forest conversion that affects all species equally.

We developed a community of 104 species (Figure 3.3) representing cryptic ($0.02 < P \leq 0.2$) and detectable species ($P > 0.2$) that are rare ($0.1 \leq \psi \leq 0.2$), common ($0.3 \leq \psi \leq 0.5$), and widespread ($0.6 \leq \psi \leq 0.8$). We then use the simulation function in PRESENCE to estimate a distribution (mean ± SD) for 500 simulated occupancy estimates based on each species' true occupancy and detection probability, and a sampling effort of 100 camera points over 30 days. For the bootstrap analysis, we first set the size (n_s) and composition of the community of species (# rare

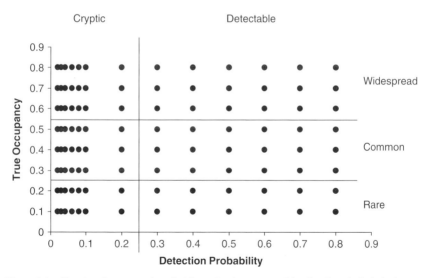

Figure 3.3 Simulated community of 104 species (represented by the closed circles) characterized by area occupied and detection probability.

cryptic species, ..., # detectable widespread species), and the number of species detected out of the total community (n_d). We then sampled with replacement for n_d times, selecting a species' true occupancy and a random estimated occupancy from a uniform distribution bounded by the 95% confidence interval based on the estimated occupancy for the species from the PRESENCE simulation. We then calculated a geometric mean based on the true occupancies and a geometric mean based on the simulated occupancy. We repeated this process 10 000 times to give us a distribution of true occupancies and simulated occupancies that could be compared for accuracy and precision.

We chose as an example a community of 60 species with four community compositions:

- 20 rare, 20 common, and 20 widespread species;
- 30 rare and 30 common or widespread species;
- 40 rare and 20 common or widespread species;
- 40 rare and 20 common species but no widespread species.

We simulated the effect of detecting 60, 50, 40, 30, and 20 species on the estimated geometric mean of occupancy and compared that to the true geometric mean of occupancy for each example (Figure 3.4). In each test, there was a small positive bias in the estimated geometric mean. This may reflect the tendency for positive bias in occupancy analysis estimates as true occupancy declines. When the number of rare species was less than or equal to the number of common and widespread species (Figure 3.4a,b) the bias was less than 5% at all levels of species detection. When the number of rare species was twice the number of common species (Figure 3.4c,d), the bias was between 5% and 10%. Precision (as measured by percentage of coefficient of variation (CV)) of the simulated geometric mean was between 5% and 17%, with precision declining (%CV increased) as the number of detected species declined. We conclude from this preliminary example that the geometric mean performs well, even in the presence of a large proportion of rare species, although accuracy and precision do decline as rare species increase.

We next asked how well the WPI would perform when a community of species was losing habitat and all species are affected equally by this loss. We examined the full community of 104 species and estimated the WPI for a simulated habitat loss equal to a loss of 1%, 2.5%, 5%, 7.5%, and 10% of occupancy (Figure 3.5). Because all species' occupancy values were reduced by the same amount, habitat loss had a greater impact on rare species than common and widespread species. When habitat loss reached 10%, 13 species disappeared from the community. We looked at two levels of species representation, 80 and 60 randomly selected species in the sample, to see how declining representativeness of the sample might affect the index. In both examples, the simulated WPI accurately tracked the true value of the WPI with a small positive bias that averaged 2.4%. Precision of the simulated WPI varied between 11%

The Wildlife Picture Index

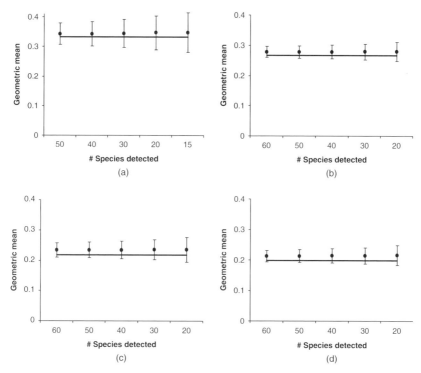

Figure 3.4 True (lines) and simulated (filled circles) geometric means (± SD) of occupancy for a community of 60 species composed of (a) 20 rare, 20 common, and 20 widespread species; (b) 30 rare and 30 common or widespread species; (c) 40 rare and 20 common or widespread species; and (d) 40 rare and 20 common species. The x-axis is the number of species from the community detected in a sample.

and 22%, increasing as the true index declined and as the number of species detected declined. With 80 species in the sample, a significant ($P < 0.05$) decline in the WPI was detected by the 5th time step. With 60 species in the sample, a significant decline was detected by the 6th time step.

Cost of implementation

To evaluate the cost-effectiveness of the WPI, we considered the implementation of a 100-point WPI survey compared to a line transect survey of 100 2-km-long transects,

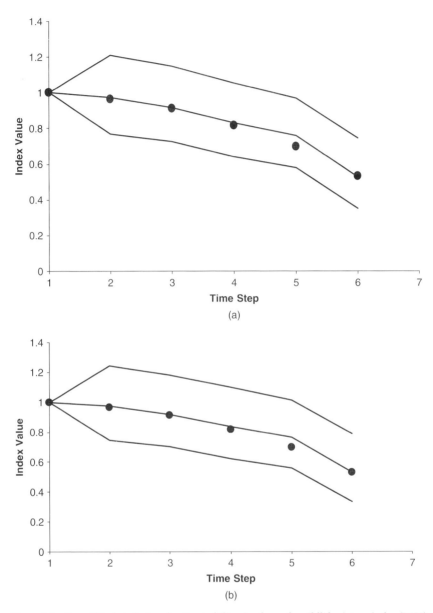

Figure 3.5 True (filled circles) and estimated (lines) values of a wildlife picture index (WPI) with 95% confidence limits when (a) 80 of 104 species are detected, and (b) 60 of 104 species are detected.

in Asian forest of Lao PDR and rangeland of Laikipia, Kenya. The WCS Lao PDR Program provided the cost data for implementing a camera trap survey versus a line transect survey in forest. We developed the rangeland budgets based on personal experience and knowledge of field expenses charged by biologists at the National Museum of Kenya. We calculated the fixed cost of 5 years of surveys in 2010 US dollars and pro-rated the equipment over that time. We ignored inflation because inflation affects all rates similarly. Equipment costs were set as one-time startup costs of US$6860 for camera traps plus accessories, and $1750 for laser rangefinders, binoculars, and digital compasses for line transect surveys. We assumed no loss or replacement of equipment during the 5 years (= five surveys). For the rangeland surveys, we used standard salaries, per diem, and minor equipment, fuel, and vehicle rental rates. For the forest surveys, we used standard transportation, per diem, field rations, camping equipment, and minor equipment. The key differences in the cost of implementation of camera trap surveys versus line transect are the startup cost (higher for camera trap surveys) and cost of keeping people in the field (higher for line transect surveys).

We estimate that the cost of implementing a single camera trap survey and line transect survey in rangeland condition averages $5750 and $7500, respectively, per year over a 5-year period. In the forest conditions of Lao PDR, we estimate the average costs at $15 900 for camera trapping and $18 400 for line transects. In savanna, a WPI survey is 30% less expensive than a line transect survey. In the forest, the cost saving for a WPI is 15% compared to a line transect survey. The similar costs in forests are due to the fact that camera trap deployment in the forest requires more time than in the savanna due to lack of roads and the rugged terrain. If we consider data returns, a line transect survey in Laikipia rangelands typically accumulates data on 28 large birds and mammals whereas a camera trap survey accumulates 59 large birds and mammals, primarily due to capturing nocturnal mammals. In the forest example, a line transect survey might encounter 20 species whereas the camera trap survey might encounter more than 40 species, again due to capturing the nocturnal community and animals that might flee humans. We expect that these figures would vary with the specifics of a deployment strategy, but the key features of startup and daily costs of conducting the surveys determine the relative cost advantage. We conclude that the WPI is a cost-effective alternative to line transect surveys, and probably other labour-intensive surveys, producing more data and more representative data for less cost.

Weaknesses and strengths

The WPI has been developed to monitor the terrestrial component of diversity represented by medium- to large-sized birds and mammals in the data-deficient tropics. The degree to which these two groups represent trends in the broader bird and mammal communities at the monitoring sites is unclear (Collen et al., 2009). There is reason to believe that larger terrestrial birds and mammals might face more threats than small birds and mammals in the same ecosystem. In areas where wildlife is hunted for food, larger body size and a terrestrial habit increase vulnerability to a wider range of hunting methods (Robinson and Bennett, 2000). Larger birds and mammals also require more habitat per unit animal (Haskell et al., 2002; Jetz et al., 2004). These same traits, however, make the larger taxa sensitive indicators of habitat and exploitation threats in forest and savanna ecosystems (Dobson et al., 2006).

We also do not know whether changes in the community of terrestrial birds and mammals are representative of changes in other vertebrate taxa, particularly reptiles, amphibians, and fish, in forest and savanna ecosystems. The long search for indicator, flagship, and umbrella species has been frustrated by the lack of congruence in species response to common threats across taxa within an ecosystem (Simberloff, 1998; Caro and O'Doherty, 1999; Andelman and Fagan 2000; Williams et al., 2000). We have some understanding of why taxa differ in response to common threats but more work is needed to determine if the lack of congruence in response to common threat is due to time lags or genuine differences in taxa responses.

The WPI targets tropical rainforests and savannas because of data deficiencies in these ecosystems. It may be subject to criticism for ignoring coverage in other tropical ecosystems and the temperate equivalent of these tropical ecosystems. To date, we do not know the extent of geographical coverage represented by current camera trap programmes, although we suspect it is biased towards the tropics and it is widespread. There is no reason why the WPI could not be applied in other ecosystems, especially in temperate forest and grasslands, and where the fauna contains a large number of nocturnal species. A global data registry similar to Ecological Archives would help track camera trap programmes and coverage.

The success of the WPI will ultimately depend on contributions and participation by a network of voluntary data providers who adopt the WPI sampling protocol in their programmes. We do not know the degree to which ongoing camera trap programmes, many of which target single species and sample accordingly, produce data that are comparable and compatible with WPI data. For example, a long-term tiger monitoring programme in Nagarahole National Park, India, is designed to maximize detection probabilities of tigers (Karanth and Nichols, 1998). The camera trap locations in this study are not systematic and not optimal for sampling chital (*Axis axis*) and several other ungulate species in the park (K.U. Karanth, personal communication). Consequently, occupancy estimates for these species may be biased low compared to true occupancy due to unrepresentative sampling. If the bias is systematic, however, the trends in occupancy estimates may still be unbiased, and a WPI based on a combination of unbiased and systematically biased occupancy estimates would still contain useful information (O'Brien, 2010b). For camera trap studies not designed with WPI objectives in mind, the sampling designs and datasets will have to be scrutinized to determine if and where biases occur in the representation of the medium- to large-sized bird and mammal community. The usefulness of this retrospective analytical approach needs to be evaluated.

Basing the WPI on occupancy rather than abundance means that the index responds to spatial distributions rather than abundance distributions. While there are many reasons to assume that occupancy and abundance should be positively associated (Royle and Nichols, 2003; Royle, 2004; MacKenzie et al., 2006), situations may arise that cause abundance and occupancy to be out of synchrony. We can imagine situations in which a population is declining uniformly across a region. Its occupancy may stay relatively constant as the population size declines. This might manifest itself as a declining point abundance but stable occupancy. We can also imagine a situation where a recovering population is growing but has not reached the size where dispersal or recolonization of old habitat becomes detectable in the occupancy estimates. In both cases, however, we should observe changes in detection probabilities associated with changing abundance (Royle and Nichols, 2003) and we expect that the WPI will respond to changing abundances though not necessarily in the same way that a relative abundance index might.

Two final weaknesses of the WPI are shared by many biodiversity indicators. First, the WPI may include many rare species at some sites and few at others. This may affect the local precision of the WPI, making it more difficult to detect changes in the trend at some sites compared to others. Preliminary bootstrap simulations on the effect of community structure (measured by true occupancy of species), detectability, and number of species detected in samples suggest that the accuracy of the WPI is fairly insensitive to rare species in the sample. The precision of WPI estimates, however, remains sensitive both to rare species and to the number of species detected. Second, the WPI weights each species in the community equally. Declining rare and endemic species are treated the same as abundant and widespread species. Although some may argue that declines in rare and endemic species are more important than declines in widespread species (loss of evolutionary distinctiveness: Mace et al., 2003; Isaac et al., 2007) or that declines of widespread, abundant, or top trophic level species are more important to ecosystem processes (Ray et al., 2005; Sergio et al., 2005; Dobson et al., 2006), these are arguments based on different, often qualitative, notions of biodiversity and there are no agreements on how weighting schemes should be assigned. Buckland et al. (2005) recommend that weighting schemes be avoided altogether.

A significant strength of the WPI is that it monitors a component of biodiversity that is seen as important to governments and the public. The carnivores, large ungulates, seed dispersers, and seed predators are not only important components of terrestrial biodiversity, they are often the target of exploitation, management (to promote or to limit populations), and tourism (Norton-Griffiths, 2007). Wildlife species occupying the top trophic levels are an important component of agricultural landscapes, where they may prey on livestock or crops. They can be important to local or national economies where wildlife viewing earns tourism revenues. They also provide important sources of protein and hunting revenues where permitted. Top carnivores and herbivores also are vulnerable to consumptive exploitation, habitat deterioration and loss, and their disappearance is often associated with trophic cascades as their ecosystem services are lost (Dobson *et al*., 2006). Because of their economic, ecological, and aesthetic importance, we believe that wildlife managers may find the WPI especially useful to their monitoring efforts.

The future

Clearly, the WPI is a work in progress. We are still a few years away from a complete evaluation of the index using real data. We need to develop a network of WPI practitioners, although the coordinated camera trap monitoring programme by TEAM and ZSL will soon provide the critical nucleus of this community. Work on a full simulation of the accuracy and precision of the WPI under a variety of community structures is underway. More data are needed on the cost of implementation of WPI surveys. The use of camera traps in ecological studies is expanding rapidly (Rowcliffe and Carbone, 2008) in all major biomes of the world. Also, ecologists are making great progress in the development of hierarchical models of abundance and occupancy (Royle, 2004; Royle and Dorazio, 2008; MacKenzie *et al*., 2006) as well as in trend estimation with imperfect detection (Kéry *et al*., 2009). We agree with Rowcliffe and Carbone (2008) that an important next step is a global databank for camera trap information, where studies from around the world can be connected and integrated. Several efforts are underway, and we look forward to the time when the WPI becomes a useful indicator of changing biodiversity.

Acknowledgements

The development of the WPI has been supported by the Wildlife Conservation Society, the Zoological Society of London, and E. McBean. C. Hallam provided data on the cost of forest surveys in Lao PDR. We thank J.E.M. Baillie, B. Collen, J.R. Ginsberg, L. Krueger, G.M. Mace, J.D. Nichols, J.G. Robinson, and M. Rowcliffe, as well as our many colleagues at the Zoological Society of London for encouragement, stimulating discussions, and useful insights.

References

Andelman, S.J. and Fagan, W.F. (2000) Umbrellas and flagships: Efficient conservation surrogates or expensive mistakes. *Proceedings of the National Academy of Science of the USA*, 97, 5954–5959.

Balmford, A., Crane, P., Dobson, A., Green, R.E., and Mace, G.M. (2005) The 2010 challenge: data availability, information needs and extraterrestrial insights. *Philosophical Transactions of the Royal Society of London, Series B*, 360, 221–228.

Buckland, S.T., Magurran, A.E., Green, R.E., and Fewster, R.M. (2005) Monitoring change in biodiversity through composite indices. *Philosophical Transactions of the Royal Society of London, Series B*, 360, 243–254.

Caro, T.M. and O'Doherty, G. (1999) On the use of surrogate species in conservation. *Conservation Biology*, 13, 805–814.

Ceballos, G. and Ehrlich, P.R. (2002) Mammal population losses and the extinction crisis. *Science*, 296, 904–907.

Collen, B., Konie, J., Rist, J., and Howard, R. (2008a) Constructing the evidence base for pygmy hippo conservation in Sapo National Park, Liberia. Report to London Zoological Society, 22 pp.

Collen, B., Ram, M., Zamin, T., and McRae, L. (2008b) The tropical biodiversity data gap: addressing disparity in global monitoring. *Tropical Conservation Science*, 1, 75–88.

Collen, B., Loh, J., Whitmee, S., McRae, L., Amin, R., and Baillie, J.E.M. (2009) Monitoring change in vertebrate abundance: the Living Planet Index. *Conservation Biology*, 23, 317–327.

Collen, B., Howard, R., Konie, J., Daniel, O., and Rist, J. (2011) Field surveys for the Endangered pygmy hippopotamus *Choeropsis liberiensis* in Sapo National Park, Liberia. *Oryx*, 45, 35–37.

Dobson, A. (2005) Monitoring global rates of biodiversity change: challenges that arise in meeting the Convention on Biological Diversity (CBD) 2010 goals. *Philosophical Transactions of the Royal Society of London, Series B*, 360, 229–241.

Dobson, A., Lodge, D., Alder, J., *et al*. (2006) Habitat loss, trophic collapse, and the decline of ecosystem services. *Ecology*, 87, 1915–1924.

Dunning, J.B. Jr., (2008) *CRC Handbook of Avian Body Masses*, 2nd edn. CRC Press, Boca Raton, FL.

Fewster, R.M., Buckland, S.T., Siriwardena, G.M., Bailey, S.R., and Wilson. J.D. (2000) Analysis of population trends for farmland birds using generalized additive models. *Ecology*, 81, 1970–1984.

Gregory, R.D., Noble, D., Field, R., Marchant, J.H., Raven, M., and Gibbons, D.W. (2003) Using birds as indicators of biodiversity. *Ornis Hungaria*, 12-13, 11–14.

Haskell, J.P., Ritchie, M.E., and Olff, H. (2002) Fractal geometry predicts varying body size scaling relationships for mammal and bird home ranges. *Nature*, 418, 527–530.

Isaac, N.J.B., Turvey, S.T., Collen, B., Waterman, C., and Baillie, J.E.M. (2007) Mammals on the EDGE: Conservation priorities based on threat and Phylogeny. *PLoS One*, 2, 1–7.

IUCN (2001) *IUCN Red List Categories: Version 3.1*. Prepared by the IUCN Species Survival Commission. IUCN, Gland, Switzerland, and Cambridge, UK.

Jetz, W., Carbone, C., Fulford, J., and Brown, J.H. (2004). The scaling of animal space use. *Science*, 306, 266–268.

Karanth, K.U. and Nichols, J.D. (1998) Estimation of tiger densities in India using photographic captures and recaptures. *Ecology*, 79, 2852–2862.

Karanth, K.U., Nichols, J.D., and Kumar, N.S. (2004) Photographic sampling of elusive mammals in tropical rainforests. In: Thompson, W.L. (ed.), *Sampling Rare or Elusive Species*. Island Press, Washington DC, pp. 229–247.

Karanth, K.U., Kumar N.S., Srinivas, V. and Gopalaswamy, A. (2008) Revised monitoring framework for Tigers Forever – Panthera sites. Report to the Wildlife Conservation Society, 17 pp.

Kéry, M., Dorazio, R.M., Soldaat, L., van Strein, A., Zuiderwijk, A., and Royle, J.A. (2009) Trend estimation in populations with imperfect detection. *Journal of Applied Ecology*, 46, 1163–1172.

Kinnaird, M.F. and O'Brien, T.G. in press. Effects of private-land use, livestock management, and human tolerance on diversity, distribution, and abundance of large African mammals. Conservation Biology.

Kinnaird, M.F., Sanderson, E.W., O'Brien, T.G., Wibisono, H.T. and Woolmer, G. (2003) Deforestation trends in a tropical landscape and implications for forest mammals. *Conservation Biology*, 17, 245–257.

Loh, J., Green, R.E., Ricketts, T., *et al.* (2005) The Living Planet Index: using species population time series to track trends in biodiversity. *Philosophical Transactions of the Royal Society of London, Series B*, 360, 289–295.

Mace, G.M. and Baillie, J.E.M. (2007) The 2010 Biodiversity Indicators: challenges for science and policy. *Conservation Biology*, 21, 1406–1413.

Mace, G.M., Gittleman, J.L., and Purvis, A. (2003) Preserving the tree of life. *Science*, 300, 1707–1709.

MacKenzie, D.I., Nichols, J.D., Royle, J.A., Pollock, K.P., Bailey, L.L. and Hines, J.E. (2006) *Occupancy Estimation and Modeling: Inferring Patterns and Dynamics of Species Occurrence*. Academic Press, New York.

McCullagh, P. and Nelder, J.A. (1989) *Generalized Linear Models*, 2nd edn. Chapman & Hall, London.

Millennium Ecosystem Assessment (2005) *Ecosystems and Human Well-Being: Biodiversity Assessment*. Island Press, Washington, DC.

Norton-Griffiths, M. (2007) How many wildebeest do you need? *World Economics*, 8, 41–64.

O'Brien, T.G. (2010a) Wildlife Picture Index: Implementation Manual Version 1.0. Wildlife Conservation Society Working Paper Series, New York.

O'Brien, T.G. (2010b) Wildlife picture index and biodiversity monitoring: issues and future directions. *Animal Conservation*, 13, 350–352.

O'Brien, TG. (2011) Abundance, density and relative abundance: A conceptual framework. In: O'Connell, A.F., Nichols, J.D., and Karanth, K.U. (eds), *Camera Traps in Animal Ecology: Methods and Analyses*. Springer Verlag, New York, pp. 71–96.

O'Brien, T.G. and Kinnaird, M.F. (2008) A picture is worth a thousand words: the application of camera trapping to the study of birds. *Bird Conservation International*, 18, S144–S162.

O'Brien, T.G., Kinnaird, M.F., and Wibisono, H.T. (2003) Crouching tigers, hidden prey: Sumatran tiger and prey populations in a tropical forest landscape. *Animal Conservation*, 6, 131–139.

O'Brien, T.G., Baillie, J.E.M., Krueger, L., and Cuke, M. (2010) The Wildlife Picture Index: monitoring top trophic levels. *Animal Conservation*, 13, 335–343.

O'Brien, T.G., Kinnaird, M.F., and Wibisono, H.T. (2011) Species richness estimation using camera traps: An Indonesian example. In: O'Connell, A.F., Nichols, J.D., and Karanth, K.U. (eds), *Camera Traps in Animal Ecology: Methods and Analyses*. Springer-Verlag, New York, pp. 223–252.

Pollock, K.H., Nichols, J.D., Simons, T.R., Farnsworth, G.L., Bailey, L.L., and Sauer, J.R. (2002) The design of large scale wildlife monitoring studies. *Environmetrics*, 13, 105–119.

Power, M.E., Tilman, D., Estes, J., et al. (1996) Challenges in the quest for keystones. *BioScience*, 46, 609–620.

Ray, J.C., Redford, K.H., Berger, J., and Steneck, R. (2005) Conclusion: Is large carnivore conservation equal to biodiversity conservation and how can we achieve both? In: Ray, J.C., Redford, K.H., Steneck, R., and Berger, J. (eds), *Large Carnivores and the Conservation of Biodiversity*. Island Press, Washington, DC, pp. 400–427.

R Development Core Team (2012). *R: A Language and Environment for Statistical Computing*. R Foundation for Statistical Computing, Vienna, Austria (www.R-project.org/).

Robinson, J.G. and Bennett E. (eds) (2000) *Hunting for Sustainability in the Tropics*. Columbia University Press, NY.

Rowcliffe, J.M. and Carbone, C. (2008) Surveys using camera traps: are we looking to a brighter future? *Animal Conservation*, 11, 185–186.

Royle, J.A. (2004) N-mixture models for estimating population size from spatially replicated count data. *Biometrics*, 60, 108–115.

Royle, J.A. and Dorazio, R.M. (2008) *Hierarchical Modeling and Inference in Ecology: The Analysis of Data from Populations, Metapopulations and Communities*. Academic Press, San Diego.

Royle, J.A. and Nichols, J.D. (2003) Estimating abundance from repeated presence absence data or point counts. *Ecology*, 84, 777–790.

Sergio, F., Newton, I., and Marchesi, L. (2005) Top predators and biodiversity. *Nature*, 436, 192.

Sibley, C.G. and Monroe, B.L. (1990) *Distribution and Taxonomy of Birds of the World*. Yale University Press, New Haven.

Simberloff, D. (1998) Flagships, umbrellas, and keystones: is single-species management passé in the landscape era? *Biological Conservation*, 83, 247–257.

Smith, F.A., Lyons, S.K., Ernest, S.K.M., et al. (2003) Body mass of late Quaternary mammals. *Ecology*, 84, 3403.

TEAM Network (2008) *Terrestrial Vertebrate Protocol Implementation Manual, v. 3.1*. Tropical Ecology, Assessment and Monitoring Network, Conservation International, Arlington.

ter Braak, C.J.F, van Strien, A.J., Meijer, R. and Verstrael, T.J. (1994) Analysis of monitoring data with many missing values: which method? In: Hagemeijer, W. and Verstrael, T. (eds), *Bird Numbers 1992. Distribution, Monitoring and Ecological Aspects*. Proceedings of the 12th International Conference of the International Bird Census Committee and European Ornithological Atlas Committee. SOVON, Beek-Ubbergen, The Netherlands, pp. 663–673.

Thompson, W.L. (ed.) (2004) *Sampling Rare or Elusive Species*. Island Press, Washington, DC.

Townsend S.E. and Baillie, J.E.M. (2009) Wildlife inventory and implementation of the Wildlife Picture Index at Myangan Ugalzat National Park and Buffer Zone, Tsetseg Soum, Khovd Aimag, Mongolia summer 2009. Report to the Zoological Society of London.

Wacher, T., Rabeil, T. and Newby, J. (2009) Monitoring survey of Termit and Tin Touma (Niger) – December 2008. Report to Zoological Society of London, 30 pp.

Williams, P.H., Burgess, N.D., and Rahbek, C. (2000) Flagship species, ecological complementarity and conserving the diversity of mammals and birds in sub-Saharan Africa. *Animal Conservation*, 3, 249–260.

Wood, S.N. (2006) *Generalized Additive Models: An Introduction with R*. Chapman & Hall/CRC, Boca Raton, FL.

Yoccoz, N.G., Nichols, J.D., and Boulinier, T. (2001) Monitoring of biological diversity in space and time. *Trends in Ecology and Evolution*, 16, 446–453.

4

Tracking Change in Abundance: The Living Planet Index

Ben Collen[1], Louise McRae[1], Jonathan Loh[2], Stefanie Deinet[1], Adriana De Palma[1], Robyn Manley[1] and Jonathan E.M. Baillie[3]

[1]Institute of Zoology, Zoological Society of London, London, UK
[2]WWF International, Gland, Switzerland
[3]Conservation Programmes, Zoological Society of London, London, UK

Introduction

Understanding change in population abundance is critical to understanding change in biodiversity. The loss of populations is a prelude to species extinction, and tends to reduce taxonomic, genetic, and functional diversity, so is integrally related to many different elements of biodiversity. Abundance trend data, that is, change in the number of individuals over time, can inform about the importance of both variability and quantity of biodiversity. For example, local populations retain local adaptations, therefore maintaining abundance is integral to maintaining the variability of biodiversity (Mace, 2008). Further, provisioning and regulating services may depend on quantity or overall abundance – for example, food, fresh water, and the long-term viability of such services. There is also a growing body of work that supports the assertion that common or abundant species contribute a disproportionately large amount to biomass, function, and resilience of ecosystems (Gaston, 2010; Gaston and Fuller, 2008). Measuring rates of change in population size is perhaps one of the most sensitive metrics for long-term measurement of biodiversity change (Balmford *et al.*, 2003; Buckland *et al.*, 2005; Pereira and Cooper, 2006) and is a trend that can be updated annually, which is important given the immediacy with which biodiversity information is needed. Abundance measures are an important proxy for biodiversity at higher levels, and can be used to infer community change (Buckland *et al.*, 2005),

---

Biodiversity Monitoring and Conservation: Bridging the Gap between Global Commitment and Local Action, First Edition. Edited by Ben Collen, Nathalie Pettorelli, Jonathan E.M. Baillie and Sarah M. Durant.
© 2013 John Wiley & Sons, Ltd. Published 2013 by John Wiley & Sons, Ltd.

allow integration of threats such as exploitation (e.g. many populations provide us with harvested goods), and are indicative of change in habitats (which provide essential services; Balmford *et al.*, 2003), as well as changes in the impact of population demographics. Here, we present an indicator measuring change in abundance of vertebrate species over a 37-year time period from 1970 to 2007.

At the 8th Convention of the Parties, the Living Planet Index (LPI) was adopted as one of the potential measures to address the Convention on Biological Diversity (CBD) headline indicator: change in abundance of selected species (UNEP, 2006). The LPI is based on what is believed to be one of the largest time-series databases on vertebrate populations, available to view online at www.livingplanetindex.org. It provides the capacity for a broad range of vertebrate population trend indicators. The LPI began life as a communications tool for a World Wildlife Fund (WWF) campaign. One of its biggest assets is that it is a simple yet powerful way of conveying information about changing trends in biodiversity to non-experts, from policy- and decision-makers to the general public. Due to its increasingly prevalent role as a policy tool for monitoring progress toward the 2010 CBD and other biodiversity targets, ever more effort is being focused on ensuring that this indicator is as robust, sensitive, and representative as possible (Collen *et al.*, 2009a).

Ideally an indicator measuring change in population abundance would track a randomly selected representative subset of taxa stratified across the main strata for which one would like information; in the context of this book that might be nations, or even regions within nations. These data do not exist. One recourse is to use available data on monitored populations to generate an indicator of population trends, and analyse the coverage of these data to understand how they deviate from the ideal indicator. Long-term data have much to offer current conservation efforts (Willis *et al.*, 2007), but like all other species-based indicators, the LPI relies on compiling data collected for a range of different purposes. To ensure a robust and meaningful indicator, discrepancies in representation must be accounted for and minimized.

The LPI measures global vertebrate abundance trends over time by calculating the average change in abundance for each year compared with the preceding year, which is then chained to the previous average annual population change to make an index, starting with an initial value of 1 in 1970. It shows that over a 37-year period, global abundance of monitored populations of vertebrates has decreased, on average, by around 30% (Figure 4.1). It can thus be thought of as a biological analogue of a stock market index that tracks the value of a set of stocks and shares traded on an exchange, or a retail price index that tracks the cost of a basket of consumer goods. In terms of biodiversity monitoring, the LPI is based on the idea that population declines and losses are a prelude to species-level extinctions. This is because populations are sensitive to short-term changes caused by anthropogenic pressures (Balmford *et al.*, 2003), and may thus represent a good indicator of the loss of biodiversity (Ceballos and Ehrlich, 2002). In addition, the relatively short delay between human

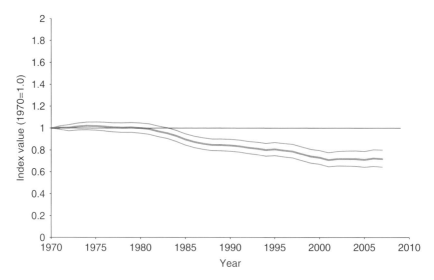

Figure 4.1 Global Living Planet Index, showing aggregated trends in the abundance of vertebrate species from across the globe ($n_{species} = 2544$; $n_{population} = 7953$).

impact and subsequent vertebrate population decline gives the opportunity for the implementation of proactive and targeted conservation action (Collen *et al.*, 2009a).

In this chapter we explore the prospect of taking this global indicator of change in species abundance, and applying it at the regional and national level, and to thematic groups. We highlight several examples where this has been done, and discuss how data coverage and data integrity are important factors in extending the utility of this index. We set out the potential strengths and limitations to using abundance indices to monitor change in biodiversity.

Methods

To produce the index, population abundance time-series data are aggregated to calculate average rates of change in a large number of populations of terrestrial, freshwater, and marine vertebrate species. LPI values relate to the proportional change in population size over time, therefore a value of 0.71 in 2007 relates to an average 29% decline in population abundance since 1970 (Figure 4.1). The LPI integrates the net impacts of populations of species gaining in number and declining in number, to result in an average proportional change in abundance between successive years.

Since the initial development of the index calculation in 2000 (Loh, 2000) the methods have been developed in a number of ways (Collen *et al.*, 2009a; Loh *et al.*, 2005), with particular reference to advancing the underlying time-series modelling. Data include time-series information for vertebrate species from published scientific literature, online databases (e.g. NERC Centre for Population Biology, 2010 (Global Population Dynamics Database); Pan-European Common Bird Monitoring, 2006), and grey non-governmental organization (NGO) and national park records literature. Data are only included if a set of criteria are met (Collen *et al.*, 2009a):

- a measure of population size is available for at least 2 years;
- information is available on how the data were collected and what the units of measurement are;
- the geographical location of the population is provided;
- the data were collected using the same method on the same population throughout the time series; and
- the data source is referenced and traceable.

Each iteration of the Living Planet Report has involved a new round of data collection, adding to the total number of populations and species in the LPI dataset (Table 4.1; Figure 4.2a). The 2010 index is currently calculated using time-series data on 7953 populations of 2544 species of mammal, bird, reptile, amphibian, and fish from around the globe (Figure 4.2b; Loh *et al.*, 2010; Secretariat of the Convention on Biological Diversity, 2010). Two complementary methods are used to generate index values, depending on time-series length: a chain method (Loh *et al.*, 2005) and a generalized additive modelling technique (Collen *et al.*, 2009a). While there are some potential issues with chaining (e.g., drift; Ehemann, 2007; Frisch, 1936), generalized additive models (GAMs) are not possible on datasets with small numbers of population estimates (Buckland *et al.*, 2005; Fewster *et al.*, 2000). We therefore determined the

Table 4.1 **Growth in number of populations and species used to calculate the Living Planet Index since 2000**

Report year	Number of populations	Number of species
2000	–	730
2002	–	694
2004	3000	1145
2006	3627	1313
2008	4645	1686
2010	7953	2544

The Living Planet Index

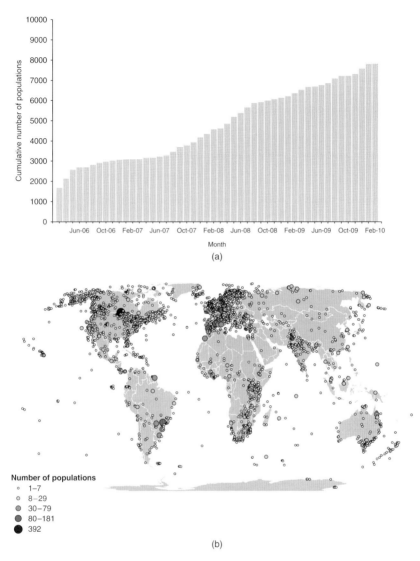

Figure 4.2 (a) **Growth in population number in the Living Planet Index database over time.** (b) **Distribution of population time-series data in the Index.**

choice of method on the length of the time-series, with time-series of $n > 6$ data points being processed using the GAM framework, and those that do not meet the criteria processed using the chain method. In order to calculate an index, the logarithm of the ratio of population measure for each species is calculated for successive years. Mean values are calculated for species with more than one population. The overall index is then calculated with the index value set to 1 in 1970. Insufficient population time-series are currently available to continue the index beyond 2007 because of a lag in publication of data. Indices are produced weighting populations equally within species, and species weighted equally within each index. Indices for terrestrial, marine, and freshwater systems are calculated as the geometric mean of tropical and temperate species (Figure 4.3).

In the context of the CBD 2010 target, it is important to evaluate significant change in the index over time, so as to be able to assess significant change in the rate of biodiversity decline. Years where the slope of the log-index value changes can identify where the annual rate of change as a proportion has significantly shifted. Using this method to assess changes in the rate of abundance is therefore vital for assessing the progress towards the CBD 2010 target (Buckland et al., 2005).

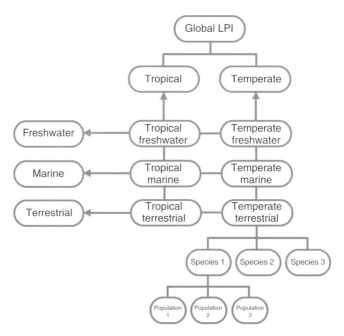

Figure 4.3 **Hierarchy of index aggregation in the Living Planet Index (LPI).**

What has the index told us about biodiversity change? Interpreting the index

When examining indices calculated using the LPI method, it is important to consider a particular characteristic of the trends produced. A decrease in the LPI represents an overall reduction of species populations, meaning that the magnitude of population decline exceeded the degree of increase in abundance. This necessarily implies that diversity will have reduced, even if none of these species populations has declined to zero, which would indicate an extinction event. A constant LPI represents no overall change in species populations, or a situation in which population gains and declines cancel each other out, implying no overall biodiversity loss. Put simply, values above 1 indicate a net increase in vertebrate abundance compared to the baseline, trend values below the baseline a net decrease.

Knowledge of the precision with which the trend curve has been estimated is central to the interpretation of the results (Fewster *et al.*, 2000), and while no traditional testing statistic is generated, the LPI offers the possibility to assess the reliability of the estimate by examining the confidence interval, which can be customized by defining the desired number of bootstraps used for resampling (Collen *et al.*, 2009a; Loh *et al.*, 2005). In addition, inflection points in the index can be identified using the bootstrap to identify time points at which the second derivative of the index differed significantly from zero (see Collen *et al.*, 2009a), that is, years in which the curvature of the index curve is statistically significant (Fewster *et al.*, 2000). A significantly positive derivative represents an increase in the rate of growth or a decrease in the rate of decline. Similarly, a set of populations is likely to have undergone a decrease in growth rate or increase in the rate of decline if the second derivative is significantly negative. As changes in the underlying species trends may not be apparent in the resulting aggregate index, it is always advisable to examine the output for the presence of change points. In conjunction with other data that can be placed temporally, such as policy changes or unusual natural events, the timing of change points can also help to identify causes of change in abundance trends (Fewster *et al.*, 2000).

Global LPI results

The global LPI shows that over a 37-year period, global population abundance of vertebrates has decreased, on average, by 29% below the baseline (95% CI: 0.64–0.80; $n = 2544$ species, 7953 populations; see Figure 4.1). The global LPI is the geometric mean of temperate and tropical indices (Figure 4.4). The temperate index, which consists of 5400 populations of 1492 species in terrestrial and freshwater species populations from the Palearctic and Nearctic realms as well as marine species

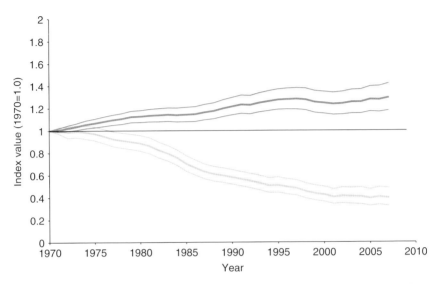

Figure 4.4 Temperate (positive slope) and tropical (negative slope) Living Planet Indices showing aggregated trends in the abundance of tropical ($n_{species} = 1216$; $n_{population} = 2543$) and temperate ($n_{species} = 1492$; $n_{population} = 5400$) vertebrate species across the globe.

populations north or south of the tropics, increases slowly over the period 1970–2007 to a value of around 29% above the baseline (95% CI: 1.18–1.42; Figure 4.4). Conversely, the tropical index, based on 2553 populations of 1216 species of terrestrial and freshwater species populations in the Afrotropical, Indo-Pacific, and Neotropical realms as well as marine species populations from the zone between the Tropics of Cancer and Capricorn, shows an average decline of 60% over the 37-year period (95% CI: 0.33–0.49; Figure 4.4).

The marked contrast in trends between temperate and tropical populations is also apparent in terrestrial, freshwater, and marine systems (not shown). The terrestrial index has decreased consistently since the mid-1970s and shows an overall 25% decline in terrestrial vertebrate populations (95% CI: 0.66–0.87). Most of this change occurred in the tropics, with tropical terrestrial species abundance halving (95% CI: 0.42–0.70) while temperate terrestrial species populations increased slightly by about 5% (95% CI: 0.97–1.14). In freshwater systems, populations of species in inland waters decreased by more than one-third from 1970 to 2007 (95% CI: 0.53–0.78), which can be attributed to freshwater tropical species decreasing in abundance by 69% (95% CI: 0.22–0.43), representing the largest decline in any of the system-based LPIs. Temperate freshwater populations appear to be faring much better with a 36% increase (95% CI: 1.12–1.66). The pronounced drop of 69% in the abundance of tropical freshwater species is particularly alarming and should be a signal for urgent action to protect life in tropical lakes, rivers, and wetlands. The tropical marine index declined by 62% (95% CI: 0.25–0.57), while the temperate marine index increased by 52% (95% CI: 1.25–1.84), resulting in a global average reduction of 24% between 1970 and 2007 in marine species abundance (95% CI: 0.6–0.95). Marine ecosystems thus show the largest discrepancy in trends between the tropical and temperate regions.

Identifying causes of decline

Any global-level index is more informative if broken down into meaningful component indices, for example of particular regions, systems, or even groups of species. Taxa do not respond in a uniform manner to anthropogenic threats (Mace *et al.*, 2003; Owens and Bennett, 2000), so making progress in conservation requires an understanding of how the action of threats might differ across groups. Meanwhile, there is geographical variation in driver intensity (Millennium Ecosystem Assessment, 2005). Therefore, any disaggregated index is only capable of supporting monitoring targets and informing policies with the help of contextual information, such as findings from other biodiversity indicators and additional information regarding changes in management and policy. Interpretation also needs to consider any intervals between censuses, which may cause changes in the trends observed. In addition, variations in the index trajectory could be attributable to delayed biological responses to external factors (McRae *et al.*, 2008). Overall, it is imperative that interpretation of an LPI takes into account changes that may have happened before the baseline year of the index in question (e.g., Lotze *et al.*, 2006; Thurstan *et al.*, 2010).

Possible reasons for the disparity between tropical and temperate systems described above are manifold but remain somewhat speculative, as the top-down approach of the LPI necessarily summarizes information and can therefore not be location-specific. In the tropics, it is clear that a combination of deforestation and other habitat destruction, driven by agricultural conversion and overexploitation from logging and hunting (Milner-Gulland and Bennett, 2003) as well as the impact of invasive species (Vitousek *et al.*, 1997; see also Chapter 7), are among the major causes of species population decreases. Declines in freshwater species abundance may be attributable to the loss and degradation of wetland areas caused by overfishing, invasive species, pollution, the creation of dams, and water diversion (Dudgeon *et al.*, 2005; Duncan and Lockwood, 2001). Of the world's oceans, 40% are estimated to be severely affected by human activities (Halpern *et al.*, 2008), so it is reasonable to assume that overexploitation, destructive fishing methods, and pollution are responsible for most of the decline in vertebrate marine fauna (Dulvy *et al.*, 2003).

Assessing data integrity

Like all indicators, the LPI has been designed to be robust, sensitive, and representative, but because it represents a summary of available vertebrate population data, it provides an overview that is necessarily only as representative as the available data feeding into it (Collen *et al.*, 2008b, 2009a; Pereira and Cooper, 2006). The legitimate worry is that due to the LPI's reliance on the published literature, some non-random selection bias results in an inaccurate index. For example, a combination of larger-bodied, commercially important, rapidly changing (either strongly positive or negative trends), or simply most easily studied populations could dominate the dataset, which would therefore not be truly representative of overall biodiversity. Monitoring schemes are more prominent in temperate regions, whereas the majority of biodiversity is undoubtedly found in the tropics (Jablonski *et al.*, 2006). Such undesirable data characteristics can be easily evaluated, and to some extent accounted for in the aggregation of the index. The effect can also be ameliorated by extracting population trends from existing widespread monitoring schemes (Gregory *et al.*, 2005), which is the approach that has been taken with the global LPI dataset (Collen *et al.*, 2009a), enabling the control and reduction of selection bias.

Data balance in the global dataset has been addressed in the short term in the index calculation by weighting tropical and temperate populations equally, and by weighting within them, the three systems equally (Figure 4.3). In the long term there are two ways in which this situation is to be further rectified:

1. Improving representation by collecting data targeting gaps.
2. The implementation of a systematic stratified design to address bias within the dataset in which weighting of different regions, taxa, and various other aspects, e.g. level of commonness or endemism, would play a crucial role (Collen *et al.*, 2009a).

While the latter approach is not currently feasible due to limited data availability, the addition of new data should achieve a dataset that is more representative geographically, taxonomically, across habitat types, and across levels of threat or degradation (Balmford *et al.*, 2005).

Another complication in the interpretation of the LPI is the fact that abundance change is measured against a baseline, which is artificially set to 1970 due to limited data availability before this point in time. It is likely that the temperate index in particular suffers from the artefact of shifting baselines (e.g., Pauly, 1995; Thurstan *et al.*, 2010). The diverging trajectories do not necessarily imply that the state of biodiversity is worse in the tropics than in temperate regions. Rather, declines occurred in temperate systems before 1970 (Millennium Ecosystem Assessment, 2005) due to the earlier onset of agricultural expansion and industrialization, with many populations now increasing in size due to the successes of concerted conservation efforts (though see Gregory *et al.*, 2005). Indeed, the choice of start year for any index may be critical, especially if it is put in a historical context and changes are compared to a pristine, historical baseline. For example, Lotze *et al.* (2006) examined historical data on relative abundance of taxonomic and functional groups from several temperate estuarine and coastal systems and found similar trajectories of long periods of slow decline followed by rapid acceleration in the last 150 to 300 years. Increases relative to the baseline may thus simply be recoveries from a state of historical depletion, as appears to be the case with many marine mammals (Lotze and Worm, 2009). For this reason, although the overall temperate trend is positive, one cannot necessarily infer that temperate populations are in a better state than tropical populations, just that they are on average increasing in abundance compared to 1970.

Implementation of the index

Indices from regions, nations, and thematic groups

In addition to implementation at the global level, however, biodiversity indicators need to be scalable so that regional, national, and other thematic indices can be produced. Changes to the method and improvements in data coverage over the past few years have now expanded the options for implementation (Collen *et al.*, 2009a). Gathering more data in a targeted manner provides a wider opportunity for disaggregating the global index and filling taxonomic and geographical gaps, and thus enabling a broader range of sub-global, regional, national, and thematic indices to be produced (Table 4.2).

Sub-global-level indices of change in abundance using the LPI approach can be implemented in two main ways: either by disaggregating data from the global index or via a more targeted approach using a data collection period for the relevant region, country, or thematic grouping. The former is constrained by the amount of data available in the existing dataset, which, despite comprising some 11 700 population time-series, is only useful for those applications for which there is already good data coverage. In order to make the index more representative rather than relying solely on data already gathered ad hoc, targeted data collection is the preferred approach. The sub-global applications of the LPI to date (Table 4.2) have used existing data from the global dataset as a base, augmented by carrying out data collection in collaboration with regional and in-country partners.

Regional, national, and thematic group examples

In response to national and international commitments to global biodiversity targets, the LPI has been used as a biodiversity indicator of global trends in biodiversity (Table 4.2), but has also been disaggregated and the method applied at a sub-global level.

Table 4.2 Use of Living Planet Index (LPI) method at global, regional, and national level, and by thematic groupings

Scale	Examples of use	References
Global	Global LPI	Collen *et al.*, 2008a; Loh 2000, 2002; Loh *et al.*, 2008; Loh and Goldfinger, 2006; Loh and Wackernagel, 2004
	Terrestrial, freshwater, and marine, tropical and temperate systems	
	Habitat (temperate and tropical forest)	Collen *et al.*, 2009a
	Taxon within habitat (grassland birds and grassland mammals)	Collen *et al.*, 2009a
Regional	Arctic	McRae *et al.*, 2010
	Regions and biogeographical realms	Collen *et al.*, 2008a; Loh *et al.*, 2006, 2008
	Mediterranean wetlands	Galewski, 2008; Galewski *et al.*, 2011
National	Norway	WWF Norway, 2005
	Uganda	Pomeroy and Tushabe, 2008
	Canada	McRae *et al.*, 2007
Thematic group	Convention on migratory species	Latham *et al.*, 2008
	Estuarine systems	Deinet *et al.*, 2010
	African Protected Areas	Craigie *et al.*, 2010

The following section outlines how the LPI method has been implemented regionally (Arctic, Mediterranean wetlands), nationally (Norway, Canada, Uganda), and thematically (migratory species) to serve the purposes of communicating biodiversity trends, in fulfilling multilateral environmental agreements, and as part of regional or national monitoring initiatives to inform national policy-makers.

Unifying national or regional monitoring initiatives can be an efficient way to harness existing data collection efforts. By communicating the results and publicizing the monitoring data used, other data providers may be encouraged to enhance the dataset from their own field of research, which can also be fed into the global database, ultimately strengthening the global index. Three initiatives focusing on Arctic species (McRae *et al.*, 2010), the Mediterranean wetlands (Galewski, 2008; Galewski *et al.*, 2011), and Uganda (Pomeroy and Tushabe, 2008) have been implemented in part for this reason. Indices based on Arctic species were published as a first report in 2010 (McRae *et al.*, 2010) and developed from the collation of data drawn from the network of eight Arctic nations and individual researchers of Arctic species and ecosystems (Figure 4.5a). This collaborative effort yielded data for 35% of Arctic vertebrate species;

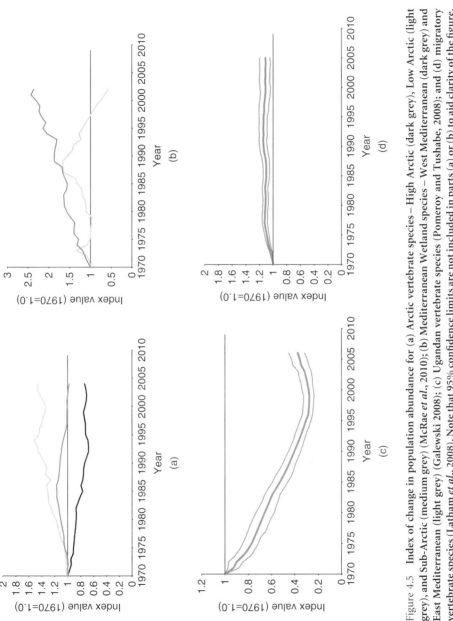

Figure 4.5 Index of change in population abundance for (a) Arctic vertebrate species – High Arctic (dark grey), Low Arctic (light grey), and Sub-Arctic (medium grey) (McRae et al., 2010); (b) Mediterranean Wetland species – West Mediterranean (dark grey) and East Mediterranean (light grey) (Galewski 2008); (c) Ugandan vertebrate species (Pomeroy and Tushabe, 2008); and (d) migratory vertebrate species (Latham et al., 2008). Note that 95% confidence limits are not included in parts (a) or (b) to aid clarity of the figure.

however, geographical distribution of data was clustered in well-studied regions and often focused at a local site scale. Encouraging further research and unifying existing data collection efforts across the pan-Arctic network was one of the recommendations of the report, so that future revisions of the Arctic indices could better represent the region geographically and taxonomically. A dedicated website (www.caff.is/asti) exists to show the results with the view to encouraging more local input and increase data coverage in the future.

The Mediterranean Wetlands Initiative promotes monitoring and awareness raising of this important ecosystem, and one of its five research centres, Tour du Valat, is using the development of an indicator to encourage data sharing and collaborative monitoring (Galewski, 2008; Figure 4.5b). This organization also has the expertise to help partners initiate new inventory and monitoring schemes thereby progressively widening the data sharing network. In a similar way, the Makerere University Institute of Environment and Natural Resources (MUIENR) in Uganda has used an application of the LPI as part of a national initiative to report on the state of biodiversity in the country to fulfil both national and international objectives (Pomeroy and Tushabe, 2008). Monitoring data are collated from different institutions in Uganda and have fed into the national index on a regular basis since 2000. However, a number of barriers exist that have impeded the expansion of this national initiative (see case study below and Figure 4.5c).

Communicating biodiversity trends to fulfil national obligations to the CBD (e.g., Xu *et al*., 2009), to inform national policy-makers, or to engage a wider public audience has also been an objective for implementing the LPI at a national level. For example, the WWF national offices of Canada and Norway have both produced national indicators using the LPI method, and communicated them via a public and policy focused report (McRae *et al*., 2007; WWF Norway, 2005).

In addition to its role as one of the indicators reporting on biodiversity trends at the global level as part of the 2010 Biodiversity Target, the index has been adapted to inform some of the other CBD themes. For example, the Circumpolar Biodiversity Monitoring Program (CBMP) is headlining a regional indicator initiative to define status and trends in Arctic biodiversity (Biodiversity Indicators Partnership, 2010). At the country level, the LPI has been used in the national reporting mechanism required by the CBD (Convention on Biological Diversity, 2010).

The LPI has also been implemented as an indicator to measure trends in migratory species (Figure 4.5d). Population trends were compared between migratory and non-migratory species, different taxonomic groups of migrants, and species grouped according to their principal migratory activity (Latham *et al*., 2008). The results were presented at the 15th Scientific Council meeting of the Convention on Migratory Species (CMS) as one of the candidate indicators for measuring progress towards the CBD 2010 target for migratory species. The decision on the set of indicators to be used for migratory species will be finalized at the 16th Scientific Council meeting and suggestions were put forward as to how the indicators can be developed in the interim.

National reporting case study: Uganda

In 2006 the Makerere University Institute of Environment and Natural Resources (MUIENR) began working with the WWF in an endeavour to assess Uganda's progression towards the CBD 2010 target by periodically producing a national Living Planet Index. This collaboration has resulted in five annual updates, and the index that results is included in MUIENR's biennial publication, the *State of Uganda's Biodiversity Report*.

The majority of population data are provided by the MUIENR National Biodiversity Data Bank, which was set up in 1990, augmented by population data from the LPI database, collected during systematic searches for the global index. The databank comprises information gathered by the MUIENR, both from ongoing data sharing with other institutions and from its own regular monitoring of selected species throughout Uganda. Though not comprehensive, the database is broad in coverage, and includes data from protected and non-protected areas and populations that are migratory and resident, utilized and unexploited, managed and unmanaged. The database is routinely updated and submitted annually to the Living Planet Index group, for the generation of a new Uganda LPI so biodiversity trends can be regularly monitored.

The most recent index (Figure 4.5c) is based on trends in 106 populations from 48 vertebrate species. It reveals that biodiversity declined rapidly in Uganda between 1970 and the late 1990s after which the decline slowed, with indication of slight recovery in recent years. Uganda's National Environment Management Authority ascribed this recovery to increasing conservation efforts and awareness (National Environment Management Authority, 2008). The result is an approximate overall loss in biodiversity of 66% between 1970 and 2006 (index value 0.34; 95% CI: 0.26–0.44; Pomeroy and Tushabe, 2010).

This decline is exemplified by a number of species but there are two particularly striking cases. The first is that of the grey-crowned crane (*Balearica regulorum*), a national bird for Uganda, which has declined in abundance by approximately 75% since 1970 (Pomeroy and Tushabe, 2010). Another dramatic decline has been observed in the straw-coloured fruit bat in Kampala, a species once common, with a population size of over 200 000 individuals (Pomeroy and Tushabe, 2008). Its population is now estimated to have been reduced to only 10% the size in 1970, due to the effects of habitat destruction at roost sites leading to the break-up of the original larger roosts (Pomeroy and Tushabe, 2008).

Although decreases in the abundance are observed in a number of taxa, the decline shown in the Ugandan National Index should be interpreted with caution. The index

Table 4.3 **Representation of numbers of species and populations by vertebrate class in the Ugandan national index of change in abundance**

Class	No. species	No. populations
Mammals	31	65
Birds	16	40
Reptiles	1	1

largely reflects the steep declines in terrestrial mammals, which contribute the majority of species to the index. Therefore, as all species are equally weighted, this group has the greatest influence (Table 4.3). This restricted dataset means that disaggregation at the system level (terrestrial and freshwater) is not yet possible. As the coverage of the dataset is improved, it will be more amenable to indicating how policy changes are affecting abundance trends – as Gregory *et al.* (2005) suggest is possible for European farmland birds – or how interventions might best be targeted (McRae *et al.*, 2008). By including a suite of indices that can complement this approach, it is possible to gain a broader picture of national biodiversity trends. For example, the *State of Uganda's Biodiversity* report (Pomeroy and Tushabe, 2008) includes an index for habitat cover, species richness, and species population index, as well as indices for individual species.

Greater population time-series representation at a national level relies on investment in local institutions that can develop the infrastructure and expertise for the implementation of long-term monitoring (Collen *et al.*, 2008b). The challenges faced, highlighted in Uganda's third national report to the CBD, include those born of financial and political insecurity, as well as limited public and stakeholder involvement. In this report, CBD Article 7 (Identification and Monitoring) was given the lowest priority rating out of 24 other CBD articles questioned (National Environment Management Authority, 2006). This outlook must change if monitoring programmes are to be effective and the indices derived from them informative.

Regardless, the Uganda LPI has had a significant impact, informing national policy at multiple levels. For example, the index has been integrated into the National Environment Management Authority's (NEMA's) *State of Environment Report for Uganda* (National Environment Management Authority, 2008), which aims to inform decision-makers and suggest new policies for the environment, in line with the Earth Summit of 1992. The Uganda LPI is also enlisted in the effort meet global commitments, to satisfy the second objective of Uganda's National Biodiversity Strategy and Action Plan, which is 'to facilitate research, information management and information exchange on biodiversity issues', and is also an important mechanism for fulfilling the requirements of CBD articles 17.1, 17.2, and 7d (National Environment Management Authority Ministry of Water and Environment, 2009).

Although there is scope for improvement and refinement, the Uganda Living Planet Index has had an impact in terms of meeting both national needs and global commitments. As long as national monitoring programmes receive the attention and funding warranted, the Living Planet Index is one mechanism that can provide the information needed by public and policy-makers to make significant changes for the benefit of Uganda's biodiversity, and ultimately its people.

Critique of strengths and weaknesses

Strengths

The extinction of species is a crude measurement for the loss of biodiversity in part due to the time lag in classifying a species as extinct and also because most species are poorly known (Mace *et al.*, 2003). Species-level extinctions are preceded by declines at the population level, so a benefit of capturing trends at this smaller scale can be a more sensitive indicator to the loss of biodiversity (Ceballos and Ehrlich, 2002). Population responses to environmental change can occur over a short time scale so using populations as the base unit means that the LPI can provide an alert to policy-makers on the decline of species populations in time to implement effective conservation measures. In addition, species population trends can be used as a proxy indicator of the state of the ecosystems that the species inhabit; they can be used as biodiversity indices in a broader sense, rather than the state of species only.

Developments of the LPI method, with input from a broad group of scientists, statisticians, conservation biologists, and population modellers, have helped optimize the use of available data and strengthened the calculation of the index. The generalized additive modelling (GAM) technique now employed is advantageous in long-term trend analysis because it allows changes in mean abundance to follow any smooth curve, not just a linear form (Collen *et al.*, 2009a; Fewster *et al.*, 2000; though see Soldaat *et al.*, 2007). In addition, the GAM approach renders the index insensitive to the effects of background noise in population fluctuation, and is more suitable for calculating significant points of change in index trajectory.

Population monitoring programmes are highly variable in terms of species and sites monitored, and methods and units used (Marsh and Trenham, 2008). By using relative measures of change over time (and ensuring consistent use of methods within a population study), it is possible to compare trends even if the method and units differ across species. This flexibility enables the collation of a wide variety of population time-series, which allows for a greater geographical coverage and representation of species within the dataset. One of the greatest strengths of the LPI is that it is applicable at many scales: global, national, regional, biome, or taxonomic class, provided sufficient data exist. The LPI can be aggregated into big-picture 'headline' indicators for

Table 4.4 **Strengths and weaknesses of the Living Planet Index method for aggregating population abundance trends**

Strengths	Weaknesses
Large and up-to-date online database	Requires large quantities of data
Long time-series data	Limited to available data
Quantitative	Restricted to vertebrate species
Responsive to environmental change over short time scale	Data are better for temperate regions than tropical regions
Insensitive to background noise in the data	Uneven coverage across groups of vertebrates (birds and mammals are better covered than amphibians, reptiles, and fishes)
Applicable at many scales (e.g. global, national, regional, taxonomic, thematic area)	Applying the method at smaller scale (e.g. national level) means index is reliant on fewer datasets
Easy to understand, present, and communicate	
Policy relevant – linked to CBD 2010 targets	

communicating broad-scale trends and also disaggregated into meaningful subsets to show underlying changes.

The LPI is easy to understand, present, and communicate, rendering it a useful reporting tool for addressing a variety of different audiences. The index shows the average change in the abundance of a large number of vertebrate species over time in a way that is analogous to well-known stock market indices. Like the FTSE it can be presented using a basic line graph, making it easy for a non-scientific audience to understand the general message without prior detailed explanation (Table 4.4).

Weaknesses and solutions

The perfect index would comprise extensive and unbiased time-series data, collected across all regions and species of the world. It is important to recognize the limitations of the LPI in order to find solutions to improve it, and to interpret the indices correctly. Weaknesses of the LPI relate to how representative the index is of changes in the parts of biodiversity in which humans are interested. Currently, trends at the species level are calculated by averaging the population trends available for that species, a criticism of which is that all changes in population size are weighted equally regardless of the size or vulnerability of the population (Pereira and Cooper, 2006). However, this doesn't appear to affect the analysis as no difference was found between indices calculated with equal and proportional weighting methods (Collen *et al.*, 2009a).

The LPI is restricted to vertebrate species, and therefore the extent to which it is a measure of global biodiversity is reliant on the extent to which trends in vertebrate species populations are representative of wider trends in all species, genes, and ecosystems (Balmford *et al.*, 2003; Loh *et al.*, 2005). It remains largely unknown how vertebrate trends relate to biodiversity as a whole (e.g., Collen *et al.*, 2009b). Plants are a major omission given their ecological importance (Pereira and Cooper, 2006); however, the LPI approach is applicable to any data that meet the selection criteria so there is potential to extend the approach to plants. Understanding the relationship between vertebrate populations and biodiversity in its broadest sense is integral to increasing the utility of the LPI for policy-making.

Data are collected according to availability, from primary literature, reports, online databases, and from specifically designed monitoring programmes. Inevitably, data are biased towards better-studied species and regions, though coverage is improving over time. Firstly, data coverage is dominated by birds and mammals, with disproportionately fewer amphibians, reptiles, and fishes. In general, temperate terrestrial systems, birds, commercially important, or large and easy-to-count species are monitored more extensively than other species (Collen *et al.*, 2008b). Secondly, geographical coverage is uneven. There are many more data from temperate regions than tropical regions in the dataset, though in the 2010 update of the index, for the first time, tropical species coverage was comparable to that of temperate species numbers (Loh *et al.*, 2010). Within temperate regions, monitoring has been more extensive in Europe and North America, and in tropical regions eastern and southern Africa have more widespread systems for tracking change in species abundance, in particular through their Protected Area systems (Craigie *et al.*, 2010). There remain some clear gaps in the data, specifically from the Neotropical and Indo-Pacific realms. Considering that the tropics are far more speciose than temperate regions (Jablonski *et al.*, 2006), and that long-term data are limited (although this is improving, e.g. Whitfield *et al.*, 2007), this area has to be the focus of future improvements.

To some extent, the geographical inequality of data is compensated for by weighting tropical and temperate species equally in the index calculation. Although it is possible to weight taxonomic classes relative to others within a realm, this is not currently advisable because of limited data when disaggregating to this level. For the LPI to be used as a tool for tracking progress towards the 2010 target, regular monitoring of a large number of species across the globe is essential to ensure biases are further reduced. Although well-designed indicator programmes do exist in some regions (e.g. Pan-European Common Bird Monitoring, 2006; Sauer *et al.*, 2008 – http://www.pwrc.usgs.gov/bbs), these are somewhat restricted at the moment (though see Chapter 10). There remains a pressing need to initiate further well-designed programmes to monitor changes in biodiversity, particularly in data-poor regions. Targeted data searches and collection have been employed in the short term to fill the data gaps highlighted here. Putting all the LPI population time-series data into a searchable interactive online database and allowing remote data input from anywhere in the world may also, it is hoped, help fill the data gaps.

A long-term solution to control for uneven representation in the dataset is to implement a systematic stratified design. For example, prior to calculating the index the data would be stratified by taxon within realm. However, this is not yet possible as the numbers of species for which data are available are too small in many realm-class combinations (e.g. amphibians in the Afrotropical realm).

Conclusions

To carry out the directives of the Convention on Biological Diversity, monitoring of biological diversity is essential (Mace and Baillie, 2007), but it is also an important component of the obligations of reporting to other international agreements. The technique of measuring change in biodiversity through time using aggregated population trends has been applied globally, regionally, and also to a variety of taxonomic groups (Buckland *et al.*, 2005; Collen *et al.*, 2009a; Gregory *et al.*, 2005; Houlahan *et al.*, 2000; Loh *et al.*, 2005). Understanding the direction, magnitude, and timing of changes in abundance over time is vital to identify populations, species, and regions of conservation concern, and to understand and address the drivers underlying such population changes. The aggregation of population trends across a set of species results in a sensitive measure of their change in abundance and allows the tracking of the impact of human pressures on that sample of species. The LPI is one of the most developed direct measures of biodiversity (Walpole *et al.*, 2009), and it can provide a general indication of how effectively policies are reducing the loss of biodiversity, and, by looking at the disaggregated indices, show where additional policies and interventions need to be targeted (McRae *et al.*, 2008). The observed overwhelming declines across many regions of the world, and the assumption that the 2010 target cannot and will not be met (Mace and Baillie, 2007), are supported by the most recent global findings (Butchart *et al.*, 2010). The global vertebrate population abundance trend and other disaggregated indices remain below the 1970s baseline, indicating that species and ecosystems are still under threat across the world.

Acknowledgements

We are grateful to Julia Latham, Mike Gill, Christoph Zockler, and Thomas Galewski for comments and their input into sub-global indices. WWF International provided funding (LM and JL). BC is supported by the Rufford Foundation.

References

Balmford, A., Green, R.E., and Jenkins, M. (2003) Measuring the changing state of nature. *Trends in Ecology and Evolution*, 18, 326–330.

Balmford, A., Bennun, L.A., ten Brink, B., *et al.* (2005) The Convention on Biological Diversity's 2010 target. *Science*, 307, 212–213.

Biodiversity Indicators Partnership (2010) National Biodiversity Indicators Portal (http://www.bipnational.net/).

Buckland, S.T., Magurran, A.E., Green, R.E., and Fewster, R.M. (2005) Monitoring change in biodiversity through composite indices. *Philosophical Transactions of the Royal Society of London, Series B*, 360, 243–254.

Butchart, S.H.M., Walpole, M., Collen, B., *et al.* (2010) Global biodiversity: indicators of recent declines. *Science*, 328, 1164–1168.

Ceballos, G. and Ehrlich, P.R. (2002) Mammal population losses and the extinction crisis. *Science*, 296, 904–907.

Collen, B., McRae, L., Kothari, G., *et al.* (2008a) *2010 and Beyond: Rising to the Bbiodiversity Challenge*. WWF, Gland, Switzerland.

Collen, B., Ram, M., Zamin, T., and McRae, L. (2008b) The tropical biodiversity data gap: addressing disparity in global monitoring. *Tropical Conservation Science*, 1, 75–88. Available at: tropicalconservationscience.org.

Collen, B., Loh, J., Whitmee, S., McRae, L., Amin, R., and Baillie, J.E.M. (2009a) Monitoring change in vertebrate abundance: the Living Planet Index. *Conservation Biology*, 23, 317–327.

Collen, B., Ram, M., Dewhurst, N., *et al.* (2009b) Broadening the coverage of biodiversity assessments. In: Vié, J.-C., Hilton-Taylor, C., and Stuart, S.N. (eds), *Wildlife in a Changing World – An Analysis of the 2008 IUCN Red List of Threatened Species*. IUCN, Gland, pp. 67–76.

Convention on Biological Diversity (2010) Convention on Biological Diversity – National Biodiversity Strategies and Action Plans. Available at: http://www.cbd.int/nbsap/.

Craigie, I.D., Baillie, J.E.M., Balmford, A., *et al.* (2010) Large mammal population declines in Africa's protected areas. *Biological Conservation*, 143, 2221–2228.

Deinet, S., McRae, L., De Palma, A., Loh, J., Manley, R., and Collen, B. (2010) *The Living Planet Index for Global Estuarine Systems*. WWF Netherlands.

Dudgeon, D., Arthington, A.H., Gessner, M.O., *et al.* (2005) Freshwater biodiversity: importance, threats, status and conservation challenges. *Biological Review*, 81, 163–182.

Dulvy, N.K., Sadovy, Y., and Reynolds, J.D. (2003) Extinction vulnerability in marine populations. *Fish and Fisheries*, 4, 25–64.

Duncan, J.R. and Lockwood, J.L. (2001) Extinction in a field of bullets: a search for causes in the decline of the world's freshwater fishes. *Biological Conservation*, 102, 97–105.

Ehemann, C. (2007) Evaluating and adjusting for chain drift in national economic accounts. *Journal of Economics and Business*, 59, 256–273.

Fewster, R.M., Buckland, S.T., Siriwardena, G.M., Baillie, S.R., and Wilson, J.D. (2000) Analysis of population trends for farmland birds using generalized additive models. *Ecology*, 81, 1970–1984.

Frisch, R. (1936) Annual survey of economic theory: The problem of index numbers. *Econometrica*, 4, 1–39.

Galewski, T. (2008) Towards an observatory of Mediterranean wetlands: Evolution of biodiversity from 1970 to the present. Tour du Valat.

Galewski, T., Loh, J., McRae, L., Collen, B., Grillas, P., and Devictor, V. (2011) Long-term trends in the abundance of Mediterranean wetland vertebrates: from global recovery to localized declines. *Biological Conservation*, 144, 1392–1399.

Gaston, K.J. (2010) Valuing common species. *Science*, 327, 154–155.

Gaston, K.J. and Fuller, R.A. (2008) Commonness, population depletion and conservation biology. *Trends in Ecology and Evolution*, 23, 14–19.

Gregory, R.D., van Strien, A., Vorisek, P., *et al.* (2005) Developing indicators for European birds. *Philosophical Transactions of the Royal Society of London, Series B*, 360, 269–288.

Halpern, B.S., Walbridge, S., Selkoe, K.A., *et al.* (2008) A global map of human impact on marine ecosystems. *Science*, 319, 948–952.

Houlahan, J.E., Findlay, C.S., Schmidt, B.R., Meyer, A.H., and Kuzmin, S.L. (2000) Quantitative evidence for global amphibian population declines. *Nature*, 404, 752–755.

Jablonski, D., Roy, K., and Valentine, J.W. (2006) Out of the tropics: dynamics of the latitudinal diversity gradient. *Science*, 314, 102–106.

Latham, J., Collen, B., McRae, L., and Loh, J. (2008) *The Living Planet Index for Migratory Species: An Index of Change in Population Abundance*. Convention on Migratory Species, United Nations Environment Programme.

Loh, J. (2000) *Living Planet Report 2000*. WWF, Gland.

Loh, J. (2002) *Living Planet Report 2002*. WWF, Gland.

Loh, J. and Goldfinger, S. (2006) *Living Planet Report 2006*. WWF, Gland.

Loh, J. and Wackernagel, M. (2004) *Living Planet Report 2004*. WWF, Gland.

Loh, J., Green, R.E., Ricketts, T., *et al.* (2005) The Living Planet Index: using species population time series to track trends in biodiversity. *Philosophical Transactions of the Royal Society of London, Series B*, 360, 289–295.

Loh, J., Collen, B., McRae, L., *et al.* (2006) The Living Planet Index. In: Loh, J. and Goldfinger, S. (eds), *Living Planet Report 2006*. WWF, Gland.

Loh, J., Collen, B., McRae, L., *et al.* (2008) The Living Planet Index. In: Hails, C. (ed.), *Living Planet Report 2008*. WWF International.

Loh, J., Collen, B., McRae, L., *et al.* (2010) The Living Planet Index. In: Almond, R. (ed.), *The Living Planet Report 2010*. WWF International, Gland.

Lotze, H.K. and Worm, B. (2009) Historical baselines for large marine mammals. *Trends in Ecology and Evolution*, 24, 254–262.

Lotze, H., Lenihan, H.S., Bourque, B.J., *et al.* (2006) Depletion, degradation, and recovery potential of estuaries and coastal seas. *Science*, 312, 1806–1809.

Mace, G.M. (2008) An index of intactness. *Nature*, 434, 32–33.

Mace, G.M. and Baillie, J.E.M. (2007) The 2010 Biodiversity Indicators: challenges for science and policy. *Conservation Biology*, 21, 1406–1413.

Mace, G.M., Gittleman, J.L., and Purvis, A. (2003) Preserving the tree of life. *Science*, 300, 1707–1709.

Marsh, D.M. and Trenham, P.C. (2008) Current trends in plant and animal population monitoring. *Conservation Biology*, 33, 647–655.

McRae, L., Loh, J., Collen, B., et al. (2007) *A Living Planet Index for Canada*. WWF, Canada.
McRae, L., Loh, J., Bubb, P.J., Baillie, J.E.M., Kapos, V., and Collen, B. (2008) *The Living Planet Index – Guidance for National and Regional Use*. United Nations Environment Programme – World Conservation Monitoring Centre, Cambridge.
McRae, L., Zockler, C., Gill, M., et al. (2010) Arctic Species Trend Index 2010: Tracking trends in Arctic wildlife. *Conservation of Arctic Flora and Fauna Circumpolar Biodiversity Monitoring Program Report No. 20*.
Millennium Ecosystem Assessment (2005) Ecosystems and human well-being: biodiversity synthesis. World Resources Institute, Washington, DC.
Milner-Gulland, E.J. and Bennett, E.L. (2003) Wild meat: the bigger picture. *Trends in Ecology and Evolution*, 18, 351–357.
National Environment Management Authority (2006) *Third National Biodiversity Report for the Convention on Biological Diversity*. National Environment Management Authority, Kampala, 166 pp.
National Environment Management Authority (2008) *State of Environment Report for Uganda*. National Environment Management Authority, Kampala, 265 pp.
National Environment Management Authority Ministry of Water and Environment (2009) *Fourth National Report to the Convention on Biological Diversity*. National Environment Management Authority, Kampala, 132 pp.
NERC Centre for Population Biology Imperial College London. (2010) The Global Population Dynamics Database. Imperial College London. Available at: http://www.imperial.ac.uk/cpb/databases/gpdd.
Owens, I.P.F. and Bennett, P.M. (2000) Ecological basis of extinction risk in birds: habitat loss versus human persecution and introduced predators. *Proceedings of the National Academy of Sciences of the USA* 97, 12144–12148.
Pan-European Common Bird Monitoring (2006) *European Common Bird Index: Population Trends of European Common Birds 2005 Update*. EBCC – European Bird Census Council.
Pauly, D. (1995) Anecdotes and the shifting baseline syndrome of fisheries. *Trends in Ecology and Evolution*, 10, 430.
Pereira, H.M. and Cooper, H.D. (2006) Towards the global monitoring of biodiversity change. *Trends in Ecology and Evolution*, 21, 123–129.
Pomeroy, D. and Tushabe, H. (2008) *The State of Uganda's Biodiversity 2008*. National Biodiversity Databank and Makerere University Institute of Environment and Natural Resources, Kampala, Uganda.
Pomeroy, D. and Tushabe, H. (2010) *The State of Ugands's Biodiversity 2010*. National Biodiversity Databank and Makerere University Institute of Environment and Natural Resources, Kampala.
Sauer, J.R., Hines, J.E., and Fallon, J. (2008) *The North American Breeding Bird Survey, Results and Analysis 1966–2007*. USGS Patuxent Wildlife Research Center, Laurel, MD.
Secretariat of the Convention on Biological Diversity (2010) *Global Biodiversity Outlook 3*. United Nations Environment Programme, Montreal, 94 pp.
Soldaat, L., Visser, H., van Roomen, M., and van Strien, A. (2007) Smoothing and trend detection in waterbird monitoring data using structural time-series analysis and the Kalman filter. *Journal of Ornithology* 148, 351–357.

Thurstan, R.H., Brockington, S., and Roberts, C.M. (2010) The effects of 118 years of industrial fishing on UK bottom trawl fisheries. *Nature Communications*, 1, 1–6.

UNEP (2006) *Report on the Eighth Meeting of the Conference of the Parties to the Convention on Biological Diversity*. Convention on Biological Diversity.

Vitousek, P.M., Mooney, H.A., Lubchenco, J., and Melillo, J.M. (1997) Human domination of Earth's ecosystems. *Science*, 277, 494–499.

Walpole, M., Almond, R., Besançon, C., *et al.* (2009) Tracking progress towards the 2010 biodiversity target and beyond. *Science*, 325, 1503–1504.

Whitfield, S.M., Bell, K.E., Philippi, T., *et al.* (2007) Amphibian and reptile declines over 35 years at La Selva, Costa Rica. *Proceedings of the National Academy of Sciences of the USA* 103, 3165–3170.

Willis, K.J., Araujo, M.B., Bennett, K.D., Figueroa-Rangel, B., Froyd, C.A., and Myers, N. (2007) How can a knowledge of the past help to conserve the future? Biodiversity conservation and the relevance of long-term ecological studies. *Philosophical Transactions of the Royal Society of London, Series B*, 362, 175–186.

WWF Norway (2005) *Naturindeks for Norge 2005 – Utfor bake med norsk nature*. WWF Norway.

Xu, H., Tang, X., Liu, J., *et al.* (2009) China's progress toward the significant reduction of the rate of biodiversity loss. *BioScience*, 59(10), 843–852.

Part II
Indicators of the Pressures on Biodiversity

5

Satellite Data-Based Indices to Monitor Land Use and Habitat Changes

Nathalie Pettorelli

Institute of Zoology, The Zoological Society of London, London, UK

Introduction

The last century was extraordinary in terms of human expansion and global changes: within 100 years, the human population grew from an estimated 1.6 billion (www.un.org) to more than 7 billions in 2012 (www.census.gov), while mean life expectancy increased from 30–40 years to more than 70 years (www.who.int). During this same period, the proportion of people living in urban areas increased from 13% to an estimated 50% (www.un.org). Meanwhile, the development of the commercial aviation and the car industry dramatically increased our ability to travel (both in terms of distance and frequency), increasing immigration and emigration, and technology and good exchanges worldwide. With the advent of the automobile and airplanes, oil became the dominant fuel during the twentieth century, and in 2005 total worldwide energy consumption was estimated to reach 500 exajoules, with 80–90% derived from the combustion of fossil fuels (www.eia.doe.gov).

Parallel to these changes, the twentieth century also witnessed a change in atmospheric gas concentrations leading to changes in climate (Easterling *et al.*, 2000; IPCC, 2007b). The global atmospheric concentration of carbon dioxide (CO_2) increased from a pre-industrial value of about 280 ppm to 379 ppm in 2005, and between 1970 and 2004 annual CO_2 emissions grew by about 80%, from 21 to 38 gigatonnes. The global atmospheric concentration of methane has increased from a pre-industrial value of about 715 ppb to 1774 ppb in 2005, while the global atmospheric nitrous oxide concentration has increased from a pre-industrial value of about 270 ppb to 319 ppb in 2005. Within 100 years, the global surface temperature increased by 0.74 $\pm 0.18°C$

Biodiversity Monitoring and Conservation: Bridging the Gap between Global Commitment and Local Action, First Edition. Edited by Ben Collen, Nathalie Pettorelli, Jonathan E.M. Baillie and Sarah M. Durant.
© 2013 John Wiley & Sons, Ltd. Published 2013 by John Wiley & Sons, Ltd.

(IPCC, 2001; IPCC, 2007). This translated into, glacier reduction, ice sea surface reduction, or sea level increase (Meehl *et al.*, 2005): since 1900, the maximum areal extent of seasonally frozen ground has, for example, decreased by about 7% in the northern hemisphere, with decreases in spring of up to 15% (IPCC, 2007b). Global average sea level rose at an average rate of 1.8 (1.3–2.3) mm per year over 1961 to 2003 and at an average rate of about 3.1 (2.4–3.8) mm per year from 1993 to 2003 (IPCC, 2007b), while 11 of the last 12 years (1995–2006) rank among the 12 warmest years in the instrumental record of global surface temperature (since 1850; IPCC, 2007b).

But climate change is only one of the environmental changes characterizing the twentieth century, as human expansion also translated into land use changes, habitat degradation, or habitat fragmentation (Reynolds *et al.*, 2002; World Meteorological Organization, 2005; Millennium Ecosystem Assessment, 2005a). In the face of such drastic environmental changes, species have started to exhibit diet change, variation in life history, changes in population growth rates, changes in temporal patterns (such as changes in migration date, egg-laying date), or changes in spatial patterns (with, e.g., some species shifting their distribution range) (Crick and Sparks, 1999; Inouye *et al.*, 2000; Stenseth *et al.*, 2002; Visser *et al.*, 2004). However, not all species were able to keep up with the rate of change imposed by human expansion, and the twentieth century is also characterized by numerous species disappearing from the surface of the Earth (Millennium Ecosystem Assessment, 2005b). But how fast is biodiversity decreasing? Are there any hotspots where it disappears faster or in greater number? To answer such questions, we need to monitor biodiversity and biodiversity change across the world and link it to measurable changes.

Satellites and their possible role in helping monitoring biodiversity through habitat change

Defining biodiversity and habitat

Biodiversity can be defined as 'variation of life at all levels of biological organization' (Gaston and Spicer, 2004): biodiversity (or biological diversity) thus encompasses all the variety of life forms at all biological system levels (i.e., molecular, organismic, population, species, and ecosystem). Biodiversity therefore refers to a concept associated with many levels, so that no single measure can fully encompass it. How can we then monitor biodiversity? We can, for example, monitor direct components of biodiversity, such as species richness, by assessing the number of species and their relative abundance per site and monitoring changes over time. But we can also estimate biodiversity change by monitoring variables that are expected to correlate with biodiversity (Hill *et al.*, 2005). Examples of variables that correlate with biodiversity loss include habitat loss, human density, abundance of invasive species, or intensity of wildlife trade (Wood *et al.*, 2000; McNeely, 2001; Hill *et al.*, 2005).

Because of the close association between biodiversity and habitat (Statzner and Moss, 2004), and because, depending on how habitat is defined, it could be seen as a component of biodiversity, habitat monitoring is a priority for several international conventions, as well as international and national organizations. The Convention on Biological Diversity for example listed the conservation of ecological regions and the decrease of habitat degradation and habitat loss as being two major targets for 2010 (www.cbd.int/). The Convention on the Conservation of Migratory Species (CMS) explicitly states that 'parties that are range states of a migratory species shall endeavour to conserve and, where feasible and appropriate, restore those habitats of the species which are of importance in removing the species from danger of extinction' (www.cms.int). Focusing on forest monitoring, the UN Programme on Reducing Emissions from Deforestation and Forest Degradation in Developing Countries (REDD; see Glossary) aims to contribute to the development of capacity for implementing REDD and to support the international dialogue for the inclusion of a REDD mechanism in a post-2012 climate regime (www.undp.org/mdtf/un-redd/overview.shtml).

But how can we define habitat? According to Morrison and colleagues (1992), habitat is defined as the resources and conditions present in an area that produce occupancy – including survival and reproduction – by a given organism. The term habitat is thus organism-specific, and relates the presence of a species, population, or individual to an area's physical and biological characteristics. A habitat is defined as an area that provides a given organism with resources that allow it to survive – meaning that migration and dispersal corridors and the land that animals occupy during breeding and non-breeding seasons are habitats (Hall *et al.*, 1997). The spatial scale at which it is defined is a function of several variables, such as the number of individuals considered or the size of the species considered. The term habitat thus encompasses several variables, of different importance according to the species considered, the location considered, and the spatio-temporal scale considered (e.g., something might be important for one species and not for another, something might be important for a species in this location but not in that one, or something might be important at that time and that place for a given species, but might become less important as time goes). Because the term habitat can encompass several, sometimes correlated variables, because it can be defined at several spatial scales, and because habitat selection is scale-dependent (see, e.g., Ciarniello *et al.*, 2007; Zweifel-Schielly *et al.*, 2009), monitoring habitat change is far from being trivial. Like any form of monitoring, habitat monitoring is thus linked to several questions such as 'At which spatial scale is it relevant to monitor? How often do we need to monitor? What should we monitor?'

Satellites: the potential 'goldmine'

For decades, habitat monitoring has been carried out by sending people into the field. The good side of such an approach is that there is little restriction regarding what can be monitored (e.g., possibility to monitor the distribution of termite mounds, caves, waterholes, or the distribution of a given tree species). The downside is that the data are traditionally collected at small spatial and temporal scales, they vary in their type and reliability, and are therefore difficult to use for assessing or predicting regional or global change. In such a context, satellite monitoring could be seen as a potential goldmine (Kerr and Ostrovsky, 2003; Turner *et al.*, 2003; Pettorelli *et al.*, 2005a; Pettorelli *et al.*, 2011). With data collected at the scale of the Earth and monitoring programmes that started in the 1970s, satellites indeed offer a great possibility to carry out relatively cheap long-term monitoring of variables that might help assess changes in habitat distribution and degradation, at spatial resolutions spanning from a few metres up to a few kilometres (see, e.g., Lahoz-Monfort *et al.*, 2010; Hortal *et al.*, 2010). Satellite data can, for example, provide valuable information regarding changes in forest distribution, forest composition, primary productivity, or phenology, or in forest and land degradation (Pettorelli *et al.*, 2005a). They can also help monitor droughts, fires, frosts, floods, ice sheet area changes, or the spatial distribution of human disturbances such as night-time light brightness or road density (Campbell, 2007; Jensen, 2007). Remote sensing can therefore help quickly identify areas of concern at a global scale, supporting managers in their efforts to design and apply adaptive management strategies. It can also provide a cost-effective way to target monitoring effort, by identifying areas with rapid changes in functional attributes of ecosystems where more intense monitoring might be required. Advantages of remote sensing particular to vegetation studies include the fact that it is contactless and non-destructive.

Satellite and remote sensing: how does it work?

All objects emit radiation, albeit in various amounts and at differing wavelengths. Radiation travels in a wave-like manner and the distance between wave peaks is known as the wavelength. When organized by wavelength and frequency, these emissions collectively form the electromagnetic spectrum. The electromagnetic spectrum spans from high-frequency, short-wavelength gamma rays to low-frequency, long-wavelength radio waves (Jensen, 2007). As well as emitting radiation, all objects reflect different wavelengths of light emitted by other objects (such as the Sun): a red carpet is, for example, red because the carpet does not absorb well the red light emitted by the Sun. Likewise, chlorophyll absorbs in the red but not in the green, meaning that objects that contain a lot of chlorophyll (such as healthy leaves) tend to be green (Jensen, 2007). Humans can only see a very small part of the electromagnetic spectrum, the visible spectrum, and unlike bees or butterflies, cannot, for example, see what is going on in the ultra-violet or near-infrared part of the spectrum. Sensors, on the other hand, are able to capture information on a much larger range of the spectrum, and remote sensing could be defined as deriving information about the Earth's surface by measuring the electromagnetic radiations it reflects (Tucker, 1979; Campbell, 2007). Remote sensing involves the use of sensors, which can be placed on board aircraft or satellites. Satellites offer several advantages over airborne-based monitoring platforms: they provide a synoptic view (observation of large areas in a single image), as well as fine detail and systematic, repetitive coverage. Such capabilities are well suited to creating and maintaining a worldwide cartographic infrastructure as well as to monitoring broad-scale environmental problems (Campbell, 2007).

Satellites: a world of possibilities

The world of aerial-picture-like information

Monitoring at very high resolution started with the Satellite Pour l'Observation de la Terre (SPOT), a system that was initiated by the Centre National d'Etudes Spatiales in the 1970s and that was developed in association with the Belgian Scientific, Technical and Cultural Services and the Swedish National Space Board. SPOT is a high-resolution, optical imaging Earth observation satellite system, which means that SPOT satellites are specifically designed to observe the Earth from an orbit, similar to reconnaissance satellites but intended for non-military uses. The first satellite, SPOT 1, was launched in 1986 and was designed to produce multispectral pictures at a 20 m resolution. The current SPOT satellite, SPOT 5, was launched in 2002 with 2.5 m to 5 m (panchromatic (black-and-white) mode) and 10 m (multispectral mode) spatial resolution capabilities. The SPOT orbit is polar, circular, and Sun-synchronous, and the satellite flies over any point on Earth within 26 days (www.spotimage.fr/web/en/172-spot-images.php).

The real explosion of 'Google Earth' type of satellite imagery occurred in the early years of the twenty-first century, with the launch of IKONOS and QuickBird. Ikonos imagery began being sold in 2000, and Ikonos was the first satellite to collect publicly available high-resolution imagery at 1 m (panchromatic mode) and 4 m (multispectral mode) spatial resolutions. Imagery from both sensors (multispectral and panchromatic) can be merged to create 1 m colour imagery (referred as pan-sharpened; see www.geoeye.com for more details). At such resolution, detail such as buildings or other infrastructure are visible and the imagery can be imported into remote sensing image-processing software, as well as into GIS packages for analysis. QuickBird, on the other hand, was launched in 2001 and collects panchromatic imagery at 60–70 cm resolution and multispectral imagery at 2.4 m and 2.8 m resolutions (for more details, see www.digitalglobe.com). The latest satellite from GeoEye Inc., GeoEye-1, is currently the world's highest resolution commercial Earth-imaging satellite, with

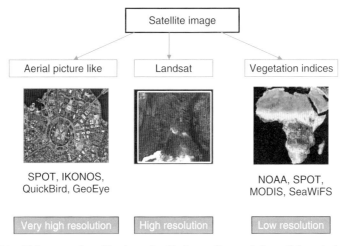

Figure 5.1 **Main types of satellite data, classified according to their spatial resolution.**

images collected at 41 cm (panchromatic mode) and 1.65 m (multispectral imagery) resolution.

Satellites such as QuickBird, IKONOS, or GeoEye-1 can be perfect for gathering high-resolution, spatially precise information, and can, for example, help monitor the distribution of human disturbances (e.g. deforestation) or the distribution of human infrastructures such as road or rail networks, or city delimitation. Because of the costs of the images (all these satellites are commercial) as well as their high spatial resolution (meaning that the number of images required to cover an area increases dramatically with the size of the area), such satellites are most useful to projects that have the required amount of funds and aim to monitor a relatively small area at a relatively low temporal scale (Figure 5.1).

The Landsat Program

The Landsat ('land satellite') Program is a series of Earth-observing satellite missions jointly managed by NASA and the US Geological Survey (http://landsat.gsfc.nasa.gov/). The Landsat system consists of spacecraft-borne sensors that observe the Earth and transmit information by microwave signals to ground stations that receive and process the data for dissemination. The first Landsat satellite was launched in 1972, while the most recent one, Landsat 7, was launched in 1999. Landsat sensors have a moderate spatial resolution (e.g. Landsat 7 spatial resolution ranges from 15 m to 60 m).

Contrary to images produced by GeoEye-1 or QuickBird, individual houses cannot be individualized on a Landsat image, but large man-made objects such as highways can be distinguished. Landsat thus provides access to an important level of spatial resolution – one that is coarse enough for global coverage, yet detailed enough to characterize human-scale processes such as urban growth. Landsat 80-, 30-, and 15-m satellite data are, moreover, the only record of global land surface conditions at a spatial scale of tens of metres spanning the last 30 years (Tucker *et al.*, 2004; Figure 5.1).

The particularity of Landsat satellites is that they collect several images at once (all of the images are obtained at the same time, and at the exact same location), with each image showing a specific section of the electromagnetic spectrum, called a band. Landsat 7, for example, collects information in seven bands, with each band maximizing the ability to differentiate particular objects or structure: Band 1 helps coastal water mapping, soil/vegetation discrimination, forest classification, and man-made feature identification. Band 2 helps vegetation discrimination and health monitoring, as well as man-made feature identification. Band 3 helps plant species identification and man-made feature identification, while Band 4 is useful when monitoring soil moisture and vegetation, as well as for water body discrimination. Band 5 is generally used for vegetation moisture content monitoring, and Band 6 for monitoring surface temperature, vegetation stress, and soil moisture. Band 6 can also help with cloud differentiation and volcanic monitoring. Band 7, finally, helps with mineral and rock discrimination, as well as with vegetation moisture content monitoring. It is the combination of the information collected in each band that allows mapping of habitats. Such mapping, however, generally involves a first phase of ground truth and calibration, where image data are related to real features and materials on the ground (see, e.g., Koy *et al.*, 2005).

Landsat is probably one of the most successful satellite programmes to date, providing pictures that can help the mapping and/or monitoring of, for example, deforestation, species suitable habitat extent within a defined area, water pollution, drought impact, fires, or mine waste pollution (Turner *et al.*, 2003; Koy *et al.*, 2005; see http://landsat.gsfc.nasa.gov for an extensive list). Thanks to long-term programmes such as this one, satellite imagery can be applied retrospectively across wide regions, creating fantastic opportunities to reanalyse old data and make use of previously unavailable information. Landsat images are free, which contributes to the success of this programme. Landsat 7 Enhanced Thematic Mapper Plus (ETM+) scan-line corrector, however, failed in 2003, leading to a reduction in data quality. Although such failure is regrettable, 78% of the data within each Landsat 7 scene are unaffected, meaning that Landsat 7 is still providing valuable information for many environmental applications (Loveland *et al.*, 2008). NASA and the USGS are, moreover, currently developing the Landsat Data Continuity Mission, which should result in the launch of Landsat 8 (Loveland *et al.*, 2008).

Radio detection and ranging (RADAR) and light detection and ranging (LIDAR)

The sensors on board the satellites introduced so far in this chapter can be all categorized as 'passive', meaning that they measure radiation that reaches a detector without the sensor first transmitting a pulse of radiation (Table 5.1). Passive sensors are all sensitive to cloud cover, and land-cover and land-use monitoring generally uses passive sensors to measure visible, near- and middle-infrared, and thermal-infrared radiation. 'Active' sensors, on the other hand, emit a pulse and later measure the energy bounced back to a detector, and such sensors include RADAR and LIDAR systems. Vegetation structure and ground surface elevations are often measured using active sensors (Turner *et al.*, 2003).

The principle of RADAR systems is based on a transmitter emitting microwaves or radio waves, which bounce off any object in their path. The object then returns a tiny part of the wave's energy to a dish or antenna (which is usually located at the same site as the transmitter) and the time it takes for the reflected waves to return to the dish enables a computer to calculate how far away the object is (Kasischke *et al.*,

Table 5.1 Non-exhaustive list of relevant satellite sensors for biodiversity monitoring and assessment. Adapted from Gillespie *et al.* (2008)

Sensor	Type	Spatial resolution
QuickBird	Passive	0.6–2.5 m
IKONOS	Passive	1–4 m
OrbView	Passive	1–4 m
Landsat (TM, ETM+)	Passive	15–120 m
IRS (LISS III)	Passive	5–70 m
EOS (ASTER)	Passive	15–90 m
SPOT	Passive	2.5–1150 m
EOS (Hyperion)	Passive	30 m
ALOS	Passive	2.5–10 m
NOAA (AVHRR)	Passive	1100 m
EOS (MODIS)	Passive	250–1000 m
SRTM	Active	30–90 m
QSCAT	Active	2500 m
Radarsat-1	Active	8–100 m
SIR-C	Active	10–200 m
TRMM (TMI)	Active	18000 m
ERS-2	Active	26 m
Envisat (ASAR)	Active	30–1000 m

1997; Goetz *et al.*, 2009). Synthetic-aperture radar (SAR) is a form of RADAR in which multiple radar images are processed to yield higher-resolution images. These multiple RADAR images are obtained either by using a single antenna mounted on a moving platform (such as a spacecraft) or by using many low-directivity small stationary antennas scattered over an area near the target area. In both cases, the many echo waveforms received at the different antenna positions are post-processed to resolve the target. SAR and RADAR are sometimes employed indifferently (Goetz *et al.*, 2009). A number of RADAR/SAR satellites are currently in operation, including RADARSAT, ALOS/PALSAR, ENVISAT/ASAR, TerraSAR-X, and SkyMed (Goetz *et al.*, 2009). SAR sensors can operate day or night, and can penetrate through haze, smoke, and clouds. Such sensors systems are associated with several capabilities useful to habitat monitoring. When it comes to forest monitoring, the main attraction of SAR systems is that they provide a possibility to monitor above-ground biomass (AGB) in forested areas, and that such monitoring remains possible even in persistently cloudy areas (Goetz *et al.*, 2009). However, extensive analyses with existing SAR sensors suggest the sensitivity of RADAR backscatter saturates around 100–150 tons/ha (Kasischke *et al.*, 1997), meaning that such systems are unable to monitor AGB variation in densely vegetated areas. Preliminary investigations using RADARSAT-1 imagery have then proven especially useful for identifying wetland habitats and flooded areas, and since vector-borne diseases are most often found in tropical environments and during rainy seasons with persistent cloud cover, RADAR/SAR sensors could become an important tool for monitoring and mapping the environmental indicators of disease (Kaya *et al.*, 2004).

In contrast to RADAR/SAR sensors, which make use of microwaves, LIDAR is an optical remote-sensing technology that determines distance to an object using laser pulses (Dubayah and Drake, 2000). Like RADAR technology, the range to an object is determined by measuring the time delay between transmission of a pulse and detection of the reflected signal. LIDAR has many applications in forestry or agriculture, providing a fantastic opportunity to access measurements of canopy height or AGB, which can then be related to wildlife. For example, data on crop and field boundary height derived from LIDAR data were used to predict the distribution of breeding skylarks (*Alauda arvensis*) in a farmed landscape (Bradbury *et al.*, 2005). In the same article, the authors presented another example where LIDAR-retrieved canopy height and structural data were used to predict the breeding success of great tits (*Parus major*) and blue tits (*Cyanistes caeruleus*) in broad-leaved woodland (Bradbury *et al.*, 2005). Topographic maps can also be generated readily from LIDAR (Turner *et al.*, 2003), and in oceanography, LIDAR has been used to estimate phytoplankton fluorescence and biomass in the surface layers of the ocean (Churnside and Ostrovsky, 2005; Churnside and Donaghay, 2009). Mean body size and species composition of forest-dwelling beetle assemblages were shown to have considerable predictability using LIDAR-derived variables (Muller and Brandl, 2009). When it comes to AGB monitoring, the main differences between LIDAR and RADAR/SAR are that (i) LIDAR does not saturate, meaning that LIDAR systems are able to monitor AGB variation in densely vegetated areas, and (ii) LIDAR is highly sensitive to aerosols and clouds, meaning that AGB monitoring using LIDAR is impossible in persistently cloudy areas (Goetz et al., 2009). There is only one LIDAR instrument currently operating from a satellite platform, and this instrument was not designed specifically for vegetation characterization – the currently operating Geoscience Laser Altimetry System (GLAS) on board ICESAT was indeed originally designed for monitoring ice dynamics (Lefsky et al., 2005; Sun et al., 2008) ICESAT was retired in February 2010.

The Normalized Difference Vegetation Index (NDVI) and its promising links with ecology

I have so far described the type of information that can be accessed by each sensor on board various satellites, such as land cover, land use, vegetation structure, and ground surface elevations. I will now focus on one type of information derived from various sensors and satellites (e.g. Landsat, AVHRR, MODIS, SeaWiFS, SPOT), namely the information encapsulated in vegetation indices. Vegetation indices are quantitative measures that attempt to measure vegetation biomass or vegetative vigour. Usually, vegetation indices are formed from combinations of several spectral values that are added, divided, or multiplied in a manner designed to yield a single value that indicates the amount or vigour of vegetation within a pixel (Campbell, 2007). Among the various vegetation indices that have been proposed, the Normalized Difference Vegetation Index (NDVI, see Box 5.1) has proven to be extremely useful to ecologists dealing with assessing ecological responses to environmental change (Pettorelli et al., 2005a, 2011).

Because the NDVI directly correlates with vegetation productivity, the possible applications of this index for ecological purposes are wide: the NDVI can help differentiate savanna, dense forest, non-forest, and agricultural fields (Achard and Blasco, 1990; Achard and Estreguil, 1995), and phenological characteristics can be used to determine evergreen forest versus seasonal forest types (Achard and Estreguil, 1995; Van Wagtendonk and Root, 2003), or trees versus shrubs (Senay and Elliott, 2002). However, differentiating between forests with, for example, different dominant species cannot be achieved using NDVI, because several assemblages of plant species can produce a similar NDVI value or a similar NDVI temporal trend. Even with data of sufficiently high spectral and spatial resolution, few plant species, if any, can be identified accurately (Nagendra, 2001). The NDVI can then be used to map, assess or predict the occurrence and impact of disturbances such as drought, fire, flood, and frost (Singh et al., 2003; Maselli et al., 2003; Wang et al., 2003; Tait and Zheng, 2003). It can help map and predict land degradation (Thiam, 2003), as shown by a study carried out by Prince and colleagues (2009) in Zimbabwe. Based on the difference

Box 5.1 **The Normalized Difference Vegetation Index (NDVI)**

The Normalized Difference Vegetation Index (NDVI) is a satellite-based vegetation index that correlates strongly with above-ground net primary productivity (Running, 1990; Myneni *et al.*, 1995). The NDVI is derived from the red:near-infrared reflectance ratio:

$$\text{NDVI} = (\text{NIR} - \text{RED}) / (\text{NIR} + \text{RED})$$

where NIR and RED are the amounts of near-infrared and red light, respectively, reflected by the vegetation and captured by the sensor of the satellite. The formula is based on the fact that chlorophyll in green leaves absorbs incoming solar radiation in the Photosynthetic Active Radiation (PAR) spectral region, drawing from this irradiance the energy needed to power photosynthesis. Green leaf molecules, however, reflect infrared radiation, since its absorption would overheat the plant and denature its proteins (Tucker, 1979; Jensen, 2007). So green leaves have low high visible light absorption and high near-infrared reflectance, which will result in NDVI values close to 1 (Tucker *et al.*, 1985). Conversely, senescing vegetation, soil, cloud, and snow will have higher near-infrared absorbance, thus driving NDVI values closer to -1 (Tucker *et al.*, 1985; Neigh *et al.*, 2008).

Several satellites collect information that can be used to calculate an NDVI value per pixel. The oldest ones are Landsat (Van Wagtendonk and Root, 2003) and the NOAA/AVHRR satellites (Tucker *et al.*, 2005), both providing more than 25 years worth of information on the temporal and spatial distributions of primary productivity across the globe. More recently, other datasets originating from other programmes have appeared such as SPOT, SeaWiFS, or MODIS (Tucker *et al.*, 2005).

The NDVI is a crude estimate of vegetation health and a means of monitoring changes in vegetation over time (Goward and Prince, 1995; Pettorelli *et al.*, 2011), and its ability to monitor variations in primary productivity can sometimes be reduced (Markon and Peterson, 2002). NDVI indeed integrates the composition of species within the plant community; the vegetation form, vigour, and structure; the vegetation density in vertical and horizontal directions; the reflection, absorption, and transmission within and on the surface of the vegetation or ground; and the reflection, absorption, and transmission by the atmosphere, clouds, and atmospheric contaminants (Markon *et al.*, 1995; Pettorelli *et al.* 2011). The quality of the information regarding primary productivity variation encompassed in NDVI values is therefore a function of the type of processing applied on raw data (Markon and Peterson, 2002; Tucker *et al.*, 2005), as well as the spatial location: the relationship between NDVI and vegetation can, for example, be biased in low vegetated areas and very dense canopies (Huete, 1988). Moreover, clouds cover about 60% of the land surface at any given time (Rossow and Schiffer, 1999), meaning that NDVI might be of limited use in areas of persistent cloudiness.

between potential and observed primary productivity, the authors were indeed able to produce a map of land degradation, at the scale of the country, which matched independent available information.

Because it provides information on vegetation phenology and biomass (Figure 5.2), the NDVI can be used to assess vegetation quantity and quality (Pettorelli *et al.*, 2007, 2009), and such information has been shown to correlate with the distribution, behaviour, life-history traits, and abundance of several species (Table 5.2). In Norway, NDVI values around the vegetation onset were shown to be a strong predictor of red deer (*Cervus elaphus*) altitudinal migration date (Pettorelli *et al.*, 2005b), while NDVI-based estimates of vegetation growth proved to be helpful in understanding wildebeest (*Connochaetes taurinus*) migration in the Serengeti (Boone *et al.*, 2006). In Kenya, the NDVI was a strong predictor of topi (*Damaliscus lunatus*) distribution (Bro-Jørgensen *et al.*, 2008), while the home ranges of the grivet monkey (*Cercopithecus aethiops*) exhibited a significantly higher average NDVI value than the average of the survey area in Eritrea (Zinner *et al.*, 2002). Birth occurrence and synchrony were reported to be driven by NDVI variation for buffaloes (*Syncerus caffer*) in South Africa (Ryan *et al.*, 2007), while NDVI variations were shown to correlate with elephant (*Loxodonta africana*) diet in Kenya (Wittemyer *et al.*, 2009). In Norway and France, NDVI values around the vegetation onset were shown to be a strong predictor of body mass for reindeer (*Rangifer tarandus*) and roe deer (*Capreolus capreolus*) (Pettorelli *et al.*, 2005c, 2006). In Africa, satellite-based primary productivity estimates were reported to correlate positively with density estimates of 13 ungulate species, although this link was more pronounced for some species than others (Pettorelli *et al.*, 2009). Interestingly, energy has been hypothesized to determine species richness through its effects on total species biomass or abundance (Hurlbert, 2004; Evans *et al.*, 2005, 2006), meaning that biodiversity might correlate with primary productivity distribution. Such results have been reported in the United States, with NDVI-based estimates of productivity being reported to explain up to 61% of avian species richness (Hurlbert and Haskell, 2003). Because climatic models can be used to predict spatio-temporal changes in the

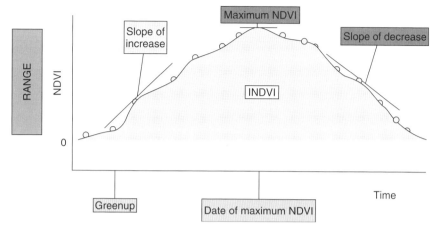

Figure 5.2 Different indices that can be derived from Normalized Difference Vegetation Index (NDVI) time-series over a year: the slopes of increase (spring) and decrease (fall); the maximum NDVI value; the integrated NDVI (INDVI, the sum of NDVI values over a year); the date when the maximum NDVI value occurs; the range of annual NDVI values; and the date of greenup.

NDVI (see, e.g., Anyamba *et al.*, 2006; Funk and Brown, 2006), such results highlight satellite-based analyses as a potentially crucial tool in assessing the population (and potentially biodiversity) consequences of future changes in primary productivity on this continent.

The NDVI is thus a vegetation index that has demonstrated its usefulness in many ecological studies, as exemplified above. However, in some particular situations, other vegetation indices might be more appropriate than the NDVI. The relationship between the NDVI and vegetation can, for example, be biased in sparsely vegetated areas and very dense canopies (Huete, 1988). In sparsely vegetated areas with a leaf area index (LAI, defined as the one-sided green leaf area per unit ground area) of less than 3, the NDVI is indeed mainly influenced by soil reflectance, whereas for LAI > 6, that is, in densely vegetated areas, the relationship between NDVI and near-infrared (NIR) saturates (Asrar *et al.*, 1984). In sparsely vegetated areas, the SAVI (soil-adjusted vegetation index; Huete, 1988) might perform better than the NDVI in indexing vegetation biomass (Pettorelli *et al.*, 2005a). Another index that has appeared with MODIS is the Enhanced Vegetation Index (EVI; Huete *et al.*, 2002). The idea behind this index is to provide complementary information on the spatial and temporal variations of vegetation, while minimizing many of the contamination problems present in the NDVI, such as those associated with canopy background and

Table 5.2 **Examples of ecological studies linking the Normalized Difference Vegetation Index (NDVI) to animal data**

Parameters	Taxa/species	References
Diet	Elephants	Wittemyer et al., 2009
Habitat use	Grivet monkey	Zinner et al., 2002
	Mongolian gazelles	Mueller et al., 2008
	Ticks	Estrada-Pena et al., 2004
	Vervet monkeys	Willems et al., 2009
	Reindeer	Hansen et al., 2009
Movement patterns	Red deer	Pettorelli et al., 2005b
	Wildebeest	Boone et al., 2006
	Mongolian gazelles	Ito et al., 2006
	Elk	Hebblewhite et al., 2008
	Elephants	Wittemyer et al., 2009
	Roe deer	Andersen et al., 2004
Body mass	Roe deer	Pettorelli et al., 2006
	Red deer	Pettorelli et al., 2005b
	Reindeer	Pettorelli et al., 2005c
	Reindeer	Couturier et al., 2009
Reproductive patterns	Buffalo	Ryan et al., 2007
	Barn swallow	Saino et al., 2004
	Red deer	Loe et al., 2005
	Elephants	Trimble et al., 2009
Survival	White stork	Schaub et al., 2005
	Alpine ungulates	Pettorelli et al., 2007
Abundance	Topis	Bro-Jørgensen et al., 2008
	Ungulates	Pettorelli et al., 2009
Species richness	Birds	Hurlbert and Haskell, 2003
	Birds	St-Louis et al., 2009
	Birds	Ding et al., 2006

residual aerosol influences. Whereas the NDVI is chlorophyll sensitive and responds mostly to red band variations, the EVI is more NIR sensitive and responsive to canopy structural variation, including leaf area index, canopy type, and canopy architecture. This index is thus meant to take full advantage of MODIS's new, state-of-the-art measurement capabilities. An important point is also that the EVI does not become saturated as easily as the NDVI when viewing rainforests and other areas of the Earth with large amounts of green material. However, the EVI has been developed on MODIS data and so does not exist before 2000.

Limits to satellite data

As we saw in the previous sections of this chapter, satellites, sensors, and satellite-derived vegetation indices such as the NDVI have their limits. Passive sensors and LIDAR are all sensitive to cloud cover, meaning that the ability for passive sensors to monitor land-cover and land-use change is limited in areas of persistent cloudiness. On the other hand, RADAR systems, which are not sensitive to cloud cover, are unable to monitor AGB variation in densely vegetated areas. Problems can then occur at high latitudes and during the winter, because reflectance resolution can deteriorate in such areas (and such deterioration will impact vegetation indices such as NDVI; Goward et al., 1991). Errors can also occur near the Equator (from 30°N to 30°S) owing to solar zenith angle variation within most of the satellite records (Tucker et al., 2005) and for pixels near water bodies (mixed pixels; Justice et al., 1985). Specific limits linked to the use of the NDVI are detailed in Pettorelli et al. (2005a). Errors can sometimes occur in back-calculating the position of the satellite at the time the image was taken, meaning that the accuracy of the downloaded data should be checked by superimposing the satellite data on known maps. Depending on the sensor and the satellite, the temporal and spatial resolutions can vary: some information can be available daily at the 8 × 8 km scale, some Landsat images are only available once a decade, some information started to be collected in the 1980s, some less than a decade ago.

Conclusions: biodiversity monitoring, satellites, and national reporting

With the increase in international treaties and conventions aimed at coordinating efforts among countries to face global environmental change and its expected consequences on biodiversity – such as the Kyoto Protocol to the United Nations Framework Convention on Climate Change (UNFCCC), the Convention on Biological Diversity (CBD), or the Convention on Migratory Species (CMS) – there is a real need to monitor the Earth using standardized methodologies, and in such a context, environmental data gathered by satellites are expected to play a significant role in the current and future management and implementation processes (Onoda, 2008). Satellites could indeed provide a relatively cheap and relatively accurate way to help monitor national change in the distribution of certain types of habitat (DeFries et al., 2007) as well as land degradation (Goetz et al., 2009; Prince et al., 2009). Programmes such as Landsat or NOAA/AVHRR have, moreover, demonstrated that satellites can provide long-term information, meaning that satellite-based monitoring programmes have a high probability of being sustainable. Satellite-based data could then allow nations to report standardized and transparent information, which can be easily verified since some of these data are already freely available and since the reported information can be downscaled to allow verification using ground-based methodologies. Satellite-based information can finally provide a cost-effective way of concentrating our monitoring efforts, by identifying areas where more intense monitoring might be required.

However, several challenges remain. First, each convention, treaty, or national agreement is linked to specific terminology and targets, and it will be important to define the role of satellite measurements in relation to the treaty terminology and management procedures, in order to make the best use of satellite systems and data (Onoda, 2008). Then, each country is linked to specific environmental and economic conditions, so that, for a given common target (e.g. monitoring deforestation), no single method, index, or satellite will be appropriate for all national circumstances. It can be anticipated that each country might choose the method it will follow according to several variables such as average level of cloud contamination, the average AGB density, the financial and human resources available, the size of the country, the accuracy and the temporal resolution at which the country decides to report. Technical and financial challenges should not be underestimated: many satellites still produce pictures that aren't free, and the amount of data gathered by modern remote-sensing systems is extremely vast. As an example, Landsat Thematic Mapper generates over 10^4 megabits of data per second, and much higher rates may be expected from the larger space platforms of the Earth Observation System (EOS; Barrett and Curtis, 1999). Satellite-based data analysis has thus a non-negligible cost, linked to access to the relevant data, access to computer and memory resources, and access to software packages for data analysis and display, as well as access to the relevant expertise in large dataset processing and analysis. Such costs can be considerable for some of the countries where such monitoring would be the most beneficial. Habitat monitoring is, moreover, not a 'one-off' process, and it will be essential within the context of national monitoring to have methods that are adapted to local conditions and are consistently applied over the years to guarantee that trends can be detected if they occur.

Glossary

AGB	Above-ground biomass
AVHRR	Advanced Very High Resolution Radiometer
CBD	Convention on Biological Diversity
CMS	Convention on the Conservation of Migratory Species
EOS	Earth Observation System
EVI	Enhanced Vegetation Index
fAPAR	Fraction of absorbed photosynthetic active radiation
GAC	Global area coverage

GIMMS	Global Inventory Modelling and Mapping Studies
INDVI	Integrative Normalized Difference Vegetation Index
IPCC	International Panel on Climate Change
LAI	Leaf area index
LIDAR	Light detection and ranging
MODIS	The Moderate Resolution Imaging Spectroradiometer
NASA	National Aeronautics and Space Administration
NDVI	Normalized Difference Vegetation Index
NIR	Near-infrared reflectance
NOAA	National Oceanic and Atmospheric Administration
PAR	Photosynthetic active radiation
RADAR	Radio detection and ranging
RED	Red reflectance
REDD	UN Programme on Reducing Emissions from Deforestation and Forest Degradation in Developing Countries
SAR	Synthetic-aperture radar
SAVI	Soil Adjusted Vegetation Index
SPOT VGT	Satellite Pour l'Observation de la Terre, Vegetation
UNFCCC	United Nations Framework Convention on Climate Change
USGS	United States Geological Survey

References

Achard, F. and Blasco, F. (1990) Analysis of vegetation seasonal evolution and mapping of forest cover in West Africa with the use of NOAA AVHRR data. *Photogrammetric Engineering and Remote Sensing*, 56, 1359–1365.

Achard, F. and Estreguil, C. (1995) Forest classification of Southeast Asia using NOAA AVHRR data. *Remote Sensing of Environment*, 54, 198–208.

Andersen, R., Herfindal, I., Saether, B.E., Linnell, J.D.C., Odden, J. and Liberg, O. (2004) When range expansion rate is faster in marginal habitats. *Oikos*, 107, 210–214.

Anyamba, A., Davies, G., Indeje, M., Ogallo, L.J., and Ward, M.N. (2006) Predictability of the Normalized Difference Vegetation Index in Kenya and potential applications as an indicator of Rift Valley fever outbreaks in the Greater Horn of Africa. *Journal of Climate*, 19, 1673–1687.

Asrar, G., Fuchs, M., Kanemasu, E.T., and Hatfield, J.L. (1984) Estimating absorbed photosynthetic radiation and leaf area index from spectral reflectance in wheat. *Agronomy Journal*, 76, 300–306.

Barrett, E.C. and Curtis, L.F. (1999) *Introduction to Environmental Remote Sensing*, 4th edn. Stanley Thornes Ltd.

Berteaux, D., Humphries, M.M., Krebs, C.J., *et al.* (2006) Constraints to predicting the effects of climate change on mammals. *Climate Research*, 32, 151–158.

Boone, R.B., Thirgood, S.J., and Hopcraft, J.G.C. (2006) Serengeti wildebeest migratory patterns modeled from rainfall and new vegetation growth. *Ecology*, 87, 1987–1994.

Bradbury, R.B., Hill, R.A., Mason, D.C., *et al.* (2005) Modelling relationships between birds and vegetation structure using airborne LiDAR data: a review with case studies from agricultural and woodland environments. *Ibis*, 147, 443–452.

Bro-Jørgensen, J., Brown, M., and Pettorelli, N. (2008) Using NDVI to explain ranging patterns in a lek-breeding antelope: the importance of scale. *Oecologia*, 158, 177–182.

Campbell, J.B. (2007) *Introduction to Remote Sensing*, 4th edn. Guilford Press.

Churnside, J.H. and Donaghay, P.L. (2009) Thin scattering layers observed by airborne lidar. *ICES Journal of Marine Science*, 66, 778–789.

Churnside, J.H. and Ostrovsky, L.A. (2005) Lidar observation of a strongly nonlinear internal wave train in the Gulf of Alaska. *International Journal of Remote Sensing*, 26, 167–177.

Ciarniello, L.M., Boyce, M.S., Seip, D.R., and Heard, D.C. (2007) Grizzly bear habitat selection is scale dependent. *Ecological Applications*, 17, 1424–1440.

Couturier, S., Côté, S.D., Otto, R.D., Weladji, R.B., and Huot, J. (2009) Variation in calf body mass in migratory caribou: the role of habitat, climate, and movements. *Journal of Mammalogy*, 90, 442–452.

Crick, H.Q.P. and Sparks, T.H. (1999) Climate change related to egg-laying trends. *Nature*, 399, 423–424.

DeFries, R., Achard, F., Brown, S., *et al.* (2007) Earth observations for estimating greenhouse gas emissions from deforestation in developing countries. *Environmental Science and Policy*, 10, 385–394.

Ding, T.S., Yuan, H.W., Geng, S., Koh, C.N., and Lee, P.F. (2006) Macro-scale bird species richness patterns of the East Asian mainland and islands: energy, area and isolation. *Journal of Biogeography*, 33, 683–693.

Dubayah, R.O. and Drake, J.B. (2000) Lidar remote sensing for forestry. *Journal of Forestry*, 98, 44–46.

Easterling, D.R., Meehl, G.A., Parmesan, C., Changnon, S.A., Karl, T.R., and Mearns, L.O. (2000) Climate extremes: observations, modelling, and impacts. *Science*, 289, 2068–2074.

Estrada-Pena, A., Quilez, J., and Sanchez Acedo, C. (2004) Species composition, distribution, and ecological preferences of the ticks of grazing sheep in North-Central Spain. *Medical and Veterinary Entomology*, 18, 123–133.

Evans, K.L., Warren, P.H. and Gaston, K.J. (2005) Species-energy relationships at the macroecological scale: a review of the mechanisms. *Biological Review*, 8, 1–25.

Evans, K.L., James, N.A., and Gaston, K.J. (2006) Abundance, species richness and energy availability in the North American avifauna. *Global Ecology and Biogeography*, 15, 372–385.

Funk, C.C. and Brown, M.E. (2006) Intra-seasonal NDVI change projections in semi-arid Africa. *Remote Sensing of Environment*, 101, 249–256.

Gaston, K.J. and Spicer, J.I. (2004) *Biodiversity: an Introduction*, 2nd edn. Blackwell Publishing.

Gibbs, H.K., Brown, S., Niles, J.O., and Foley, J.A. (2007) Monitoring and estimating tropical forest carbon stocks: making REDD a reality. *Environmental Research Letters*, 2, 1–13.

Gillespie, T.W., Foody, G.M., Rocchini, D., Giorgi, A.P., and Saatchi, S. (2008) Measuring and modelling biodiversity from space. *Progress in Physical Geography*, 32, 203–221.

Goetz, S.J., Baccini, A., Laporte, N.T., *et al.* (2009) Mapping and monitoring carbon stocks with satellite observations: a comparison of methods. *Carbon Balance and Management*, 4, 1–7.

Goward, S.N. and Prince, S.D. (1995) Transient effects of climate on vegetation dynamics: satellite observations. *Journal of Biogeography*, 22, 549–564.

Goward, S.N., Markham, B., Dye, D.G., Dulaney, W., and Yang, J. (1991) Normalized difference vegetation index measurements from the Advanced Very High Resolution Radiometer. *Remote Sensing of the Environment*, 35, 257–277.

Hall, L.S., Krausman, P.R., and Morrison, M.L. (1997) The habitat concept and a plea for standard terminology. *Wildlife Society Bulletin*, 25, 173–182.

Hansen, B.B., Aanes, R., Herfindal, I., Saether, B.E., and Henriksen, S. (2009) Winter habitat-space use in a large arctic herbivore facing contrasting forage abundance. *Polar Biology*, 32, 971–984.

Hebblewhite, M., Merrill, E., and McDermid, G. (2008) A multi-scale test of the forage maturation hypothesis in a partially migratory ungulate population. *Ecological Monographs*, 78, 141–166.

Hill, D., Fasham, M., Tucker, G., Shewry, M., and Shaw, P. (2005) *Handbook of Biodiversity Methods: Survey*, Evaluation and Monitoring. Cambridge University Press, UK.

Hortal, J., Roura-Pascual, N., Sanders, N.J., and Rahbek C. (2010) Understanding (insect) species distributions across spatial scales. *Ecography*, 33, 51–53.

Huete, A.R. (1988) A soil-adjusted vegetation index (SAVI). *Remote Sensing of Environment*, 25, 295–309.

Huete, A., Didan, K., Miura, T., Rodriguez, E.P., Gao, X., and Ferreira, L.G. (2002) Overview of radiometric and biophysical performance of the MODIS vegetation indices. *Remote Sensing of the Environment*, 83, 195–213.

Hurlbert, A.H. (2004) Species-energy relationships and habitat complexity in bird communities. *Ecology Letters*, 7, 714–720.

Hurlbert, A.H. and Haskell, J.P. (2003) The effect of energy and seasonality on avian species richness and community composition. *American Naturalist*, 161, 83–97.

Inouye, D.W., Barr, B., Armitage, K.B., and Inouye, B.D. (2000) Climate change is affecting altitudinal migrants and hibernating species. *Proceedings of the National Academy of Sciences of the USA*, 97, 1630–1633.

IPCC (Intergovernmental Panel on Climate Change) (2001) *Climate Change 2001*. Synthesis Report. Cambridge University Press, Cambridge.

IPCC (Intergovernmental Panel on Climate Change) (2007) *Climate Change 2007: Synthesis Report*. Fourth Assessment Report of the Intergovernmental Panel on Climate Change. IPCC. Available at: http://www.ipcc.ch/pdf/assessment-report/ar4/syr/ar4_syr.pdf.

Ito, T.Y., Miura, N., Lhagvasuren, B., *et al.* (2006) Satellite tracking of Mongolian gazelles (*Procapra gutturosa*) and habitat shifts in their seasonal ranges. *Journal of Zoology*, 269, 291–298.

Jensen, J.R. (2007) *Remote Sensing of the Environment: An Earth Resource Perspective*, 2nd edn. Prentice Hall.

Justice, C.O., Townshend, J.R.G., Holben, B.N., and Tucker, C.J. (1985) Analysis of the phenology of global vegetation using meteorological satellite data. *International Journal of Remote Sensing*, 6, 1271–1318.

Kaya, S., Sokol, J., and Pultz, T.J. (2004) Monitoring environmental indicators of vector-borne disease from space: a new opportunity for RADARSAT-2. *Canadian Journal of Remote Sensing*, 30, 560–565.

Kasischke, E.S., Melack, J.M., and Dobson, M.C. (1997) The use of imaging radars for ecological applications – a review. *Remote Sensing of Environment*, 59, 141–156.

Kerr, J.T. and Ostrovsky, M. (2003) From space to species: ecological applications for remote sensing. *Trends in Ecology and Evolution*, 18, 299–305.

Koy, K., McShea, W.J., Leimgruber, P., Haack, B.N., and Aung, M. (2005) Percentage canopy cover – using Landsat imagery to delineate habitat for Myanmar's endangered Eld's deer (*Cervus eldi*). *Animal Conservation*, 8, 289–296.

Lahoz-Monfort, J.J., Guillera-Arroita, G., Milner-Gulland, E.J., Young, R.P., and Nicholson, E. (2010) Satellite imagery as a single source of predictor variables for habitat suitability modelling: how Landsat can inform the conservation of a critically endangered lemur. *Journal of Applied Ecology*, 47, 1094–1102.

Lefsky, M.A., Harding, D.J., Keller, M., *et al.* (2005) Estimates of forest canopy height and above-ground biomass using ICESat. *Geophysical Research Letters*, 32; doi:10.1029/2005GL023971.

Loe, L.E., Bonenfant, C., Mysterud, A., *et al.* (2005) Climate predictability and breeding phenology in red deer: timing and synchrony of rutting and calving in Norway and France. *Journal of Animal Ecology*, 74, 579–588.

Loveland, T.R., Cochrane, M.A., and Henebry, G.M. (2008) Landsat still contributing to environmental research. *Trends in Ecology and Evolution*, 23, 182–183.

Markon, C.J. and Peterson, K.M. (2002) The utility of estimating net primary productivity over Alaska using baseline AVHRR data. *International Journal of Remote Sensing*, 23, 4571–4596.

Markon, C.J., Fleming, M.D., and Binnian, E.F. (1995) Characteristics of vegetation phenology over the Alaskan landscape using AVHRR time-series data. *Polar Records*, 31, 179–190.

Maselli, F., Romanelli, S., Bottai, L., and Zipoli, G. (2003) Use of NOAA-AVHRR NDVI images for the estimation of dynamic fire risk in Mediterranean areas. *Remote Sensing of Environment*, 86, 187–197.

McNeely, J. (2001) Invasive species: a costly catastrophe for native biodiversity. *Land Use and Water Resources Research*, 1, 1–10.

Meehl, G.A., Washington, W.M., Collins, W.D., *et al.* (2005) How much more global warming and sea level rise? *Science*, 307, 1769–1772.

Millennium Ecosystem Assessment (2005a) *Ecosystems and Human Well-being: Synthesis*. Island Press, Washington, DC.

Millennium Ecosystem Assessment (2005b) *Ecosystems and Human Well-being: Biodiversity Synthesis*. World Resources Institute, Washington, DC.

Morrison, M.L., Marcot, B.G., and Mannan, R.W. (1992) *Wildlife-Habitat Relationships: Concepts and Applications*. University of Wisconsin Press, Madison, 343 pp.

Mueller, T., Olson, K.A., Fuller, T.K., Schaller, G.B., Murray, M.G., and Leimgruber, P. (2008) In search of forage: predicting dynamic habitats of Mongolian gazelles using satellite-based estimates of vegetation productivity. *Journal of Applied Ecology*, 45, 649–658.

Muller, J. and Brandl, R. (2009) Assessing biodiversity by remote sensing in mountainous terrain: the potential of LiDAR to predict forest beetle assemblages. *Journal of Applied Ecology*, 46, 897–905.

Myneni, R.B., Hall, F.G., Sellers, P.J., and Marshak, A.L. (1995) The interpretation of spectral vegetation indexes. *IEEE Transactions on Geoscience and Remote Sensing*, 33, 481–486.

Nagendra, H. (2001) Using remote sensing to assess biodiversity. *International Journal of Remote Sensing*, 22, 2377–2400.

Neigh, C.S.R., Tucker, C.J., and Townshend, J.R.G. (2008) North American vegetation dynamics observed with multi-resolution satellite data. *Remote Sensing of Environment* 112, 1749–1772.

Onoda, M. (2008) Satellite observation of greenhouse gases: Monitoring the climate change regime. *Space Policy*, 24, 190–198.

Pettorelli, N., Vik, J.O., Mysterud, A., Gaillard, J.-M., Tucker, C.J., and Stenseth N.C. (2005a) Using the satellite-derived Normalized Difference Vegetation Index (NDVI) to assess ecological effects of environmental change. *Trends in Ecology and Evolution*, 20, 503–510.

Pettorelli, N., Mysterud, A., Yoccoz, N.G., Langvatn, R., and Stenseth N.C. (2005b) Importance of climatological downscaling and plant phenology for red deer in heterogeneous landscapes. *Proceedings of the Royal Society of London (B)*, 272, 2357–2364.

Pettorelli, N., Weladji, R., Holand, Ø., Mysterud, A., Breie, H., and Stenseth, N.C. (2005c) The relative role of winter and spring conditions: linking climate and landscape-scale plant phenology to alpine reindeer body mass. *Biology Letters*, 1, 24–26.

Pettorelli, N., Gaillard, J.-M., Mysterud, A., *et al.* (2006) Using a proxy of plant productivity (NDVI) to track animal performance: the case of roe deer. *Oikos*, 112, 565–572.

Pettorelli, N., Pelletier, F., Von Hardenberg, A., Festa-Bianchet, M., and Cote, S. (2007) Early onset of vegetation growth versus rapid green-up: impacts on juvenile mountain ungulates. *Ecology*, 88, 381–390.

Pettorelli, N., Bro-Jørgensen, J., Durant, S.M., Blackburn, T., and Carbone, C. (2009) Energy availability and density estimates in African ungulates. *American Naturalist*, 173, 698–704.

Pettorelli, N., Ryan, S., Mueller, T., *et al.* (2011) The Normalized Difference Vegetation Index (NDVI): unforeseen successes in animal ecology. *Climate Research*, 46, 15–27.

Prince, S.D., Becker-Reshef, I., and Rishmawi, K. (2009) Detection and mapping of long-term land degradation using local net production scaling: application to Zimbabwe. *Remote Sensing of Environment*, 113, 1046–1057.

Reed, B.C., Brown, J.F., VanderZee, D., Loveland, T.R., Merchant, J.W., and Ohlen, D.O. (1994) Measured phenological variability from satellite imagery. *Journal of Vegetation Science*, 5, 703–714.

Reynolds, J.F. and Stafford Smith, M. (eds) (2002) *Global Desertification: Do Humans Create Deserts?* Dahlem University Press, Berlin.

Rossow, W.B. and Schiffer, R.A. (1999) Advances in understanding clouds from ISCCP. *Bulletin of the American Meteorological Society*, 80, 2261–2287.

Running, S.W. (1990) Estimating primary productivity by combining remote sensing with ecosystem simulation. In: Hobbs, R.J. and Mooney, H.A. (eds), *Remote Sensing of Biosphere Functioning*. Springer-Verlag, New York, pp. 65–86.

Ryan, S.J., Knechtel, C.U., and Getz, W.M. (2007) Ecological cues, gestation length, and birth timing in African buffalo. *Behavioral Ecology*, 18, 635–644.

Saino, N., Szep, T., Ambrosini, R., Romano, M., and Moller, A.P. (2004) Ecological conditions during winter affect sexual selection and breeding in a migratory bird. *Proceedings of the Royal Society of London B* 271, 681–686.

Schaub, M., Kania, W., and Koppen, U. (2005) Variation in primary production during winter indices synchrony in survival rates in migratory white storks *Ciconia ciconia*. *Journal of Animal Ecology*, 74, 656–666.

Senay, G.B. and Elliott, R.L. (2002) Capability of AVHRR data in discriminating rangeland cover mixtures. *International Journal of Remote Sensing*, 23, 299–312.

Singh, R.P., Roy, S., and Kogan, F.N. (2003) Vegetation and temperature condition indices from NOAA AVHRR data for drought monitoring over India. *International Journal of Remote Sensing*, 24, 4393–4402.

Statzner, B. and Moss, B. (2004) Linking ecological function, biodiversity and habitat: a mini-review focusing on older ecological literature. *Basic and Applied Ecology*, 5, 97–106.

Stenseth, N.C., Mysterud, A., Ottersen, G., Hurrell, J.W., Chan, K.S., and Lima, M. (2002) Ecological effects of climate fluctuations. *Science*, 297, 1292–1296.

St-Louis, V., Pidgeon, A.M., Clayton, M.K., Locke, B.A., Bash, D., and Radeloff, V.C. (2009) Satellite image texture and a vegetation index predict avian biodiversity in the Chihuahuan desert of New Mexico. *Ecography*, 32, 468–480.

Sun, G., Ranson, K.J., Kimes, D.S., Blair, J.B., and Kovacs, K. (2008) Forest vertical structure from GLAS: An evaluation using LVIS and SRTM data. *Remote Sensing of Environment*, 112, 107–117.

Tait, A. and Zheng, X.G. (2003) Mapping frost occurrence using satellite data. *Journal of Applied Meteorology*, 42, 193–203.

Thiam, A.K. (2003) The causes and spatial pattern of land degradation risk in southern Mauritania using multitemporal AVHRR-NDVI imagery and field data. *Land Degradation and Development*, 14, 133–142.

Trimble, M.J., Ferreira, S.M., and van Aarde, R.J. (2009) Drivers of megaherbivore demographic fluctuations: inference from elephants. *Journal of Zoology*, 279, 18–26.

Tucker, C.J. (1979) Red and photographic infrared linear combinations for monitoring vegetation. *Remote Sensing of the Environment*, 8, 127–150.

Tucker, C.J., Townshend, J.R.G., and Goff, T.E. (1985) African land-cover classification using satellite data. *Science*, 227, 369–375.

Tucker, C.J., Grant, D.M., and Dykstra, J.D. (2004) NASA's global orthorectified Landsat data set. *Photogrammetric Engineering and Remote Sensing*, 70, 313–322.

Tucker, C.J., Pinzon, J.E., Brown, M.E., *et al.* (2005) An extended AVHRR 8-km NDVI data set compatible with MODIS and SPOT vegetation NDVI data. *International Journal of Remote Sensing*, 26, 4485–4498.

Turner, W., Spector, S., Gardiner, N., Fladeland, M., Sterling, E., and Steininger, M. (2003) Remote sensing for biodiversity science and conservation. *Trends in Ecology and Evolution*, 18, 306–314.

Van Wagtendonk, J.W. and Root, R.R. (2003) The use of multi-temporal Landsat Normalized Difference Vegetation Index (NDVI) data for mapping fuel models in Yosemite National Park, USA. *International Journal of Remote Sensing*, 24, 1639–1651.

Visser, M.E., Both, C., and Lambrechts, M.M. (2004) Global climate change leads to mistimed avian reproduction. *Advances in Ecological Research*, 35, 89–110.

Wang, Q., Watanabe, M., Hayashi, S., and Murakami, S. (2003) Using NOAA AVHRR data to assess flood damage in China. *Environmental Monitoring Assessment*, 82, 119–148.

Willems, E.P., Barton, R.A., and Hill, R.A. (2009) Remotely sensed productivity, regional home range selection, and local range use by an omnivorous primate. *Behavioral Ecology*, 20, 985–992.

Wittemyer, G., Cerling, T.E., and Douglas-Hamilton, I. (2009) Establishing chronologies from isotopic profiles in serially collected animal tissues: An example using tail hairs from African elephants. *Chemical Geology*, 267, 3–11.

Wood, A., Stedman-Edwards, P., and Mang, J. (eds) (2000) *The Root Causes of Biodiversity Loss*. Earthscan, London, 399 pp.

World Meteorological Organization (2005) *Climate and Land Degradation*. WMO, Geneva.

Zinner, D., Pelaez, F., and Torkler, F. (2002) Distribution and habitat of grivet monkeys in eastern and central Eritrea. *African Journal of Ecology*, 40, 151–158.

Zweifel-Schielly, B., Kreuzer, M., Ewald, K.C., and Suter, W. (2009) Habitat selection by an Alpine ungulate: the significance of forage characteristics varies with scale and season. *Ecography*, 32, 103–113.

6

Indicators of Climate Change Impacts on Biodiversity

Wendy B. Foden[1,2], Georgina M. Mace[3] and Stuart H.M. Butchart[4]

[1]International Union for the Conservation of Nature, Cambridge, UK
[2]Animal, Plant and Environmental Sciences, University of the Witwatersrand, Johannesburg, South Africa
[3]Imperial College London, Centre for Population Biology, Ascot, UK
[4]BirdLife International, Cambridge, UK

Introduction

Climate change is expected to have profound impacts on biodiversity, with some estimates suggesting that up to one-third of assessed species may become 'committed to extinction' during this century (Fischlin *et al.*, 2007; Thomas *et al.*, 2004), though such extinctions may be delayed for decades to centuries (Heywood *et al.*, 1994). Currently, most population declines and species extinctions are caused by habitat change and loss (IUCN, 2009), but climate change is expected to become an increasingly dominant pressure during this century (Jetz *et al.*, 2007; Pereira *et al.*, 2010; Sala *et al.*, 2000).

Despite the relatively limited amount of climatic change that has been observed so far, a range of effects on biodiversity have already been observed (Figure 6.1). For example, species' geographical ranges are documented to be shifting and phenologies are changing, mostly in the way that would be expected if they were tracking their recent climates (Chen *et al.*, 2011; Parmesan and Yohe, 2003; Walther *et al.*, 2002). There is already evidence of community effects as species respond individually to climate change and disrupt existing interactions (Bale *et al.*, 2002; Thackeray *et al.*, 2010; Traill *et al.*, 2010).

Biodiversity Monitoring and Conservation: Bridging the Gap between Global Commitment and Local Action, First Edition. Edited by Ben Collen, Nathalie Pettorelli, Jonathan E.M. Baillie and Sarah M. Durant.
© 2013 John Wiley & Sons, Ltd. Published 2013 by John Wiley & Sons, Ltd.

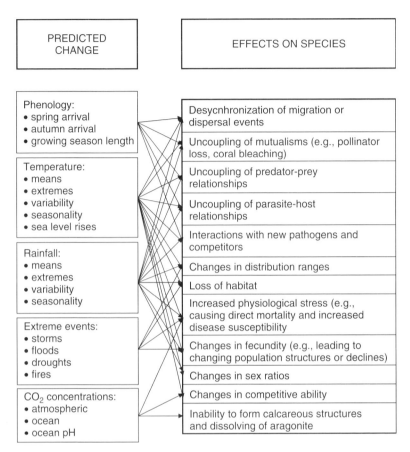

Figure 6.1 **A summary of some of the observed impacts of climate change on species (from Foden *et al.*, 2008).**

Given the current trends in greenhouse gas emissions, the world is already committed to future warming that will have global impacts for centuries to come (Solomon *et al.*, 2009). Immediate actions are therefore needed to both anticipate and manage the resulting impacts on biodiversity. Tracking climate change, its impacts on biodiversity, and the success or otherwise of our conservation actions is therefore an essential task that will require new kinds of indicators.

Indicators are measures, usually derived from some kind of monitoring, that are used to report on the condition of a resource of interest. They need to be designed with a purpose in mind. There are three common roles for indicators in environmental management (Failing and Gregory, 2003):

1. to track performance (results-based management);
2. to discriminate among competing hypotheses (scientific exploration); and
3. to discriminate among alternative policies (decision analysis).

These three more or less map onto a dichotomy recognized by Jones *et al.* (2010), who distinguished objectives for monitoring biodiversity ranging on a continuum from those that are knowledge-focused (the information collected has no direct link to management actions) to those that are action-focused (the information collected can be applied directly to management action). Whatever the classification used, indicators should be designed and tested with their primary role in mind (Green *et al.*, 2005).

The features of successful indicators go beyond ensuring that they are fit for purpose. In the case of climate change indicators, the aspect of greatest importance is determining how climate change (the pressure) is affecting biodiversity (the state) and how various management and policy interventions (the response) might be effective in reducing deleterious consequences. It is therefore useful to consider designing indicators for monitoring pressure, state, response, and benefit, and measures of the benefits that people derive from intact biodiversity may provide a valuable addition (Sparks *et al.*, 2011). Other considerations for indicator design are ensuring that the chosen indicator can be effectively communicated to relevant audiences and maximizing the cost-effectiveness of indicator development and compilation (Jones *et al.*, 2010; Mace and Baillie, 2007). Funds used for monitoring and indicators should also not unnecessarily detract from resources for conservation or environmental management (Field *et al.*, 2004).

Climate change indicators are a little different from other indicators of biodiversity change. Widely used indicators of species and population change such as the Living Planet Index and the Red List Index are based on observed trends during recent decades. In extant, well-monitored populations, consequences of climate change of the type anticipated for the future have so far been documented largely as range shifts and changes in phenologies. Evidence of population decline and extinction due solely to climate change has, to date, been limited. The main causes of biodiversity decline globally are currently habitat loss and degradation, over-exploitation, and the impacts of invasive alien species (IUCN, 2009). Attributing species decline and extinction to climate is complicated by these multiple and interacting threatening processes (Parmesan *et al.*, 2011) and the fact that climate change impacts to date are generally less severe than other threatening processes and compared to what they will be in future. Even in apparently well-documented cases such as the golden toad, the causation remains disputed (Anchukaitis and Evans, 2010).

Emerging climate change indicators tend to focus on tracking climatic changes and their likely effects, rather than measuring actual impacts on species or the effectiveness of responses. Secondly, because climate change is a relatively new pressure (at least in the context of the current biodiversity crisis), we understand less about exactly how and when its consequences will be greatest for biodiversity. The reliability of current predictions is still under debate (Botkin *et al.*, 2007), and the extent to which species will be able to cope, adapt, or evolve is the subject of continuing research (Gienapp *et al.*, 2008; Visser, 2008). Even with better general understanding of the significance of different climate variables and their interactions with species ecologies, the uncertainties in current climate predictions (Hulme *et al.*, 2009) and the evidence that many species will respond idiosyncratically to climate change (Doak and Morris, 2010; Johnson *et al.*, 2010; Ozgul *et al.*, 2010) means that making reliable predictions for longer-term consequences is difficult.

To date, no global-scale indicators of climate change impacts on biodiversity have emerged, but it is clear that putting in place pressure, state, and response indicators now is necessary to provide a reliable baseline for interpreting future impacts. Before designing an indicator of change, we need to be able to establish a link between changes in climate and those in communities, species, or populations. Measures related to extinction risk, population trends, or range decline will be most useful and resonate with users familiar with such measures in the current indicator set for monitoring biodiversity change (Butchart *et al.*, 2010). In this chapter, we discuss four ways in which climate change impacts on species are assessed, and evaluate the merits of each for development of a global indicator framework. We describe two indicators that have been put into practice at a regional scale, and finally, we discuss the role that monitoring can and should play in validating both assessments and indicators of climate change impacts on biodiversity.

Attributing and measuring climate change impacts on species

Given the limited empirical information on species' responses to climate change, it is necessary to consider alternative ways for measuring how species are being or will be affected, and the degree to which these are positive or negative effects. To date, four different kinds of methods have been used to determine the impact of climate change on species:

1. Comparing observed population and distribution changes with those predicted under climate change

This approach is based on recent observations of species' distribution, population, or phenological changes in response to climate. In individual studies, attribution to climate change is often highly uncertain, but across multiple studies and in formal meta-analyses there is a systematic bias to shifts that are in the direction that would be expected if climate change were the driving force (Chen *et al.*, 2011; Parmesan and Yohe 2003; Walther *et al.*, 2002). To date there has been more evidence for climate-driven range expansion than contraction. But this does not necessarily mean that such species are coping. They may have reduced fitness or show low rates of recruitment in the new locations; following an extinction time-lag, local extinction may be inevitable. Also, due to both sampling effects and lags, declines may be harder to detect than expansions (Thomas *et al.*, 2006).

These meta-analyses of recent observed trends have great value in demonstrating the existence of a climate driver of change, but because the sampling of species for such studies tends to be opportunistic, they are often biased towards well-resourced, northern hemisphere, temperate regions (Figure 6.2) and well-studied taxa, and they

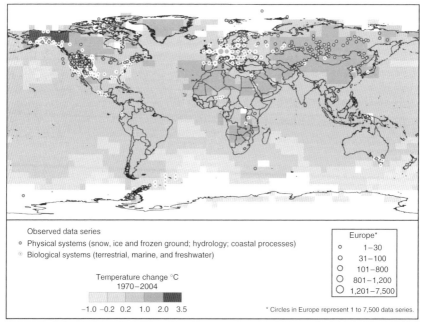

Figure 6.2 **Locations of observations of significant changes in physical and biological systems, together with surface air temperatures changes over the period 1970–2004. The data series met the following criteria: (i) ending in 1990 or later; (ii) spanning a period of at least 20 years; and (iii) showing a significant change in either direction, as assessed by individual studies. White areas do not contain sufficient observational climate data to estimate a temperature trend. From Rosenzweig, C. *et al.*, (2007).**

are less useful for empirically determining the most vulnerable species, communities, or places globally. For example, some comparative studies of birds have highlighted the most affected groups as migrants (Both *et al.*, 2010) or high-latitude species (Jetz *et al.*, 2007) but the information for other taxonomic groups is far poorer. In single-species studies, attribution to climate change remains problematic except in cases where long-term, intensive field studies permit good understanding of causal processes (Doak and Morris, 2010; Ozgul *et al.*, 2010). Lastly, publication bias, or the tendency of researchers to publish only significant findings, further decreases confidence in global trends derived from this approach. While meta-analyses of emerging observed climate change impacts can provide a good source of information on impacts, particularly where they involve new or unexpected impact mechanisms, this approach is unlikely to provide a reliable overall picture of potential climate change impacts.

In some well-studied species, the way that climate impacts species viability has been studied empirically in the field or through controlled experiments of ecophysiology, ecology, behaviour, and genetic change. In such cases it is possible to develop mechanistic or process-based models for how future climate change might affect particular species (Chuine and Beaubien, 2001; Deutsch *et al.*, 2008; Keith *et al.*, 2008; Morin *et al.*, 2008; Sinervo *et al.*, 2010). These methods offer great potential for predictive modelling (Brook *et al.*, 2009), but require many data and much parameterization before they can be useful. While clear conclusions have resulted in some cases, so far they have not supported broader generalizations from null models (Travis, 2003).

2. Climate envelope models

The most widely applied method for predicting and attributing climate change effects is through the use of so-called 'climate envelope models', whereby associations between species' observed distribution ranges and various climate and environmental variables (e.g. temperature and precipitation) are used to describe a climate space for each species. This is then used to predict their distributions under current and future climate scenarios, as predicted by General Circulation Models, providing a basis for estimating extinction risk. The change in location or area of suitable climate space can then be used to assess species' vulnerability (Guisan and Thuiller, 2005; Thomas *et al.*, 2004) and implications for site-based conservation approaches (Hole *et al.*, 2009, 2011). Large-scale studies using a bioclimatic approach include those for European plants (Thuiller *et al.*, 2006b), African mammals (Thuiller *et al.*, 2006a) and birds (Hole *et al.*, 2009, 2011), European breeding birds (Huntley *et al.*, 2008), and marine mammals (Kaschner *et al.*, 2011). Thomas *et al.* (2004) synthesized a number of these assessments to produce the first global-scale prediction of climate change impacts on biodiversity, and following from this, in the IPCC's Fourth Assessment Report, Fischlin *et al.* (2007)

created what remains the largest and most comprehensive meta-analysis of predictions, the majority of which is based on bioclimatic model assessments.

Envelope models have been the basis for many large-scale projections of climate change impacts (Guisan and Thuiller, 2005; IPCC, 2007; Thomas *et al.*, 2004). They are, however, subject to a wide range of criticisms, which highlight drawbacks that are often ignored when interpreting results. Firstly, models assume that species' niches are bioclimatically determined, discounting factors such as competition and resource availability. Ignoring the potential for behavioural or evolutionary adaptation (Chevin *et al.*, 2010), they assume that species' niche requirements will remain static in the future under a changing climate that may introduce novel combinations of bioclimatic variables (Williams *et al.*, 2007). Even if assumptions about bioclimatic niches hold, such models fail to account for many important mechanisms of climate change impacts (e.g., phenological mismatches, disruption of environmental triggers, and changing interspecific interactions including at community and habitat scales), as well as for the biological traits that make individual species more or less susceptible to impacts (Dawson, 2011). The quantity and quality of species' distribution data needed means that models tend to focus on well-studied northern hemisphere temperate regions and better-studied taxa such as mammals and birds. For the many poorly known and geographically restricted species that are often of greatest conservation concern, the use of climate envelope models is considered inadvisable and unreliable if there are too few (typically 10–30) known localities of 'presence'. As a result, the vulnerability of habitats containing disproportionate numbers of such species (e.g. tropical montane forests) may be underestimated, leading to their under-prioritization during conservation priority setting (N. Burgess *et al.*, in preparation).

Nevertheless, climate envelope model predictions have been shown to be at least generally consistent with historical (Araujo *et al.*, 2005) and observed population trends (Jiguet *et al.*, 2010; Tingley *et al.*, 2009), including both species projected to increase in numbers and those projected to decline under climate change (Gregory *et al.*, 2009). Because climate envelope models require only climate and distributional information in order to derive predictions, they are the most widely used approach and have been extremely influential in highlighting vulnerable areas and species for conservation, and in quantifying the likely scale of climate change impacts on species and ecosystems in coming decades.

3. Criteria-based methods

The concern that many existing methods emphasize the negative impacts of climate change, rather than the possibility that there will be many winners as well as losers, has led to the development and application of a framework to identify the threats and benefits of climate change for individual species, based on climate-attributed decline in the recent area of distribution and also observed or potential increases outside the recently occupied historical range. Species are classified into categories according to the degree of risk or benefit that climate change may afford them. The method uses the available trend information and some expert opinion, and has been successfully applied to a wide range of UK species (Thomas *et al.*, 2010).

4. Susceptibility frameworks

Susceptibility or vulnerability frameworks are used in a range of contexts as a way to assess risk and also to support the design of effective mitigation and adaptation measures (Turner *et al.*, 2003). Susceptibility is generally assessed according to three components: sensitivity, exposure, and adaptability. In the context of climate change impacts on species, sensitivity reflects the intrinsic characteristics of a species' ecology and life history that make it more or less at risk from climate change; exposure is the degree of climate change that it will face; and adaptability is the potential for the species' own coping mechanisms, such as phenotypic plasticity or dispersal to mitigate impacts (Williams *et al.*, 2008). The advantage of susceptibility frameworks is that they incorporate all aspects of susceptibility in an explicit manner that readily translates to the design of management interventions (Dawson *et al.*, 2011; W. Foden *et al.*, in preparation). The drawback is that, as discussed above, we do not yet have sufficient empirical information about how and when different species will be at risk from climate change, nor the potential for microevolution, phenotypic plasticity, dispersal, or conservation actions to be effective in reducing the threat. Nevertheless, this information is now becoming increasingly available, permitting the application of susceptibility frameworks for climate change, albeit involving a degree of extrapolation among related species and requiring explicit recognition of where major uncertainties lie (Young *et al.*, 2011; W. Foden *et al.*, in preparation; Chin *et al.*, 2010).

Developing a theoretical framework for climate change indicators

Indicators are needed to track the intensity of climate change as a pressure on biodiversity, its positive or negative impacts on the state of biodiversity (at all levels, including genes, populations, species, and ecosystems), the effectiveness of responses taken to reduce these impacts, and the consequence of all three of these on the benefits that humans derive from biodiversity. Ideally these indicators should be linked together in a framework that allows the measures to be reported consistently at a range of spatial scales (particularly regional to global) (Sparks *et al.*, 2011). The methods discussed above for projecting or measuring climate change impacts on species are based on fundamentally disparate approaches, ranging from empirical observations to experts' knowledge and statistical modelling. Some are not even intended for use in tracking the magnitude of impacts over time (i.e. as indicators), but rather to help shape conservation strategies in future. Nevertheless, they are highly relevant. Can the various approaches be integrated into a coherent climate change indicator framework?

Observational and empirical studies have been widely used in meta-analyses to document climate change as an influential driver of change. However, the generally opportunistic selection of target species and the unsystematic and non-random nature of sampled studies mean that analyses are seldom replicable or representative. As a result, impacts cannot be tracked over time or in different places, and species' responses to management interventions cannot be monitored over time. For an indicator framework, sampling needs to be designed systematically across areas, taxa, or climate change extents of interest.

The prospects for mechanistic or process-based models are much better. Because these are based on documented relationships between climate and aspects of the species' life histories, predictions about future climates can be incorporated into species-specific spatially explicit models, to derive both predictions of future trends under climate change, and effective interventions (Sinervo *et al.*, 2010). In terms of their application in an indicator framework, the main drawback is that a large number of models would need to be developed and parameterized if outcomes for any more than a handful of species are to be reflected, and the manner in which these species are selected will dictate the generality of conclusions drawn. One approach to this is to move to models of medium complexity (M. Dickinson *et al.*, in preparation), which reflect broad ecological and life history types rather than being either species-specific (Keith *et al.*, 2008) or null models (Travis, 2003). Such models could then be used to represent the components of the biota in particular areas or ecosystems, and the consequences of different kinds of climate change incorporated to represent indicators of change and response options.

Envelope models have been widely used to predict future impacts of climate change on biodiversity in order to assess the magnitude of these and to inform conservation planning. Their use as an indicator is clearly limited by the fact that the predictions will only change in response to changes in climate projections. Because interactions between climate change and population viability are not currently incorporated into these models, they are blind to species' differences, non-linear effects, and species interactions; are susceptible to inaccuracies resulting from uncertainties in either species' ranges or modelled predictions of future climates; and tend to be less useful at fine spatial scales.

Criteria-based methods offer an alternative that could, along with a well-designed sampling programme, provide measures of the impact of climate change on the state of biodiversity. This would mirror the approach now used for tracking trends in species' extinction risk through the IUCN Red List (Butchart *et al.*, 2005). However, the validation of thresholds and the weighting of different criteria need to be resolved, a process that took many years in the case of the Red List criteria (Mace *et al.*, 2008). An additional challenge in this case is the application to an unprecedented and projected threat rather than to observed scenarios of current or recent threat impacts on species.

Susceptibility frameworks can be based on information from all of the above methods and might be a useful way to measure climate change pressure (exposure) and state (susceptibility based on sensitivity, adaptability, and exposure), while close links with the Red List Index facilitate the possible inclusion of a response measurement (e.g. down-listing of species into less or not threatened IUCN Red List categories in response to conservation actions to address climate change threats). Recent work to define susceptibility traits (W. Foden *et al.*, in preparation) can start the process to establish indicators (Figure 6.3). At large scale, the susceptibility framework could be applied to identify areas where species are of greatest current

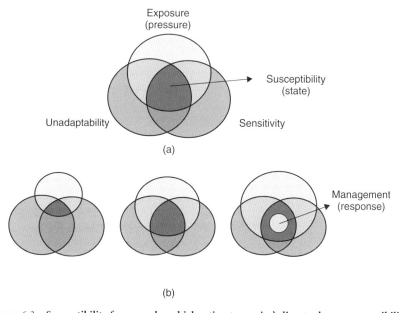

Figure 6.3 **Susceptibility frameworks, which estimate species' climate change susceptibility (state) based on independent assessments of their climate change exposure (pressure), sensitivity, and unadaptability to climatic change, might be a useful basis for a global or regional climate change indicator framework (a). They could allow us to measure and track increasing exposure and overall climate change susceptibility, while links with the Red List Index facilitate measurement of the effectiveness of conservation management responses (b).**

concern (high exposure, high sensitivity, high unadaptability/low adaptability), of high latent risk (high sensitivity, high unadaptability, and low exposure), or of high potential risk (high exposure, low sensitivity, low unadaptability).

Climate change indicators in practice

The most well-established indicator developed so far is the Climatic Impact Indicator for birds in Europe (Gregory *et al.*, 2009), developed for the Streamlining European 2010 Biodiversity Indicators (SEBI) programme (European Environment Agency, 2010). The approach distinguishes species that are expected to respond to climate change positively versus negatively depending on whether climate envelope models project range expansion or contraction in Europe (with the latter set outnumbering the former by about 3:1). The average recent (post-1980) population trends for each of these two sets of species are calculated using data from systematic population monitoring across the region, and an indicator is calculated as the ratio of these two values. Since 1990 there has been a sharp increase in the indicator, showing that there has been an increasing signal of climate change impacts on bird population trends over the last two decades, a period of rapid warming in Europe (Gregory *et al.*, 2009; Figure 6.4). The method provides a statistically sensitive and interpretable indicator,

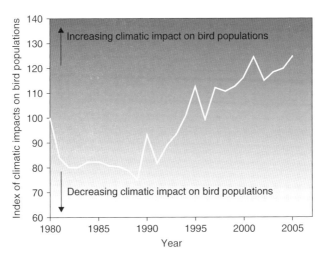

Figure 6.4 Index of the impact of climatic change on populations of European birds (1980–2005). The indicator is the ratio of the index for the 30 bird species whose potential geographical ranges are projected to expand in Europe, to that for those 92 bird species projected to contract their distribution ranges under climatic change. © 2009 Gregory *et al*. This is an open-access article distributed under the terms of the Creative Commons Attribution License.

although it can be challenging to explain its basis to policy-makers. This method was also tested on a subset of European butterflies, but trends for climate positive versus climate negative species were difficult to differentiate, possibly due to the low number of climate positive species and possible inaccuracies in modelled ranges (van Swaay et al., 2008). The method relies on being able to develop good climate envelope models and having good recent population trend information, both of which (particularly the latter) are likely to be challenging to obtain for many taxa outside better-studied regions. Nevertheless, appropriate data are now being collected in a number of tropical countries for birds, so there is potential to expand this approach in future.

The Community Temperature Index, first used for French breeding birds (Devictor et al., 2008), was adapted by van Swaay et al. (2008; Figure 6.5) for use as a SEBI indicator for tracking changes in butterfly communities in response to climate change. For each butterfly species, the average temperature across its distribution range (its Species Temperature Index (STI)) was calculated using distribution and climate data. At each recording site and every year, a Community Temperature Index (CTI) was calculated as the average of the STI of all species present in the assemblage. van Swaay et al. (2008) found that overall, the CTI for butterflies in Spain (Catalonia), The Netherlands, the United Kingdom, and Finland increased significantly between 1990 and 2005, reflecting turnover in the community from cool-adapted species to those historically dwelling at higher temperatures. As the indicator relies on presence/absence data rather than abundance data, it is likely to be less sensitive than the Climatic Impact Indicator, but potentially less data-demanding, although it still

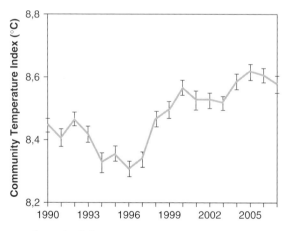

Figure 6.5 Temporal trend of the Community Temperature Index (± standard error) for butterflies in Europe – represented by Spain (Catalonia), The Netherlands, the United Kingdom, and Finland – weighted by area of the country. The trend was found to be significant for the periods 1990–2007 and 1999–2007 (From van Swaay et al., 2008. Reproduced with permission.).

requires intensive sampling to determine presence/absence at a representative suite of sites across the geographical area of interest, as well as accurate global maps of the distribution of each species considered.

Validating climate change vulnerability assessments and indicators

Because few expert knowledge and model-based studies have attempted on-the-ground evaluations of the accuracy of their predictions, our confidence in their predictions is handicapped and essential feedback for potential model improvement remains scarce. The establishment and coordination of a systematic species monitoring framework is a clear and urgent need both for evaluating and improving climate change impact assessments and indicators, as well as for establishing a baseline against which future changes can be measured. In theory, such a system would not be hard to design, though attributing cause and effect requires careful consideration, especially with multiple extrinsic threats interacting in their effects on species in most places.

To date, global monitoring efforts for terrestrial ecosystems include the Tropical Ecology Assessment and Management network (TEAM; www.teamnetwork.org) and the Global Observation Research Initiative in Alpine Environments (GLORIA; www.gloria.ac.at) network, both of which track changing climates and their impacts. Using global networks of long-term field stations, TEAM aims to monitor trends in tropical terrestrial biodiversity while GLORIA focuses on plants in alpine environments. These initiatives lead the way in monitoring global impacts of climate change on their respective focal ecosystems, and regions and have begun to establish baselines and the systematic long-term datasets needed for climate change research. Additionally, BirdLife International are expanding systematic bird population monitoring programmes worldwide, and have established standard protocols to track trends in the state, pressure, and responses for biodiversity at the a network of 11 000 Important Bird Areas (BirdLife International, 2010) globally. Strong and strategic coordination between such initiatives, and the addition of new networks and complementary approaches covering other ecosystems and regions, could provide the monitoring information needed to underpin a sound global climate change indicator framework.

Conclusions

A wide variety of approaches are used to assess and project climate change impacts on biodiversity, but few are explicitly designed to deliver indicators showing trends over time. A global climate change indicator framework would help to coordinate and structure efforts to track trends in the pressure that climate change exerts on biodiversity, its impact on the state of biodiversity, and the implementation and effect of responses to mitigate these impacts. Currently available measures and projections of climate change impacts on species are insufficient to provide such indicators at a global scale for a range of reasons.

At a broad scale, we suggest that a susceptibility framework is a useful approach for identifying regions where species are of greatest concern (high exposure, high sensitivity, low adaptability), of high latent risk (high sensitivity, low adaptability, and low exposure), or of high potential risk (high exposure, low sensitivity, high adaptability). Within these regions, mechanistic and demographic models are useful for further examining species' susceptibility. Selected species within these regions could then be systematically monitored to track trends in their occurrence, abundance, and phenology.

A successful indicator framework to measure climate change impacts on biodiversity will clearly require multiple approaches. Perhaps the most challenging to implement will be the development of sustainable, unbiased, systematic on-the-ground monitoring schemes for tracking species' population trends, in order to detect and quantify climate change impacts on these. This will require innovative partnerships between governments and civil society groups, and crucially, the involvement of local stakeholders, in order to generate robust, long-term datasets on climate change impacts on biodiversity. In addition, considerably improved communication of the trends detected, and consequences of these for the benefits and services that humans derive from biodiversity, will be essential in order to persuade governments and society more broadly of the urgent imperative to mitigate climate change and help biodiversity adapt to it.

Acknowledgements

We thank Guy Dutson for his comments and valuable suggestions.

References

Anchukaitis, K.J. and Evans, M.N. (2010) Tropical cloud forest climate variability and the demise of the Monteverde golden toad. *Proceedings of the National Academy of Sciences of the USA*, 107, 5036–5040.

Araujo, M.B., Pearson, R.G., Thuiller, W., and Erhard, M. (2005) Validation of species-climate impact models under climate change. *Global Change Biology*, 11, 1504–1513.

Bale, J.S., Masters, G.J., Hodkinson, I.D., *et al.* (2002) Herbivory in global climate change research: direct effects of rising temperature on insect herbivores. *Global Change Biology*, 8, 1–16.

BirdLife International (2010) State of the world's birds. Available at: http://www.birdlife.org/datazone/sowb

Both, C., Van Turnhout, C.A.M., Bijlsma, R.G., Siepel, H., Van Strien, A.J., and Foppen, R.P.B. (2010) Avian population consequences of climate change are most severe for long-distance migrants in seasonal habitats. *Proceedings of the Royal Society B: Biological Sciences*, 277, 1259–1266.

Botkin, D.B., Saxe, H., Araújo, M.B., et al. (2007) Forecasting the effects of global warming on biodiversity. *Bioscience*, 57, 227–236.

Brook, B.W., Akcakaya, H.R., Keith, D.A., Mace, G.M., Pearson, R.G., and Araujo, M.B. (2009) Integrating bioclimate with population models to improve forecasts of species extinctions under climate change. *Biology Letters*, 5, 723–725.

Butchart, S.H.M., Stattersfield, A.J., Baillie, J., et al. (2005) Using Red List Indices to measure progress towards the 2010 target and beyond. *Philosophical Transactions of the Royal Society B: Biological Sciences*, 360, 255–268.

Chen, I.-C., Hill, J.K., Ohlemüller, R., Roy, D.B., and Thomas, C.D. (2011) Rapid range shifts of species of climate warming. *Science*, 333, 1024–1026.

Chevin, L.-M., Lande, R., and Mace, G.M. (2010) Adaptation, plasticity, and extinction in a changing environment: towards a predictive theory. *PLoS Biology*, 8(4), e1000357.

Chin, A., Kyne, P.M., Walker, T.I., and McAuley. R.B. (2010) An integrated risk assessment for climate change: analysing the vulnerability of sharks and rays on Australia's Great Barrier Reef. *Global Change Biology*, 16, 1936–1953.

Chuine, I. and Beaubien, E.G. (2001) Phenology is a major determinant of tree species range. *Ecology Letters*, 4, 500–510.

Dawson, T.P. (2011) Beyond predictions: biodiversity conservation in a changing climate. *Science*, 332, 53–58.

Devictor, V., Julliard, R., Couvet, D., Jiguet, F. (2008) Birds are tracking climate warming, but not fast enough. *Proceedings of the Royal Society of London B, Biological Sciences*, 275, 2743–2748.

Deutsch, C.A., Tewksbury, J.J., Huey, R.B., et al. (2008) Impacts of climate warming on terrestrial ectotherms across latitude. *Proceedings of the National Academy of Sciences of the USA* 105, 6668–6672.

Doak, D.F. and Morris, W.F. (2010) Demographic compensation and tipping points in climate-induced range shifts. *Nature*, 467, 959–962.

European Environment Agency (2010) Assessing biodiversity in Europe – the 2010 report. *EEA Report no. 5/2010*. EEA, Copenhagen.

Failing, L. and Gregory, R. (2003) Ten common mistakes in designing biodiversity indicators for forest policy. *Journal of Environmental Management*, 68, 121–132.

Field, S.A., Tyre, A.J., Jonzen, N., Rhodes, J.R. and Possingham, H.P. (2004) Minimizing the cost of environmental management decisions by optimizing statistical thresholds. *Ecology Letters*, 7, 669–675.

Fischlin, A., Midgley, G.F., Price, J.T., et al. (2007) Ecosystems, their properties, goods and services. In: Parry, M.L., Canziani, O.F., Palutikof, J.P., van der Linden P.J., and Hanson, C.E. (eds), *Climate Change 2007: Impacts, Adaptation and Vulnerability. Contribution of Working Group II to the Fourth Assessment Report of the Intergovernmental Panel on Climate Change*. Cambridge University Press, pp. 211–272.

Foden, W., Mace, G.M., Vié, J.-C., *et al.* (2008) Species susceptibility to climate change impacts. In: Vié, J.-C., Hilton-Taylor, C., and Stuart, S.N. (eds), *Wildlife in a Changing World*. Lynx, Barcelona, pp. 77–88.

Gienapp, P., Teplitsky, C., Alho, J.S., Mills, J.A., and Merila, J. (2008). Climate change and evolution: disentangling environmental and genetic responses. *Molecular Ecology*, 17, 167–178.

Green, R.E., Balmford, A., Crane, P.R., Mace, G.M., Reynolds, J.D., and Turner, R.K. (2005) A framework for improved monitoring of biodiversity: Responses to the World Summit on Sustainable Development. *Conservation Biology*, 19: 56–65.

Gregory, R.D., Willis, S.G., Jiguet, F., *et al.* (2009) An indicator of the impact of climatic change on European bird populations. *PLoS ONE*, 4, e4678.

Guisan, A. and Thuiller, W. (2005) Predicting species distribution: offering more than simple habitat models. *Ecology Letters*, 8, 993–1009.

Heywood, V.H., Mace, G.M., May, R.M., and Stuart, S.N. (1994) Uncertainties in extinction rates. *Nature*, 368, 105.

Hole, D.G., Huntley, B., Pain, D.J., *et al.* (2009) Projected impacts of climate change on a continental-scale protected area network. *Ecology Letters*, 12, 420–431.

Hole, D.G., Huntley, B., Collingham, Y.C., *et al.* (2011) Towards a management framework for protected area networks in the face of climate change. *Conservation Biology*, 25, 305–315.

Hulme, M., Pielke, R., and Dessai, S. (2009) Keeping prediction in perspective. *Nature Reports Climate Change*, 3, 126–127.

Huntley, B., Collingham, Y.C., Willis, S.G., and Green, R.E. (2008) Potential impacts of climatic change on European breeding birds. *PLoS ONE*, 3(1), e1439.

IPCC (2007) *Climate Change 2007: Impacts, Adaptation and Vulnerability. Contribution of Working Group II to the Fourth Assessment Report of the Intergovernmental Panel on Climate Change*. Cambridge University Press, Cambridge, UK.

IUCN (2009) *Wildlife in a Changing World – An Analysis of the 2008 IUCN Red List of Threatened Species*. IUCN, Gland, Switzerland.

Jetz, W., Wilcove, D.S., and Dobson, A.P. (2007) Projected impacts of climate and land-use change on the global diversity of birds. *PLoS Biology*, 5, 1211–1219.

Jiguet, F., Gregory, R.D., Devictor, V., *et al.* (2010) Population trends of European common birds are predicted by characteristics of their climatic niche. *Global Change Biology*, 16, 497–505.

Johnson, D.M., Büntgen, U., Frank, D.C., *et al.* (2010) Climatic warming disrupts recurrent Alpine insect outbreaks. *Proceedings of the National Academy of Sciences of the USA*, 107, 20576–20581.

Jones, J.P.G., Collen, B., Atkinson, G., *et al.* (2011) The why, what and how of global biodiversity indicators beyond the 2010 target. *Conservation Biology*, 25, 450–457.

Kaschner, K., Tittensor, D.P., Ready, J., Gerrodette, T., and Worm, B. (2011) Current and future patterns of global marine mammal biodiversity. *PLoS ONE*, 6(5), e19653.

Keith, D.A., Akcakaya, H.R., Thuiller, W., *et al.* (2008) Predicting extinction risks under climate change: coupling stochastic population models with dynamic bioclimatic habitat models. *Biology Letters*, 4, 560–563.

Mace, G.M. and Baillie, J.E.M. (2007) The 2010 biodiversity indicators: challenges for science and policy. *Conservation Biology*, 21, 1406–1413.

Mace, G.M., Collar, N.J., Gaston, K.J., et al. (2008) Quantification of extinction risk: IUCN's system for classifying threatened species. *Conservation Biology*, 22, 1424–1442.

Morin, X., Viner, D., and Chuine, I. (2008) Tree species range shifts at a continental scale: new predictive insights from a process-based model. *Journal of Ecology*, 96, 784–794.

Ozgul, A., Childs, D.Z., Oli, M.K., et al. (2010) Coupled dynamics of body mass and population growth in response to environmental change. *Nature*, 466, 482–485.

Parmesan, C. (2006) Ecological and evolutionary responses to recent climate change. *Annual Review of Ecology, Evolution, and Systematics*, 37, 637–669.

Parmesan, C. and Yohe, G. (2003) A globally coherent fingerprint of climate change impacts across natural systems. *Nature*, 421, 37–42.

Parmesan, C., Duarte, C., Poloczanska, E., Richardson, A.J., and Singer, M.C. (2011). Overstretching attribution. *Nature Climate Change*, 1, 2–4.

Pereira, H.M., Leadley, P.W., Proença, V., et al. (2010). Scenarios for global biodiversity in the 21st century. *Science*, 330, 1496–1501.

Rosenzweig, C., Casassa, G., Karoly, D.J., et al. (2007) Assessment of observed changes and responses in natural and managed systems. In: Parry, M.L., Canziani, O.F., Palutikof, J.P., van der Linden P.J., and Hanson, C.E. (eds), *Climate Change 2007: Impacts, Adaptation and Vulnerability. Contribution of Working Group II to the Fourth Assessment Report of the Intergovernmental Panel on Climate Change*. Cambridge University Press, Cambridge, UK, pp. 79–131.

Sala, O.E., Chapin, F.S., Armesto, J.J., et al. (2000) Global biodiversity scenarios for the year 2100. *Science*, 287, 1770–1774.

Sinervo, B., Mendez-de-la-Cruz, Miles, D.B., et al. (2010) Erosion of lizard diversity by climate change and altered thermal niches. *Science*, 328, 894–899.

Solomon, S., Plattner, G.K., Knutti, R., and Friedlingstein, P. (2009) Irreversible climate change due to carbon dioxide emissions. *Proceedings of the National Academy of Sciences of the USA*, 106, 1704–1709.

Sparks, T.H., Butchart, S.H.M., Balmford, A., et al. (2011) Linked indicator sets for addressing biodiversity loss. *Oryx*, 45, 1–9.

Thackeray, S.J., Sparks, T.H., Frederiksen, M., et al. (2010) Trophic level asynchrony in rates of phenological change for marine, freshwater and terrestrial environments. *Global Change Biology*, 16, 3304–3313.

Thomas, C.D., Cameron, A., Green, R.E., et al. (2004) Extinction risk from climate change. *Nature*, 427, 145–148.

Thomas, C.D., Franco, A.M.A., and Hill, J.K. (2006) Range retractions and extinction in the face of climate warming. *Trends in Ecology and Evolution*, 21, 415–416.

Thomas, C.D., Hill, J.K., Anderson, B.K., et al. (2010) A framework for assessing threats and benefits to species responding to climate change. *Methods in Ecology and Evolution*, 2, 125–142.

Thuiller, W., Broennimann, O., Hughes, G., Alkemade, J.R.M., Midgley, G.F., and Corsi, F. (2006a). Vulnerability of African mammals to anthropogenic climate change under conservative land transformation assumptions. *Global Change Biology*, 12, pp. 424–440.

Thuiller, W., Lavorel, S., Sykes, M.T., and Araújo, M.B. (2006b) Using niche-based modelling to assess the impact of climate change on tree functional diversity in Europe. *Diversity and Distributions*, 12, 49–60.

Tingley, M.W., William, W.B., Belsisinger, S.R., and Moritz, C. (2009) Birds track their Grinnellian niche through a century of climate change. *Proceedings of the National Academy of Sciences of the USA*, 106, 19637–19643.

Traill, L.W., Lim, M.L.M., Sodhi, N.S., and Bradshaw, C.J.A. (2010) Mechanisms driving change: altered species interactions and ecosystem function through global warming. *Journal of Animal Ecology*, 79, 937–947.

Travis, J.MJ. (2003) Climate change and habitat destruction: a deadly anthropogenic cocktail. *Proceedings of the Royal Society of London Series B – Biological Sciences*, 270, 467–473.

Turner, B.L., II, Kasperson, R.E., Matson, P.A., *et al.*, (2003) A framework for vulnerability analysis in sustainability science. *Proceedings of the National Academy of Sciences of the USA*, 100, 8074–8079.

van Swaay, C.A.M., van Strien, A.J., Julliard, R., *et al.* (2008) *Developing a Methodology for a European Butterfly Climate Change Indicator* Report VS2008.040. De Vlinderstichting, Wageningen.

Visser, M.E. (2008) Keeping up with a warming world; assessing the rate of adaptation to climate change. *Proceedings of the Royal Society B – Biological Sciences*, 275, 649–659.

Walther, G.-R., Post, E., Convey, P., *et al.* (2002) Ecological responses to recent climate change. *Nature*, 416, 389–395.

Williams, J.W., Jackson, S.T., and Kutzbach, J.E. (2007) Projected distributions of novel and disappearing climates by 2100 AD. *Proceedings of the National Academy of Sciences of the USA*, 104, 5738–5742.

Williams, S.E., Shoo, L.P., Isaac. J.L., Hoffmann, A.A., and Langham, G. (2008) Towards an integrated framework for assessing the vulnerability of species to climate change. *PLoS Biology*, 6, 2621–2626.

Young, B.E., Hall, K.R., Byers, E., *et al.* (2012) Rapid assessment of plant and animal vulnerability to climate change. In: Brodie, J., Post, E., and Doak, D. (eds), *Conserving Wildlife Populations in a Changing Climate*. University of Chicago Press, Chicago, IL (in press).

7

Monitoring Trends in Biological Invasion, its Impact and Policy Responses

Piero Genovesi[1], Stuart H.M. Butchart[2], Melodie A. McGeoch[3] and David B. Roy[4]

[1]IUCN SSC Invasive Species Specialist Group, ISPRA, Rome, Italy
[2]BirdLife International, Cambridge, UK
[3]Centre for Invasion Biology and Cape Research Centre, South African National Parks, Steenberg, South Africa
[4]Centre for Ecology and Hydrology, Wallingford, UK

Introduction

Biological invasions are a major threat to biodiversity, second only to habitat change (Millennium Ecosystem Assessment, 2005), and have been a primary cause of species extinction over the last 500 years (Donlan *et al.*, 2008). For example, invasive alien species are cited as a factor in over 50% of animal extinctions where the cause is known, and for 1 out of every 5 (20%) of extinctions, invasions were the only cited cause (Clavero and García-Berthou, 2005). Invasive alien species (IAS) are also among the most important threats to globally threatened species. Invasives are the second most important threat to birds, impacting 52% of Critically Endangered species and 51% of all threatened species (BirdLife International, 2008a, 2008b). They are the fourth most important threat to threatened amphibians (possibly even underestimated owing to uncertainty over the origin of chytridiomycosis), and third most important for threatened mammals (Hilton-Taylor *et al.*, 2009). Furthermore, invasions cause huge economic losses – e.g. costs in Europe exceed 12 billion euros per year (Kettunen *et al.*, 2009) – they are responsible for the spread of many diseases, and can disrupt ecosystem services of crucial importance for human well-being, such as food security and access to water (Vilà *et al.*, 2010).

Biodiversity Monitoring and Conservation: Bridging the Gap between Global Commitment and Local Action,
First Edition. Edited by Ben Collen, Nathalie Pettorelli, Jonathan E.M. Baillie and Sarah M. Durant.
© 2013 John Wiley & Sons, Ltd. Published 2013 by John Wiley & Sons, Ltd.

The number of invasive alien species appears to be increasing in all environments and among all taxonomic groups, with numbers in Europe increasing 76% over the period 1970–2007 (Butchart *et al.*, 2010) – an upward trend that has proven difficult to halt, let alone reverse (DAISIE, 2009; Millennium Ecosystem Assessment, 2005). Prevention of further unwanted introductions is a high priority, and by far the most efficient approach to limiting further invasions. The development of global early warning and rapid response policies (G8, 2009), effective documentation and monitoring of trends, and assessment of the factors correlated with introduction and establishment of invasive alien species are all required to improve our ability to respond to alien species invasions.

The global community has committed to prevent and mitigate the impacts of invasive alien species, and to monitor trends in invasions. In particular, the Convention on Biological Diversity (CBD) has called upon parties to identify invasive alien species, assess the history and ecology of invasion, the origin of alien species, the pathway of arrival, the timing of introduction, the biology of the invaders, and the impacts upon the environment and human well-being, and to monitor temporal trends in these parameters (Decision VI/23 CBD COPVI, The Hague, April 2002). This large body of information is crucial not only for the development of more effective policy and implementation, but is also valuable for testing the efficacy of efforts invested in response to the invasion problem.

Review of indicators of invasions

Following increasing recognition of the impacts of invasive alien species, there have been several attempts to develop indicators of invasion over the last decade (see review in McGeoch *et al.*, 2006). Indicators have been developed across a range of spatial scales and have been based on a range of measures, including the percentage of area covered by alien plant species (Heinz Center, 2002), the density of managed weeds (Natural Heritage Trust, 2006), the distribution and abundance of selected alien species in parks (Parks Canada, 2005), the increase in aquaculture-related introduced species in European marine environments (EEA, 2003), and the percentage of invasive alien species in selected groups (UNEP, 2003). However, there have until recently been very few examples of indicators of invasion that are based on a range of taxa, cover large spatial scales, assess temporal trends in invasions, or consider impacts of invasive species. We discuss attempts to develop such indicators at global, regional, and national scales.

Development of global indicators for biological invasion

At the global scale, development of indicators followed the CBD Decision VI/26 of 2002, which included a commitment to achieve by 2010 a significant reduction of the rate of biodiversity loss. To assess progress against these targets, the CBD initially identified two potential indicators of the threats to biodiversity: nitrogen deposition and 'numbers and cost of alien invasions'. At the 10th meeting of the CBD's Subsidiary Body on Scientific, Technical and Technological Advice (SBSTTA), held in 2005, it was agreed to use 'trends in invasive alien species', in addition to nine fully developed biodiversity indicators and an additional seven considered ready for testing (Walpole et al., 2009). As a follow-up, the CBD secretariat in 2006 asked the Global Invasive Species Program (GISP) and the International Union for the Conservation of Nature-Species Survival Commission (IUCN-SSC) to produce a global indicator for biological invasions. For this purpose the CBD's definition of an alien invasive species was used, namely 'a species outside of its [indigenous geographical] range whose introduction and/or spread threatens biodiversity' (UNEP, 2002). Following a review of a range of candidates, the development of four indicators was prioritized in light of data availability and the practicality of reporting against the 2010 target (McGeoch et al., 2009). These comprise a measure of the state of the problem (the number of IAS per country), a measure of the pressures IAS place on biodiversity (the Red List Index of impacts of IAS), and two measures of responses to the problem (trends in the number of international agreements relevant to reducing threats to biodiversity from invasive alien species, and trends in the adoption of national legislation relevant to the control of IAS) (McGeoch et al., 2010).

The number of invasive alien species per country

One of the primary purposes for an invasive alien species indicator is to measure the status of invasion and to monitor change therein. In principle, the number of IAS is a direct and simple 'problem-status' indicator (McGeoch et al., 2006). The rationale for using numbers of IAS is that the identification and designation of species as invasive is often critical for designing and prioritizing control and prevention efforts. There is, however, considerable variation in the amount of available information on IAS in different countries (Figure 7.1). Plants make up over 40% of documented IAS, followed by insects and other invertebrates (over 30%), fish, mammals, and birds (Pyšek et al., 2008). However, there are also large differences in data adequacy between species groups. For example, invasive alien birds and mammals are comparatively well known, whereas invasive invertebrates are much more poorly known, and marine invader diversity is thought to be underestimated by up to 90% in many parts of the world (Carlton, 2008).

The number of documented IAS is undoubtedly an underestimate for many countries globally. Reasons for this include under-sampling, the time delay between identification of a new IAS and publication in national species lists, as well as inadequate information availability, and little investment in research on IAS in many instances (McGeoch et al., 2010). To assess the levels of information available, countries (those signatory to the CBD and members of the United Nations) were classified as data

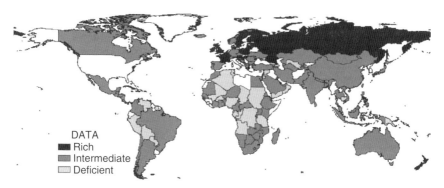

Figure 7.1 **Status of Invasive Alien Species information for countries signatory to the Convention on Biological Diversity. With permission from John Wiley & Sons.**

deficient if they were found to have comparatively little published work on IAS (Pyšek et al., 2008), and National Reports to the CBD stating that the country had not identified IAS, or assessed the risks posed by IAS, in the country. Data-rich countries, on the other hand, were those with comparatively substantial published work on IAS and/or those countries that had listed and in some cases also assessed the risks of IAS. Countries assessed to be data deficient in this way do in fact have significantly fewer listed and reported IAS than data-rich countries (McGeoch et al., 2010). Less than 11% of countries are considered to have adequate data on the IAS present, with almost one-third considered to be data deficient (31%) (see Figure 7.1). This means that the size of the global IAS problem is currently significantly underestimated quantitatively. Nonetheless, controlling for these data biases, island-based countries were shown to have more IAS than their continental counterparts and, as expected, numbers of IAS are positively related to country size (McGeoch et al., 2010).

Although the number of IAS forms a necessary basis for monitoring the status of the invasion problem, trend data are currently not available at a global scale. In addition to the shortage of information in several parts of the world for populating the indicator, variation in criteria used to designate species as invasive as well as delays and biases in species discovery and reporting mean that the construction of a global trend in the size of the problem is currently not possible (Costello and Solow, 2003; Crooks, 2005).

Nevertheless, there are several efforts to monitor trends in invasions at regional and national scales. In Europe, the European Union Council adopted a more ambitious target than the CBD 2010, committing to halt the biodiversity decline in Europe by 2010. In parallel, a European programme was developed to measure the achievements in regard to this commitment, and the European Environment Agency coordinated the development of an indicator-based assessment of European biodiversity – SEBI2010 – that includes invasive alien species as a key variable (EEA,

2009a, 2009b). The SEBI2010 invasive species indicator is the only regional indicator developed to date, and is the 'cumulative number of alien species in Europe since 1900'. This is also the first dataset (containing data for five countries) to contribute to the development of a general indicator of trends in IAS for Europe (ECCHM, 2005). The indicator is based on 163 species identified by a group of experts as causing severe impacts to biological diversity, as well as to health or economy (EEA, 2009b) and illustrates the most invaded countries of the region (Figure 7.2).

The European response to measure the impact of IAS has been further enhanced through the completion of the DAISIE (Delivering Alien Invasive Species Inventories for Europe) project funded by the European Commission (2005–2008). Although not specifically aimed at monitoring trends in invasions, DAISIE provides an inventory of alien species (invasive and non-invasive) recorded in European terrestrial, freshwater, and marine environments. DAISIE was carried out by a team of 18 European scientific institutions and has established the most comprehensive database worldwide on introduced species. The DAISIE database has collated information on over 11 000 introduced species of fungi, plants, vertebrates, and invertebrates, and of more than 50 000 records of introductions. All data have been assembled and verified by leading experts. The database includes information on year of introduction, pathway and vectors of introductions, and documented environmental or socio-economic impacts.

The major findings of the DAISIE project have been synthesized (DAISIE, 2009) to demonstrate that precise and detailed trends in invasions across a range of taxonomic

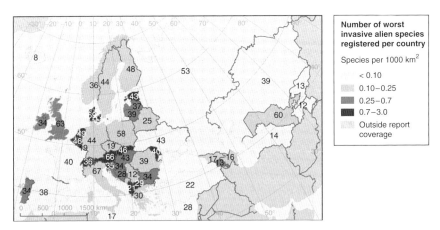

Figure 7.2 **Number of 'worst' terrestrial and freshwater invasive alien species (IAS) threatening biodiversity in Europe. List identified through an assessment process carried on by the Streamlining European Biodiversity Indicators (SEBI) 2010 expert group. From European Environment Agency (2009). Reproduced with permission.**

groups can be measured. This is made possible by three key aspects of the DAISIE dataset: the existence of a comparatively good history of taxonomic record-keeping dating back several decades (and in some cases centuries), the coverage of *all* alien species, not only those that are described as 'invasive', and the inclusion of a date or period of introduction to enable trends over time to be derived.

To provide some examples of the results that this approach permitted, Figure 7.3 reports the trend in established alien invertebrate species in Europe since 1492, expressed as the mean number of alien invertebrates recorded per year.

DAISIE (2009) reports even more detailed trend information for other taxonomic groups such as mammals (Figure 7.4), where it was possible to distinguish between trends of introduction events, and cumulative numbers of new species for Europe.

An integrated analysis of invasion trends in Europe – produced by combining the DAISIE datasets on alien mammals, alien species in European freshwaters, and alien species in the Mediterranean basin – estimated that the numbers of aliens in Europe increased 76% during 1970–2007 (Figure 7.5; Butchart *et al.*, 2010).

Another example of the potential of a retrospective approach is shown in Figure 7.6, which highlights the rate of arrival of new mammal species in Europe over four periods (based on the DAISIE dataset), providing a clear indicator of the patterns of invasion in this taxonomic group.

The DAISIE dataset also allows for much more detailed information on key correlates of introductions, crucial for enforcing response measures, such as the pathways

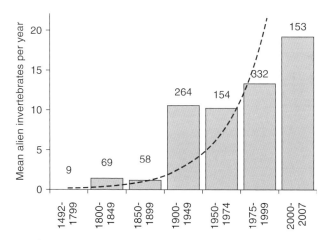

Figure 7.3 **Trends in established alien terrestrial invertebrates in Europe since 1492. Calculations made on 995 species for which precise estimates are available for the first record. The numbers above the bars correspond to the number of new species recorded per period (source: Roques *et al.*, 2009) With kind permission of Springer Science+Business Media.**

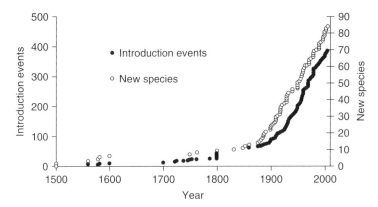

Figure 7.4 Trends of mammal invasions in Europe: number of introduction events and of new species for Europe recorded since 1500 (source: Genovesi *et al.*, 2009) With kind permission of Springer Science+Business Media.

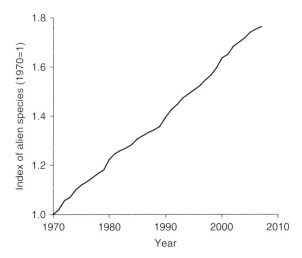

Figure 7.5 **Index of trends in number of alien species in Europe during 1970–2007.** From Butchart *et al.*, 2010.

of arrival (i.e. Figure 7.7). These data show that in recent times the intentional release or escape from captivity of IAS are by far the most common causes of introduction for mammals, and thus demonstrate the importance of more stringent rules on the import of species and on containment facilities.

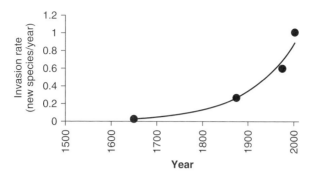

Figure 7.6 Rate of arrival of alien mammals per year in Europe in four periods: 1500–1800; 1800–1950; 1950–2000; and 2000–2005 (source: Genovesi *et al.*, 2009) With kind permission of Springer Science+Business Media.

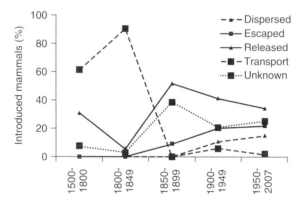

Figure 7.7 Trends in the modes by which mammal IAS were introduced to Europe (source: Genovesi *et al.*, 2009) With kind permission of Springer Science+Business Media.

Furthermore, the DAISIE approach, based on retrospective counts, appears to effectively minimize the biases due to the level of available information. For example, if we look at the number of alien marine species for Europe (Figure 7.8), the retrospective count approach highlights the trough in the 1980s in the number of alien species introduced per year in the Mediterranean. This was due to the temporary closure of the Suez Canal, and to the Arab oil embargo, that much reduced the number of vessels transiting the Mediterranean.

The pan-European inventory of alien species created through DAISIE provides a platform for European reporting on biodiversity indicators and highlights areas where

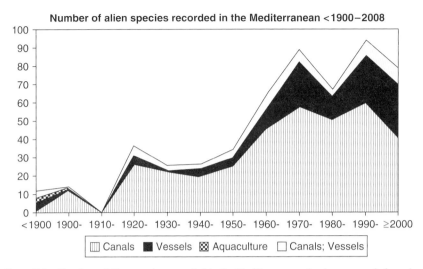

Figure 7.8 Number of alien species recorded in the Mediterranean Sea (new recorded species by decade), 1900–2007, and their means of introduction (source: Galil, 2009) With kind permission of Springer Science+Business Media.

Europe will need to direct resources to manage biological invasions. As such it can act as a template for the future development of global indicators.

At a smaller spatial scale, a number of countries reporting to the CBD have established national monitoring programmes for one or more IAS, and have developed national indicators. The United Kingdom, for example, has a number of monitoring programmes, many of which detect the abundance and distribution of alien species, although none are specifically targeted towards monitoring them. Data from five schemes have been combined to measure trends in the proportion of alien (non-native) species in survey samples for birds, mammals, plants, and marine organisms. The index covers trends in Great Britain; data for Northern Ireland were not available. The index (Figure 7.9) shows an increase in the proportion of non-native species (samples from populations of birds, mammals, plants, and marine organisms) in the period 1990–2007, suggesting that alien species are becoming more widespread and/or relatively more abundant.

A second indicator has been developed in the United Kingdom to measure the change in the extent of invasive species. Out of 3500 alien species identified, the 49 with the greatest potential impact on native biodiversity were identified using expert judgment based upon a set of standardized assessment criteria (adapted by the Belgian Forum on Invasive Species; http://ias.biodiversity.be/ias/definitions#harmonia). The number of species identified as highly invasive (i.e. of greatest threat to native biodiversity)

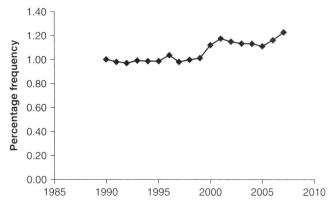

Figure 7.9 Overall proportion of non-native species in samples of birds, mammals, plants, and marine organisms, 1990–2007 in the United Kingdom (data for Great Britain only). Redrawn from http://www.jncc.gov.uk/page-4246; From the Centre for Ecology & Hydrology, British Trust for Ornithology, Marine Biological Association and the National Biodiversity Network Gateway. © Crown copyright 2011.

established across more than 10% of the land area of Britain has increased in all ecosystems (freshwater, marine, and terrestrial) (Figure 7.10).

Although such an approach is currently not feasible in many regions of the world, the UK indicators demonstrate the value of standardized monitoring schemes for measuring the status of biodiversity and the adoption of objective systems for classifying alien species as invasive.

The Red List Index of impacts of IAS on biodiversity

The second global indicator for IAS shows trends over time in the impact of IAS on biodiversity using the Red List Index (RLI). Figure 7.11 shows an example for birds, illustrating the overall rate at which birds worldwide are moving towards or away from extinction owing to the balance between the negative impacts of IAS on species and the positive impacts of conservation actions tackling IAS. It is based on repeated assessments of all birds for the IUCN Red List (which have been carried out by BirdLife International for IUCN five times during 1988–2008). Red List categories are assigned to species based on application of quantitative data (relating to the size, structure, and trend of both the population and distributional range) to explicit criteria with quantitative thresholds. Assessments require parameter estimates to be fully documented with sources and explicit estimates of uncertainty (IUCN, 2001, 2010). Only those changes to Red List categorizations resulting from genuine

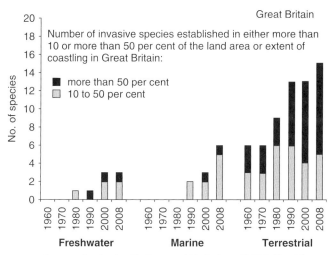

Figure 7.10 Changes in the number of moderately and very widely established IAS in freshwater, marine, and terrestrial environments, 1960 to 2008 (http://www.jncc.gov.uk/page-4246). From the Centre for Ecology & Hydrology, British Trust for Ornithology, Marine Biological Association and the National Biodiversity Network Gateway. © Crown copyright 2012.

improvement or deterioration are included in the RLI (category changes driven by improved knowledge or revised taxonomy are excluded). For all genuine category changes, the primary driver (i.e., threat leading to deterioration in status, or threat overcome by conservation action leading to improvement in status) is identified, and the overall decline in the RLI is then apportioned to different primary drivers, with the thickness of the 'slice' indicating the importance of each particular driver (see Butchart, 2008, and McGeoch et al., 2010, for further details). Determining the primary driver of category changes is facilitated by the fact that the magnitude of each threat to each species on the Red List is calculated according to its estimated scope (i.e. proportion of the population affected by the threat) and severity (rate of population decline over three generations driven by the threat within the scope), plus the fact that detailed documentation is associated with each genuine status change.

The RLI illustrates the relative importance of IAS compared to other threats to biodiversity. It shows that since 1988 the number of species improving in status as a consequence of successful eradications or control of IAS has been outweighed by the number of species deteriorating in status owing to negative impacts of IAS, leading to a negative overall trend in the RLI for impacts of IAS (Figure 7.11). Comparison with

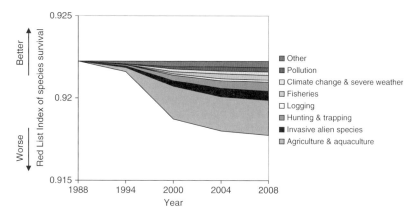

Figure 7.11 Red List Index (RLI) for birds showing trends driven by the impacts of invasive alien species (IAS) compared with trends driven by other factors, for the proportion of species expected to remain extant in the near future without additional conservation action; $n = 9785$ non-data-deficient extant bird species at start of period. An RLI value of 1.0 equates to all species being categorized as Least Concern, and hence that none are expected to go extinct in the near future. An RLI value of zero indicates that all species have gone Extinct. The coloured slices show the contribution of different drivers to the overall deterioration in the status of species over the time period. With permission from John Wiley & Sons.

the magnitude of declines driven by other factors shows that while IAS are among the most important threats to birds, agriculture and aquaculture have had a larger overall negative impact since 1988 (see McGeoch et al., 2010 for results for mammals and amphibians).

This approach can be applied to any group of species on the Red List that has been comprehensively assessed (i.e. all species have been categorized) at least twice. In due course, similar RLIs will be available for corals, cycads, conifers, and a number of other groups, plus for representative samples of a suite of additional taxonomic groups (for which comprehensive reassessments will be challenging owing to the large number of poorly known species) to provide indices more representative of all biodiversity.

Trends in the number of countries party to international agreements relating to invasive alien species

In addition to the CBD, several international agreements either explicitly consider IAS or are relevant to preventing IAS introductions and controlling exiting IAS. Examples of such agreements include the United Nations Convention on the Law of the Sea, the International Plant Protection Convention, and the Protocol on Environmental

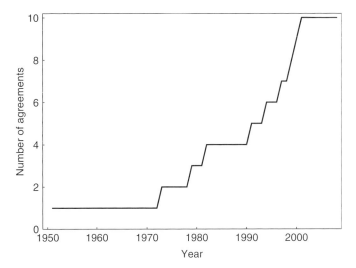

Figure 7.12 **Trends in the number of international agreements relevant to reducing threats to biodiversity from invasive alien species (excluding the Convention on Biological Diversity)** (modified from McGeoch *et al.*, 2010, with permission from John Wiley & Sons.).

Protection to the Antarctic Treaty (McGeoch *et al.*, 2010). None of these agreements were promulgated with the primary intention of reducing the rate of biodiversity impacts from IAS. There has, however, been a significant increase in the number of such IAS-relevant international agreements since the 1950s (Figure 7.12).

Trends in the adoption of national legislation relevant to the control of IAS

There has also been an increase in the development and adoption of policy to control IAS at the national scale over the last two decades (Figure 7.13). Among countries reporting to the CBD, over 80% have national biodiversity strategies and action plans, whereas far fewer have legislation pertaining to the prevention or control of IAS (Figure 7.13). Countries with IAS-relevant policies are assumed more likely to achieve the Convention on Biological Diversity (CBD) Framework Goal of controlling invasive alien species than countries without such policies. However, policy does not necessarily translate into management effectiveness, prevention, or control of IAS (McGeoch *et al.*, 2006). It is therefore insufficient for evaluating the degree to which countries have managed to control pathways for major potential IAS introductions, or whether countries have management plans in place for major alien species that threaten biodiversity. There is thus a need to collate information more directly relevant

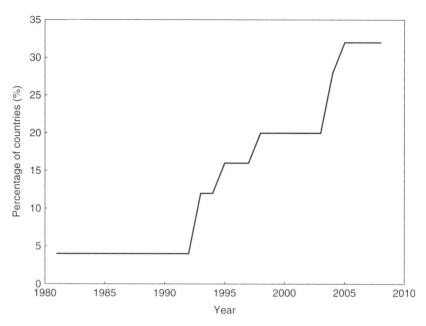

Figure 7.13 **Trend in the adoption of national legislation relevant to the control of invasive alien species by a stratified random subset of countries reporting to the Convention on Biological Diversity. With permission from John Wiley & Sons.**

to management effectiveness of IAS, such as numbers of species successfully controlled and reductions in extent of IAS (McGeoch et al., 2010).

While policy progress towards management of the IAS problem has apparently been significant at both international and national levels, it is currently not possible to assess the efficacy of such policy at reducing the rate at which new IAS are being introduced or establishing worldwide. In fact, trends in the impact of invasive species demonstrate that, in an important way, policy interventions have not been successful. Future efforts by nations to document progress towards achieving IAS targets may shed light on the current mismatch between significant improvements in international and national policy adoption, and ongoing increases in the threat to biodiversity from IAS.

Technical challenges to the development of invasion indicators

Developing indicators to track trends in invasions poses several challenges. Of primary importance is the need for indicators that encompass a range of environments (from marine, to freshwater and terrestrial) and taxonomic groups (from plants to animals; from vertebrates to invertebrates, etc.) that are affected by biological invasions. In particular, global indicators need to be developed further to incorporate information from a broad suite of taxonomic groups to reduce potential bias. One of the features of the substantial variation in the amount and reliability of available information on alien species (Pyšek *et al.*, 2008; Hulme, 2007) is that some taxonomic groups are much better known than others (e.g. vertebrates vs invertebrates). To be more representative, indicators should ideally take such bias into account when collating information on alien species in different taxonomic groups. To illustrate the uneven representation of alien species between taxonomic groups (acknowledging differences in the total species richness between groups), the alien species reported for Europe include 2260 alien terrestrial invertebrates, 737 alien multicellular marine species, and only 45 alien bryophytes (DAISIE, 2009).

Initiatives to develop indicators of invasion started only recently, and until recently (McGeoch *et al.*, 2006, 2010) in most cases have been developed at a local scale. Some of the key technical challenges to developing a global indicator of the status of invasions include the difficulty of designating alien species as invasive, treatment of geographical and taxonomic bias in data availability, the use of a wide range of definitions and criteria for designating species as invasive, the accessibility of data, and problems associated with expert opinion (McGeoch *et al.*, 2009; 2010).

The problem of classification of an alien species as invasive is one of the most crucial issues in the development of indicators. The Convention on Biological Diversity defines an invasive alien species as: a species outside of its native range whose introduction and/or spread threatens biodiversity. This definition does not explicitly include alien species affecting economies and human health, such as those included in the development of a European indicator (EEA, 2009b), even if the definition of biodiversity adopted by the CBD does make reference to the ecosystem services, and could thus cover the effects of invasives on human livelihood. This unclear definition of invasiveness could limit the comparability of future indicators from the regional to the global scale. But the problem of definition is indeed wider than this. The confusion in invasions terminology, and the lack of agreed concepts, have often been highlighted (e.g. Occhipinti-Ambrogi and Galil, 2004; Richardson *et al.*, 2000; Valéry *et al.*, 2008). The different terminologies proposed by various authors are primarily due to the different concepts of invasiveness, based on either biogeographical or impact criteria, that only consider environmental effects, or also include impacts on economy or other non-biological parameters. The effects of the inconsistency in the terms and concepts adopted in different contexts explain the large difference in the number of species listed as invasive within Europe. It must be stressed that the lack of a unanimously agreed definition of IAS does affect the development of reliable indicators. For example, for the same geographical region, EEA (2009b) identified 163 species as invasive, while Vilà and co-authors (2010) reported 1094 alien species of

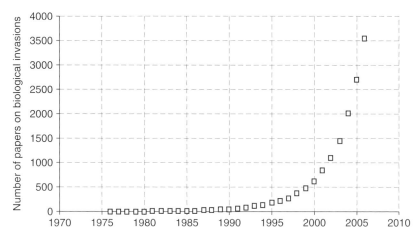

Figure 7.14 Growth in the number of papers in invasion ecology published up to 2006 and registered on the Web of Science (Richardson and Pyšek, 2008). With permission from John Wiley & Sons.

Europe known to cause some impacts to biological diversity, and 1347 known to cause some economic impacts.

Lack of data has several effects on indicators, and one of these is that alien species' invasiveness tends to be underestimated because of a lack of data on both invasive and native species, difficulty in the detection of impacts, and unclear criteria for interpreting what constitutes an impact (Vitule et al., 2009; McGeoch et al., 2010). A lack of data on invasiveness of a species could be misinterpreted as that species not being invasive, but may well be misleading (Leprieur et al., 2009). As a consequence, interpretation of the number of IAS for an area only partly depends on the number of invasive species present, but depends much more on the available information – although this is rapidly increasing in some cases (Figure 7.14). Therefore, a crucial challenge in the development of indicators is to distinguish between a real increase in invasions as opposed to improvement in knowledge of invasions (Costello and Solow, 2003).

Another aspect to be considered is that the problem of definitions applies not only to the term invasive, but also to the criteria adopted for selecting the lists of considered species. For example, the use of terms such as 'casual', 'occasional', 'cryptogenic', and 'established' differ in different projects, and this can affect the results in terms of number of species (e.g. over 30% of marine species are of cryptogenic origin; Carlton, 1996).

If indicators are also meant to support more effective responses to invasion, then monitoring efforts need to cover not only the biological patterns of invasions (number of invaders and geographical patterns of spread), but also the impacts they cause (e.g., on threatened species or economic losses), as well as the drivers and correlates of invasion (e.g., trade volumes, pathways of arrival, geographical origin of invaders, correlates of establishment success, etc.) (Jeschke and Genovesi, 2011; Hulme *et al.*, 2008; Pyšek *et al.*, 2010). Understanding the latter permits more effective responses.

Discussion and conclusions

Biological invasions are a major threat to biological diversity and to human well-being (Vilà *et al.*, 2010; McGeoch *et al.*, 2010), affecting all regions of the world, and all ecosystems (Millennium Ecosystem Assessment, 2005; Vié *et al.*, 2009). The number of documented IAS is increasing rapidly, and this is unlikely to be solely because IAS are becoming better documented. The number of alien species appears to be rapidly growing in all taxonomic groups, with no signs of slowing down (Butchart *et al.*, 2010; DAISIE, 2009). Encouragingly, there has been an increase over time in the adoption of international and regional conventions and agreements related to IAS, and of national legal frameworks to address this issue. Furthermore, the technical ability to manage (control or eradicate) IAS has significantly grown over time (i.e. Genovesi, 2007; Howald *et al.*, 2008), and there are now numerous examples of conservation successes resulting from such actions, including species brought back from the brink of extinction (e.g. Butchart *et al.*, 2006). However, IAS are continuing to drive declines in global biodiversity (McGeoch *et al.*, 2010). The results presented here, showing that the status of the world's birds continues to deteriorate owing to the impacts of IAS, are likely to be replicated when RLIs for IAS impacts on other taxa are available. More stringent responses are needed, perhaps in particular through more effective implementation of existing legislation.

To track progress in, and impacts of, such improved responses to IAS, further development of indicators and the underlying datasets is needed. This poses several complex technical challenges that need to be addressed:

1. Non-standardized terminology is still a limit to the development of indicators, and it is therefore important to continue work towards an agreed set of terms for this purpose, in particular on the criteria used to define invasiveness.
2. The availability of information about alien species remains very low in many areas of the world, and for many taxonomic groups, and it is important to address these deficiencies in order to produce comprehensive inventories covering all environments and taxa.
3. Such inventories should include all alien species, not only IAS, and should include information on the year of introduction, pathway of spread, and documented impacts, as implemented by the DAISIE project. At the same time, however, a focus on those species with the greatest negative biodiversity impacts will be most expedient for achieving biodiversity conservation objectives. Developing such comprehensive lists of alien species will require substantial resources. For example,

DAISIE, cost in total €3 450 131 – with a European Union (EU) contribution of €2 400 000 – while the species listing for the project to populate the global indicator of alien invasion (McGeoch *et al.*, 2009) cost approximately €84 000 (57 countries). However, this would be an effective investment, because it informs the development of more stringent policies and interventions to control the spread and impacts of IAS.
4. The RLI showing impacts of IAS should be extended to other taxonomic groups in which all species have been assessed for the IUCN Red List. More taxonomic groups should be completely assessed to expand taxonomic coverage, and where this is not feasible, the sampled approach of Baillie *et al.* (2008) should be implemented.
5. Data on the adoption of national policy and legislation tackling IAS needs to be expanded and kept up to date, in order to update this indicator. In addition, more direct measures of IAS management effectiveness should be adopted once such information becomes more readily available.
6. Additional indicators should be developed when feasible to track trends in the costs (economic consequences) of IAS, and their impacts on ecosystem services, and on human health and well-being, as well as more direct measures of management effectiveness. A better understanding of these will help to convince decision-makers to invest the substantial resources that are needed to tackle adequately the threats from IAS.

Acknowledgements

We thank Dian Spear, Elrike Marais, Elizabeth Kleynhans, and Riccardo Scalera for information and data collation; Andy Symes and Shyama Pagad for help in coding impacts of IAS on birds; Sarah Simons and the Global Invasive Species Programme (GISP) for advice and support; members of GISP's Expert Working Group on IAS Indicators who provided advice – Georgina Mace, Tim Blackburn, Michael Browne, Richard Smith, Jean-Christophe Vié, Tristan Tyrrell, Martin Parr, and Geoffrey Howard; and finally Phil Roberts, Ben Collen, and Jonathan Baillie for additional input. The Global Environment Facility funded the 2010-Biodiversity Indicators Partnership that supported some of the work presented here. The work on non-native indicators has been funded by the Department for Food and Rural Affairs (Defra). The DAISIE inventory was funded by the European Commission under the Sixth Framework Programme (Contract Number: SSPI-CT-2003-511202).

References

Baillie, J.E.M., Collen, B., Amin, R., *et al.* (2008) Toward monitoring global biodiversity. *Conservation Letters*, 1, 18–26.

BirdLife International (2008a) *Critically Endangered Birds: A Global Audit*. BirdLife International, Cambridge, UK.
BirdLife International (2008b) *State of the World's Birds*. BirdLife International, Cambridge, UK.
Butchart, S.H.M. (2008) Red List Indices to measure the sustainability of species use and impacts of invasive alien species. *Bird Conservation International*, 18, 245–262.
Butchart, S.H.M., Stattersfield, A.J., and Collar, N.J. (2006) How many bird extinctions have we prevented? *Oryx*, 40, 266–278.
Butchart, S.H., Walpole, M., Collen, B., *et al.* (2010) Global biodiversity: indicators of recent declines. *Science*, 328, 1164–1168.
Carlton, J.T. (1996) Biological invasions and cryptogenic species. *Ecology*, 77, 1653–1655.
Carlton, J.T. (2008) The inviolate sea? A thousand years of human-mediated global interchange of shore and shelf biota. Abstract of invited keynote lecturer. Symposium on "Fifty years of invasion ecology – the legacy of Charles Elton", 12–14 November 2008, Stellenbosch.
Clavero, M. and Garcia-Berthou, E. (2005) Invasive species are a leading cause of animal extinctions. *Trends in Ecology and Evolution*, 20, 110.
Costello, C.J. and Solow, A.R. (2003) On the pattern of discovery of introduced species. *Proceedings of the National Academy of Sciences of the USA*, 100, 3321–3323.
Crooks, J.A. (2005) Lag times and exotic species: the ecology and management of biological invasions in slow-motion. *Ecoscience*, 12, 316–329.
DAISIE (2009) *Handbook of Alien Species in Europe*. Springer, Dordrecht, The Netherlands.
Donlan, C.J. and Wilcox, C. (2008) Diversity, invasive species and extinctions in insular ecosystems. *Journal of Applied Ecology*, 45, 1114–1123.
ECCHM (European Biodiversity Clearing House Mechanism) (2005) Biodiversity monitoring and indicators. SEBI2010 Expert Group 5 – invasive species. Convention on Biological Diversity, Montreal. Available at: biodiversity-chm.eea.europa.eu/information/indicator/. Accessed July 2006.
EEA (2003) Europe's water: An indicator-based assessment. EEA *Topic Report No 1/2003*. European Environment Agency, Copenhagen. Available at: www.eea.europa.eu/.
EEA (2009a) Progress towards the European 2010 biodiversity target. *EEA Report 4*/2009. European Environment Agency, 52 pp.
EEA (2009b) Progress towards the European 2010 biodiversity target – Indicator fact sheets. *EEA Technical Report No 05*/2009. European Environment Agency, Copenhagen. Available at: www.eea.europa.eu/publications/progress-towards-the-european-2010-biodiversity-target-indicator-fact-sheets/.
G8 (2009) Carta di Siracusa on Biodiversity, adopted by the G8 Environment ministers at their meeting in Siracusa, 22–24 April 2009. Available at: www.g8.utoronto.ca/environment/env090424-biodiversity.pdf. Accessed 9 August 2009.
Galil, B.S. (2009) Taking stock: inventory of alien species in the Mediterranean sea. *Biological Invasions*, 11, 359–372.
Genovesi, P. (2007) Limits and potentialities of eradication as a tool for addressing biological invasions. In: Nentwig, W. (ed.), *Biological Invasions. Ecological Studies*, vol. 193, pp. 385–400.
Genovesi, P., Bacher, S., Kobelt, M., Pascal, M., and Scalera, R. (2009) Alien mammals of Europe. In: DAISIE (ed.), *The Handbook of Alien Species in Europe*. Springer, Berlin.

Heinz Center (2002) *The State of the Nation's Ecosystems*. Cambridge University Press, Cambridge, UK. Available from: http://www.heinzctr.org/Ecosystem_Management.html. Accessed July 2006.

Hilton-Taylor, C., Pollock, C., Chanson, J., Butchart, S.H.M., Oldfield, T. and Katariya, V. (2009) Status of the world's species. In: Vié, J.-C. Hilton-Taylor, C., and Stuart, S.N. (eds), *The 2008 Review of the IUCN Red List of Threatened Species*. IUCN, Gland, Switzerland, pp. 15–42.

Howald, G., Donlan, C.J., Galván, J.-P., *et al.* (2008) Invasive rodent eradication on islands. *Conservation Biology*, 21, 1258–1268.

Hulme, P.E. (2007) Biological invasions in Europe: drivers, pressures, states, impacts and responses. In: Hester, R. and Harrison, R.M. (eds). *Biodiversity under Threat: Issues in Environmental Science and Technology*. Royal Society of Chemistry, Cambridge, 25:56–80.

Hulme, P.E., Bacher, S., Kenis, M., *et al.* (2008) Grasping at the routes of biological invasions: a framework to better integrate pathways into policy. *Journal of Applied Ecology*, 45, 403–414.

IUCN (2001) *IUCN Red List Categories and Criteria: Version 3.1*. Species Survival Commission, IUCN, Gland, Switzerland, and Cambridge, UK.

IUCN (2010) Guidelines for using the IUCN Red List categories and criteria. Available at: www.iucnredlist.org/technical-documents/categories-and-criteria/2001-categories-criteria.

Jeschke, J.M. and Genovesi, P. (2011) Do biodiversity and human impact influence the introduction or establishment of alien mammals? *Oikos*, 120, 57–64.

Kettunen, M., Genovesi, P., Gollasch, S., Pagad, S., and Starfinger, U. (2009) *Technical support to EU strategy on invasive alien species (IAS). Assessment of the impacts of IAS in Europe and the EU*. Institute for European Environmental Policy (IEEP), Brussels.

Leprieur, F., Brosse, S., García-Berthou, E., Oberdorff, T., Olden, J.D., and Townsend, C.R. (2009) Scientific uncertainty and the assessment of risks posed by non-native freshwater fishes. *Fish and Fisheries*, 10, 88–97.

McGeoch, M.A., Chown, S.L., and Kalwij, J.M. (2006) A global indicator for biological invasion. *Conservation Biology*, 20, 1635–1646.

McGeoch, M.A., Spear, D., and Marias, E. (2009) Status of alien species invasion and trends in invasive species policy. Summary report for the Global Invasive Species Programme. http://academic.sun.ac.za/cib/IASI/archive/PoC_Summary_Report_Jan2009.pdf.

McGeoch, M.A., Butchart, S.H.M., Spear, D., *et al.* (2010) Global indicators of biological invasion: species numbers, biodiversity impact and policy responses. *Diversity and Distributions*, 16, 95–108.

Millennium Ecosystem Assessment (2005) *Ecosystems and Human Well-being: A Framework for Assessment*. Island Press, Washington, DC.

Natural Heritage Trust (2006) National land and water resources audit. Australian Government, Canberra. Available from http://www.nlwra.gov.au. Accessed July 2006.

Occhipinti-Ambrogi, A. and Galil, B. (2004) A uniform terminology on bioinvasions: a chimera or an operative tool? *Marine Pollution Bulletin*, 49, 688–694.

Parks Canada (2005) Inventory and monitoring. National Parks of Canada, Gatineau. Available from: ww.pc.gc.ca/eng/progs/np-pn/eco/eco3.aspx. Accessed August 2012.

Pyšek, P., Richardson, D.M., Pergl, J., Jarosik, V., Sixtova, Z., and Weber, E. (2008) Geographical and taxonomic biases in invasion ecology. *Trends in Ecology and Evolution*, 23, 237–244.

Pyšek, P., Jarosík V., Hulme P.E., *et al.* (2010) Disentangling the role of environmental and human pressures on biological invasions across Europe. *Proceedings of the National Academy of Sciences of the USA*, 107, 12157–12162.

Richardson, D.M. and Pyšek, P. (2008) Fifty years of invasion ecology – the legacy of Charles Elton. *Diversity and Distributions*, 14, 161–168.

Richardson, D.M., Pyšek, P., Rejmánek, M., *et al.* (2000) Naturalization and invasion of alien plants: concepts and definitions. *Diversity and Distributions*, 6, 93–107.

Rodrigues, A.S., Pilgrim, J.D., Lamoreux, J.F., Hoffmann, M., and Brooks, T.M. (2006) The value of the IUCN Red List for conservation. *Trends in Ecology and Evolution*, 21, 71–76.

Roques, A., Rabitsch, W., Rasplus, J.-Y., Lopez-Vamonde, C., Nentwig, W., and Kenis, M. (2009) Alien terrestrial invertebrates of Europe. In: DAISIE (ed.), *The Handbook of Alien Species in Europe*. Springer, Berlin.

UNEP (2002) COP 6 Decision VI/23. Alien species that threaten ecosystems, habitats or species. The Hague, 7–19 April 2002. United Nations Environment Programme. Available from: http://www.cbd.int/decisions/?id=7197= (accessed September 2008).

UNEP (2003) Monitoring and indicators: designing national level monitoring programmes and indicators. *UNEP/CBD/SBSTTA/9/A0*. United Nations Environment Programme, Montreal.

Valéry, L., Fritz, H., Lefeuvre, J., and Simberloff, D. (2008) In search of a real definition of the biological invasion phenomenon itself. *Biological Invasions*, 10, 1345–1351.

Vié, J.-C., Hilton-Taylor, C., and Stuart, S.N. (eds), (2009) *Wildlife in a Changing World – An Analysis of the 2008 IUCN Red List of Threatened Species*. IUCN, Gland, Switzerland, 180 pp.

Vilà, M., Basnou, C., Pysek, P., *et al.* (2010) How well do we understand the impacts of alien species on ecosystem services? A pan-European, cross-taxa assessment. *Frontiers in Ecology and the Environment*, 8, 135–144.

Vitule, J.R.S., Freire, A.C., and Simberloff. D., (2009) Introduction of non-native freshwater fish can certainly be bad. *Fish and Fisheries*, 10, 98–108.

Walpole, M., Almond, R., Besançon, C., *et al.* (2009) Tracking progress towards the 2010 biodiversity target and beyond. *Science*, 325, 1503–1504.

8

Exploitation Indices: Developing Global and National Metrics of Wildlife Use and Trade

Rosamunde E.A. Almond[1], Stuart H.M. Butchart[2], Thomasina E.E. Oldfield[3], Louise McRae[4] and Steven de Bie[5]

[1]United Nations Environment Programme World Conservation Monitoring Centre (UNEP-WCMC), Cambridge, UK
[2]BirdLife International, Cambridge, UK
[3]TRAFFIC International, Cambridge, UK
[4]Institute of Zoology, Zoological Society of London, London, UK
[5]Wageningen University, Wageningen, The Netherlands

Introduction

Wild animals and plants are essential for human livelihoods and well-being, and people harvest millions of individual plants and animals from tens of thousands of species every year. As the world's human population increases and demand for biological resources grows, humans will exert an ever greater pressure on both the species being targeted for exploitation, and the ecosystems in which they live. In order that wild species meet our present needs without compromising the needs of future generations, it is vital that these species are used in a biologically sustainable way. Indicators that track changes in the impact of use on wild populations and wildlife trade are useful tools in assessing how well we are balancing the conservation of species and ecosystems with the needs of people.

According to the United Nations Food and Agriculture Organisation, 40% of the world's economy is based directly and indirectly on the use of biological resources. These wild species provide a range of benefits for people, ranging from food to medicine, clothing, building materials, and transport, as well as forming a critical

---

Biodiversity Monitoring and Conservation: Bridging the Gap between Global Commitment and Local Action, First Edition. Edited by Ben Collen, Nathalie Pettorelli, Jonathan E.M. Baillie and Sarah M. Durant.
© 2013 John Wiley & Sons, Ltd. Published 2013 by John Wiley & Sons, Ltd.

part of their cultural and spiritual heritage, and being important for their well-being. A broad range of different animals and plants are used in some way, and global assessments of all species of birds, mammals, and amphibians for the International Union for the Conservation of Nature (IUCN) Red List (www.iucnredlist.org) indicate that people harvest and use products from 42.3% of the world's bird species, 25.2% of the world's mammals, and 8.1% of the world's amphibians (Hilton-Taylor *et al.*, 2009).

The trade in wild species is becoming increasingly globalised, and for many people their use of wild species now extends far beyond the resources found within the countries where they live. Timber and seafood are the most important categories of international wildlife trade, in terms of both volume and value. The UN Food and Agriculture Organisation (FAO) publishes annually a report on global fisheries – the State of the World's Fisheries and Aquaculture (SOFIA). In the latest report, SOFIA 2010, production from global capture fisheries in 2008 was just around 90 million tonnes, including just under 80 million tonnes from marine waters and a record 10 million tonnes from inland waters (FAO, 2010b). The estimated first-sale value of this catch was US$93.9 billion. The FAO also assesses the global timber trade, and in the latest Forest Resources Assessment in 2010, the global timber trade alone was conservatively estimated to be worth just over US$100 billion per year between 2003 and 2007 (FAO, 2010a). To put these figures into perspective, in 2009 the global trade value of tea, coffee, and spices all together was US$24.3 billion (for more information on the value of wildlife trade see TRAFFIC, 2012).

People benefit further from forests by extracting non-timber forest products (also labelled by the FAO as 'non-wood forest products') harvested from forested regions. These include food and food additives (such as edible nuts, mushrooms, fruits, herbs, spices and condiments, aromatic plants, game), fibres (used in construction, furniture, clothing, or utensils), resins, gums, and plant and animal products used for medicinal, cosmetic, or cultural purposes. Some of these are primarily used locally, and millions of people depend on wild products they find themselves for their livelihoods, health, and nutrition. Other products are important export commodities. The FAO collects national-level data on a broad range of wild products that are significantly traded internationally, including honey, gum arabic, rattan, bamboo, cork, nuts, mushrooms, resins, essential oils, and plant and animal parts for pharmaceutical products (for more information and national-level datasheets see http://www.fao.org/forestry/nwfp/en/). In 2005, the reported annual value of these non-timber forest products was US$18.5 billion (FAO, 2010a). However, the FAO conceded that this assessment is likely to be an underestimate as the reported statistics probably cover only a fraction of the true total economic value of such products. This is also true of any study that looks only at data on the value of the legal wildlife trade – reconciling legal and illegal wildlife trade is discussed further in the second part of this chapter.

Although a species is often associated with its most common or well-known use, it is very rare that individual species have only one use or product. According to the IUCN

Red List, the leopard (*Panthera pardus*), for example, is used for food, in medicine, in making clothing and household goods, as a pet, and is also hunted for sport (Henschel *et al.*, 2008; IUCN, 2011). Some are internationally traded and threatened, so trade is restricted under the Convention on International Trade in Endangered Species of Wild Fauna and Flora (CITES). These range from the yellow crested cockatoo (*Cacatua sulphurea*: captured in high numbers for the pet trade and listed on CITES Appendix II) to the tiger tail seahorse (*Hippocampus comes*: used for medicine and also caught for the pet trade) and the bobcat (*Lynx rufus*: primarily harvested for its fur and hunted for sport and listed on CITES Appendix I). Other species, such as mackerel (*Scomber scombrus*, used as food for people and animals, and for oil) and Canada geese (*Branta canadensis* hunted for food and as sport) are common throughout their range and are harvested and traded in high volumes.

The trade in wildlife may take place at a variety of scales ranging from local barter and exchange to international trade across borders and between continents. Although some details are known about the volume and the value of trade for specific commodities such as timber and marine fish, an accurate overview of the scale of the wildlife trade more generally is difficult to gauge. In the early 1990s, TRAFFIC, the wildlife trade monitoring network, estimated the value of legal wildlife products imported globally was around US$160 billion. In 2009, the estimated value of global imports was over US$323 billion (TRAFFIC, 2012). As part of a specific study on trade flowing into the European Union (EU), TRAFFIC estimated the legal trade of wildlife products into the EU alone was worth an estimated €93 billion in 2005, and this climbed to nearly €100 billion in 2009 (for a more detailed overview of the role of the EU in global wildlife trade see TRAFFIC, 2007). The increasing demand for some species can be met by captive breeding, ranching, or artificial propagation. However, the bulk of the trade in many species and their products comes from the wild, some such as brazil nuts and pangolins exclusively so, and therefore has the potential to impact both wild populations and the ecosystems in which they live.

Part of the challenge of the Convention on Biological Diversity 2010 target was to develop indicators that explicitly measure the link between trends in biodiversity and the benefits biodiversity provides to people, in particular under the focal area of sustainable use (CBD, 2004: COP 7 Decision VII/30). In the recently adopted 2020 Strategic Plan (CBD, 2010), the notion is maintained in Strategic Goal B: 'Reduce the direct pressures on biodiversity and promote sustainable use', within target 7, stating explicitly that 'by 2020 areas under agriculture, aquaculture, and forestry are managed sustainably, ensuring conservation of biodiversity'. The Millennium Ecosystem Assessment framed this link as ecosystem services – the services and benefits that biodiversity provides to people (MEA, 2005). As has already been discussed, wild animals and plants provide a myriad of different services, ranging from provisioning services such as food or medicine, to the cultural or spiritual services associated with a particular species.

In producing indicators for any ecosystem service, it is essential to consider both the supply of a particular resource and its use. In the case of wild species, this means establishing direct links between status and trends of species in the wild, and what, why, and how people are using them. Indicators of species use and trade are made more complicated because threats rarely occur independently, and there are often a multitude of drivers of change in population size or status of wild species, making it difficult to track the effects of exploitation alone. For example, at a species level, the IUCN Red List of Threatened Species lists unsustainable exploitation as one of the top three drivers in the change in conservation status of mammals, amphibians, and birds (Vie et al., 2009 see Figure 8.1). However, other important threats include the expansion of agriculture and aquaculture, pollution, invasive alien species, and climate change. Among individual populations of a species, the drivers of changes in population size in one area may differ markedly from those in another. Some species also have specific uses in certain parts of their range. For example, the house sparrow (*Passer domesticus*) is used in Chinese medicine, so may be harvested intensively in some parts of Asia but is unlikely to be harvested at a similar level in other parts of the world. Other species may be used sustainably in one area but overexploited in another, making it challenging to make generalisations about the sustainability of use beyond the local scale.

An additional complication for measuring trends in utilised species is that sustainability is not a fixed state, nor is the social, biological, or economic context in which sustainable off-take levels must be calculated. Instead, it is a dynamic process – the consequence of balancing a range of social, economic, political, and biological factors. The impact of each of these factors varies widely from species to species and even from place to place. Some may be based on biological characteristics of a species such as longevity or reproductive rate. Some factors may affect sustainability at a local level, such as who holds the tenure to harvest the resource and how actively any stakeholders are involved in the management of it. Others such as the effectiveness of enforcement or trade regulations may act at multiple points in the trade chain between the harvester and the end consumer.

A recommended framework for using and communicating biodiversity indicators is the Pressure-State-Response-Benefit framework (Sparks et al., 2011) as used by Butchart et al. (2010) to assess whether the CBD 2010 target had been met (Butchart et al., 2010). Indicators of exploitation can be placed in this framework and used to assess trends in: the status of wild species (state), the benefits they provide (benefit), the importance of exploitation in driving such trends (pressure), and the actions being taken to mitigate such pressures and reduce the impacts (response indicators).

Despite the known importance of wild species to human economies and livelihoods, there are relatively few indicators specifically developed to monitor the species that people use and rely upon. There are a number of well-developed sustainable use indicators for certain sectors such as forestry and marine fisheries. However, the use of wild species extends far beyond these resources, and millions of individual plants and animals from tens of thousands of species are traded every year. A suite of new indicators are currently being developed, which focus on the changes in status and the impact of use on a broad range of utilised terrestrial, freshwater, and marine species. At a global scale, these will give a broad picture of changes in these species, and will add information on drivers and pressures to existing population- and species-based indicators. In parallel with these global overviews, it is also essential that a picture is built of what, where, and how people are using wild species at a national level. Indicators of exploitation are primarily population- or species-based, and therefore are ideally suited to being scaled to a regional or national level. Such indicators have the potential to provide information that allows countries to both prioritise efforts to address threats and incorporate information on use and trade into policy- and decision-making processes.

In this chapter we first present an overview of indicators that are currently available to track trends in the status and trade in exploited wild animals and plants. These are primarily global in scope but the majority of them can be applied to a national or regional level. Once an organism is harvested, there are many steps before it reaches the end consumer. In an increasingly globalised world, this journey – commonly referred to as the trade chain – may involve tens of people and cross national boundaries, seas, and even continents. In the second part of this chapter, we use data from various points in this journey from source to consumer to explore three potential areas of future indicator development:

- indicators of the sustainability in use and trade in wild animals and plants.
- economic-based indicators; and
- indicators that measure responses to these impacts such as certification and wildlife trade interventions.

Indicators currently available

Global status indicators

Two indicators have so far been developed that can provide an overview of the status of utilised species on a global scale: the Living Planet Index (LPI: Loh *et al.*, 2005; Collen *et al.*, 2009, see also Chapter 4) and the IUCN Red List Index (RLI: Butchart *et al.*, 2004; see also Chapter 2). The LPI shows trends in population abundance of vertebrates. The RLI shows trends in survival probability of species (the inverse of extinction risk). The RLI can be used to look at both changes in the status of utilised species and trends in the contribution of use as a driver of those changes. Data on the utilisation of species are included in Red List assessments, so it is possible to apply these data to the RLI and LPI in order to show trends in the status of utilised populations and species.

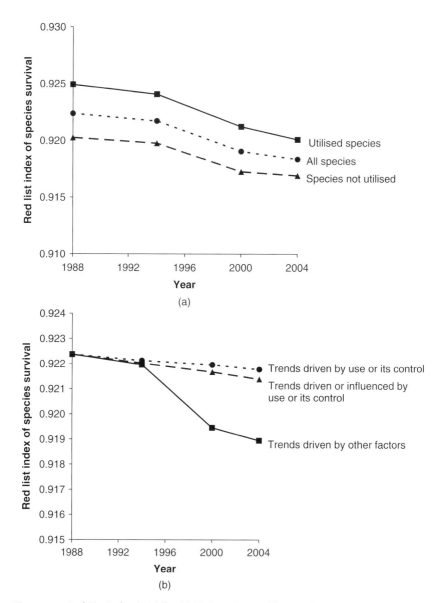

Figure 8.1 Red List Index (RLI) for (a) bird species used by people (solid line: 4481 species) and birds species not used by people (dashed line: 5402 species), showing the proportion of species expected to remain extant in the near future without additional conservation action; and (b) showing trends driven by use (or its control and management) compared to trends driven by other factors combined (e.g. habitat loss), showing the proportion of species expected to remain extant in the near future without additional conservation action ($n = 9883$ species). Sources: Butchart (2008) and BirdLife International.

Species-level indicators: changes in the survival probability of utilised species

The RLI is based on data from the IUCN Red List, on which species are placed into categories of extinction risk using five criteria with quantitative thresholds for the size, structure, and trends in population and range (IUCN, 2001). The RLI is based on the number of species in each category and the number changing categories between assessments owing to genuine improvement or deterioration in status. RLI values relate to the proportion of species expected to remain extant in the near future without additional conservation action. An RLI value of 1.0 equates to all species being categorized as Least Concern, with none expected to go extinct in the near future. An RLI value of zero indicates that all species have gone extinct. A declining trend in the index (i.e. decreasing RLI values) means that the expected rate of species extinctions is increasing, that is, that the rate of biodiversity loss is increasing (Butchart *et al.*, 2005, 2006, 2007). The IUCN Red List is becoming increasingly comprehensive in terms of its taxonomic breadth, with all species now assessed in several major taxa, including birds, mammals, amphibians, corals, sharks, conifers, and cycads, with a number of other groups close to completion. To date, repeated RLIs are available for birds (1988–2008), mammals (1996–2008), amphibians (1980–2004), and corals (1996–2008), with RLIs under development for other groups.

Each Red List assessment includes data on utilisation, specifically, the purpose to use (coded against the IUCN Utilisation Classification Scheme; Table 8.1), source (wild, captive breeding, or ranching), primary form taken (whole individual or non-lethal removal of parts), life stage (adult, chicks, or eggs), and scale (international/regional, national/sub-national, or local/subsistence).

Birds have been assessed for the Red List most frequently (five times since 1988) and are commonly used by people, in particular for food, pets, medicine, sport hunting, and apparel/ornamentation (Butchart, 2008; data available at http://www.birdlife.org/datazone/species/index.html). Among utilised species of birds, two purposes of use dominate: 3649 species (37.0% of extant species, 87.4% of utilised species) were recorded as being used as pets, and 1398 species (14.2% of extant species; 33.5% of utilised species) were recorded as being hunted for food (Butchart, 2008). Less significant uses include sport hunting, wearing apparel or ornamentation, and medicine (usually traditional), with trivial numbers of species being recorded as used for handicrafts, fuel (from oil or fat, principally from seabirds), and household goods (e.g. down for mattresses), etc. Many species are used for multiple purposes; for example, 68.9% of species that are hunted for food are also kept as pets (Butchart, 2008).

As can be seen in Figure 8.1a, the RLI for utilised birds shows a negative slope, indicating that these species are on average increasing in extinction risk, potentially because human use of birds is currently unsustainable. This negative slope occurs because although some species have been downlisted to lower categories of threat

Table 8.1 **IUCN 2007. Utilisation categories used in the Use and Trade Classification Scheme (version 1.0) for the IUCN Red List of Threatened Species (IUCN 2007). Available from http://intranet.iucn.org/webfiles/doc/SSC/RedList/AuthorityF/utilization.rtf.**

IUCN use category	End use
1	Food for people
2	Food for animals
3	Medicine – human and veterinary
4	Poisons (includes pesticides, herbicides, fish poisons)
5	Manufacturing chemicals (includes resins and dyes)
6	Other chemicals (includes incense, perfumes, and cosmetics)
7	Fuels
8	Fibre
9	Construction or structural materials
10	Wearing apparel, accessories (including leather and fur)
11	Household goods
12	Handicrafts, jewellery, decorations, and curios
13	Pets
14	Research
15	Sport hunting and specimen collecting
16	Establishing an *ex situ* population

owing to successful conservation efforts (in some cases through reducing levels of exploitation), many more have been uplisted to higher categories of threat owing to increasing threats, including over-exploitation, driving faster population declines and leading to smaller population and distribution sizes. The RLI shows that the net balance has been an overall reduction in the proportion of species expected to remain extant in the absence of additional conservation action; in other words, that this set of species is slipping further towards extinction.

Overall, utilised bird species are less threatened with extinction than non-utilised species (i.e. higher RLI values; Figure 8.1a). This is perhaps not surprising as people tend to use those species that are easiest to utilise, which are often more abundant and hence less threatened. Exceptions are species whose rarity increases their value to collectors and their price on the market. There is strong evidence that people value rare species more than they value common ones (Angulo and Courchamp, 2009). In extreme cases, economic theory predicts that people may place such a high value on owning or using products from rare and endangered species that the increase in economic incentives to exploit the last few individuals could eventually lead to their extinction (Brook and Sodhi, 2006). This is discussed further in the second part of this chapter in relation to the development of future indicators.

To examine whether the scale of trade (international vs national/local) affected trends in the status of species, Butchart (2008) compared RLIs for internationally or nationally traded bird species A total of 3337 species (33.9% of extant species) were recorded as being traded internationally, all for the pet trade, although some are also internationally traded for additional purposes. At least 46.5% of internationally traded bird species (1552 species) were also recorded in national/local trade, but the remainder probably reflect lack of data rather than being genuinely targeted only for international trade. Internationally traded species have declined in status since 1988, although they are, on average, less threatened than utilised species that are not internationally traded (Figure 8.2a). One possible reason for this difference relates to the purpose of use, as internationally traded species tend to be common and attractive species that are used as cagebirds, whereas locally used or nationally traded species tend to be larger-bodied species that are hunted for food and are more sensitive to exploitation.

Utilisation as a driver of changes in Red List status

Information on the main factors driving changes in Red List categories between assessments of species allows determination of the role that use, and more specifically international trade, plays in driving overall trends. Using data from the World Bird Database (which is summarised in the published species factsheets at http://www.birdlife.org/datazone/species/index.html), Butchart (2008) determined the primary driver of changes in status for each species that had undergone a genuine change in Red List category during the period 1988–2008. For those status changes driven by exploitation, the primary purpose and scale of use were also determined.

Among utilised species of birds, there were 137 genuine changes in Red List category during the period. Factors related to use drove just under one-third (30.9%) of these changes, being outweighed in importance by habitat loss and invasive species impacts. This is shown in Figure 8.1b where the RLI for trends driven by other factors (solid black line) shows a steeper decline than the index where changes were primarily driven by use (dashed line). Nevertheless, the net balance was negative: although some species have improved in status through successful control of unsustainable trapping and trade and/or improved harvest and trade management (e.g. Lear's macaw *Anodorhynchus leari*, imperial Amazon *Amazona imperialis*), these improvements have been outweighed by the number of species that have deteriorated in status owing to inadequate trade management or implementation of trade controls (e.g. yellow-crested cockatoo, *Cacatua sulphurea*). Therefore, the RLI showing trends driven by international trade or its management and control shows a negative slope (Figure 8.2b) indicating that the trade in these species is not sustainable. However, international trade issues drove only 16% of 93 genuine status changes among internationally traded species, indicating that the overall trend was largely driven by factors other than international trade.

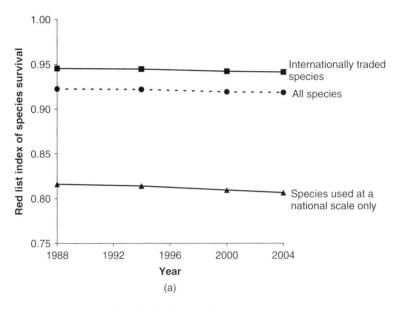
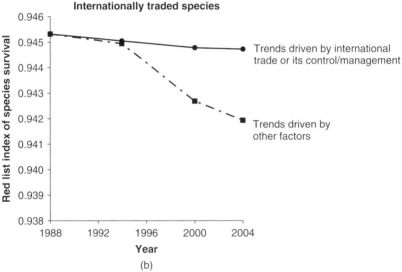

Figure 8.2 Red List Index (RLI) for (a) internationally traded bird species (3337 species), species utilised at a national scale only (745 species), and all species (9883 species), in each case showing the proportion of species expected to remain extant in the near future without additional conservation action; and (b) all bird species, showing trends driven by international trade (or its control and management) compared to trends driven by a combination of other factors (e.g. habitat loss, pollution, or climate change). The trend in the Index shows the proportion of species expected to remain extant in the near future without additional conservation action ($n = 3736$ species). Sources: Butchart (2008) and BirdLife International.

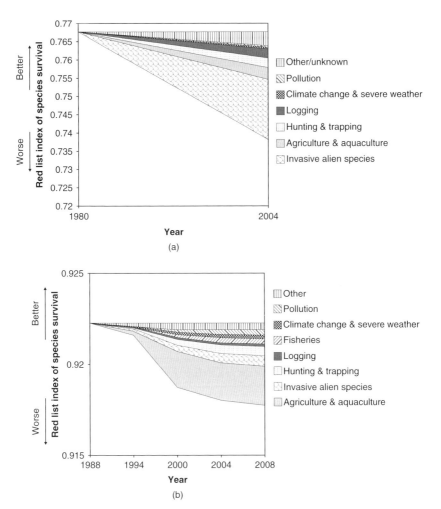

Figure 8.3 Red List Index (RLI) for (a) amphibians, (b) birds, and (c) mammals showing trends driven by the impacts of utilisation (shown here as 'hunting and trapping') compared with trends driven by other factors, for the proportion of species expected to remain extant in the near future without additional conservation action; $n = 9785$ non-Data Deficient extant bird species at start of period, 4555 mammal and 4417 amphibian species. Note that for mammals, the lines show RLIs for trends driven by each factor, rather than the proportional contribution of each factor, because it was not possible to distinguish between the importance of different drivers and/or assign a primary driver for many species.

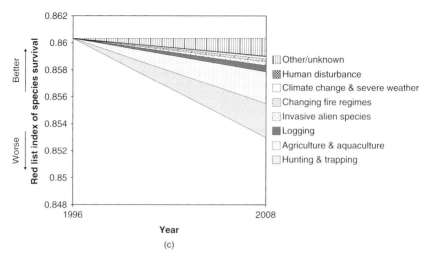

Figure 8.3 (*continued*)

A comparative analysis has been completed for birds, mammals, and amphibians, which examines species utilisation (coded as 'hunting and trapping') with other drivers of changes in Red List status. For amphibians and birds (Figures 8.3a and 8.3b), hunting and trapping is the third strongest driver of category change. For mammals, it was not possible to distinguish between primary and secondary drivers. However, hunting and trapping is still shown to be a strong driver of changes in Red List status (Figure 8.3c).

Linking state and benefits: the Red List Index for food and medicine

Many terrestrial animal and plant species are used by humans for food and medicine. These species make significant contributions to diet and healthcare, particularly in developing countries. An estimated 50 000–70 000 plant species are used in traditional and modern medicine. Many of the wild species used for food and medicine are threatened with extinction, some due to over-exploitation, or to different pressures such as habitat loss, disease, or a combination of factors. Regardless of the causes, the diminishing availability of these resources threatens the income from wild collection, and the health and well-being of the people who depend on them.

Of the extant bird species, 14% (over 1400 species) are believed to be used for food and/or medicinal purposes, while 22% of all known mammal species are used

EXPLOITATION INDICES 171

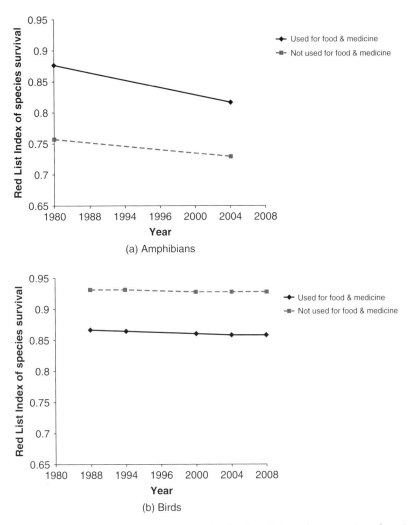

Figure 8.4 Red List Index for species used for food and medicine. The proportion of species expected to remain extant in the near future without additional conservation action for all species, species used for food and/or medicine, or species not used for these purposes, for (a) amphibians, (b) birds, and (c) mammals. Source: RLIs produced using IUCN Red List data with assistance from IUCN Species Programme and BirdLife International.

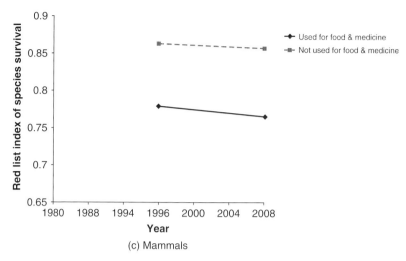

Figure 8.4 (*continued*)

in this way. For both birds and mammals the RLI shows that species used for food and medicine are in general more threatened and their conservation status is also deteriorating at a faster rate (Figure 8.4). For birds, 23% of species used for food and/or medicine are threatened, compared with 12% for all bird species (Figure 8.4). In contrast, amphibians used for food and medicine appear to be less threatened overall than species not used for these purposes, although they have declined faster.

Comparable data are not available for plants as just 3% of the world's well-documented medicinal plants have been evaluated for global conservation status. The current limited assessments suggest that medicinal plants face a high level of extinction risk in those parts of the world where people are most dependent on them for healthcare and income from wild collection – namely Africa, tropical Asia, the Pacific, and South America.

Population-based indicators: the Living Planet Index (LPI) of utilised species

Indicators based on population trends can provide a more sensitive complement to the species-level assessments of the RLI. The LPI shows trends in vertebrate abundance by calculating a chain of average annual changes in abundance to make an index, starting with an initial value of 1 in 1970 (Collen *et al.*, 2009; Loh *et al.*, 2005, also see Chapter 4). In order to track trends in the status of utilised species, each of the 2500 species in the LPI database was classified using the IUCN utilisation classification scheme (Table 8.1). This information came from a wide variety of published and unpublished sources, including: the IUCN Red List (www.iucnredlist.org), the World Bird Data base (www.birdlife.org/datazone/species/index.html), the CITES trade database (www.cites.org/eng/resources/trade.shtml), FAO forestry country profiles (www.fao.org/forestry/nwfp/en/ and www.fao.org/forestry/country/en/), the International Timber Trade Organisation (ITTO: www.itto.int), publications by the Center for International Forestry Research (CIFOR: www.cifor.cgiar.org), the University of British Columbia (UBC) Sea Around Us Project (www.seaaroundus.org), and the FishBase online database (www.fishbase.org/search.php).

The LPI of utilised species is based on time-series data from 5448 populations of 1394 species of mammals, birds, amphibians, reptiles, sharks, and bony fish. The purpose of use of these species ranged from food to sport, pets, clothing, and medicines. As expected, most species were utilised for multiple purposes. The index has declined by around 15% between 1970 and 2006, meaning that populations of utilised species fell by nearly a sixth over that time period (Figure 8.5), with concomitant declines in the species' availability to be used by people. Multiple factors are likely to have driven these declines in addition to unsustainable exploitation, and a limitation of this analysis is that although all the species are used by people in some way, there is no information on whether the specific population included within the index has had individuals harvested from it. Determining the contribution of exploitation in driving trends is problematic, as drivers of trends tend to vary both in time and in space (see Corey and Wait, 2008; Issac and Cowlishaw, 2004; Kotiaho et al., 2005;

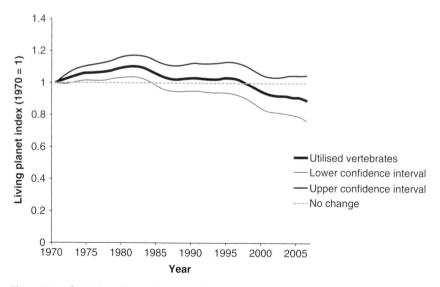

Figure 8.5   The Living Planet Index for utilised species; $n = 5448$ populations of 1394 species.

Owens and Bennett, 2000; Price and Gittleman, 2007; Purvis *et al.*, 2000; Thomas, 2008). Even if drivers are known, how populations may respond to them may vary and interact with each other, resulting in non-linear population responses (Lomolino and Channell, 1995; Rodriguez, 2002; Thomas, 2008). Adding such information to the LPI is discussed later in this chapter.

Indicators of sustainable use

Together, the RLI and LPI paint a pessimistic picture of global declines in the status and population abundance of utilised species in the last few decades. The negative trend in both of these indicators implies that our use of those species may not be sustainable. However, the RLI shows that utilised species are, in general, less threatened than non-utilised species. In addition to such global overviews, it is essential to look directly at the sustainability of use and trade in wild species and the impact that use may be having on both the species being harvested and the people who rely on them for subsistence, their wealth, or their well-being. The most widely available indicators measuring the sustainability of use have been developed for the forestry and fisheries sectors, as discussed below.

Timber: the area of sustainably managed forest certified by the Forest Stewardship Council

One way of assessing sustainable use is to focus on the management of areas where important resources are produced in a sustainable manner. In the case of forests, sustainably managed areas may include natural or semi-natural forests that are used to produce timber and non-timber forest products, and forest plantations. Currently the best way of assessing the extent of sustainable management of forests on a large scale is through certification systems that have been developed to assess the management of individual forests according to agreed criteria in order to enable forest products to be sold as coming from sustainable sources.

Several forest certification schemes exist, including Forest Stewardship Council (FSC), Pan-European Forest Council (PEFC), Canadian Standards Association (CSA), Sustainable Forest Initiative (SFI), ISO 14001, and Malaysian Timber Certification Council (MTCC). One of the indicators the CBD used to track progress towards the 2010 Target was 'area of certified forest under sustainable management'. The United Nations Environment Programme World Conservation Monitoring Centre (UNEP-WCMC) conducted a thorough assessment of the different certification schemes available and their contribution to biodiversity conservation, and concluded that the area of forest certified by the FSC can be used as a proxy for the 'area of certified forest under sustainable management' (information on FSC indicators is updated monthly at www.fsc.org).

The area of forest under FSC certified management globally has increased from 3.24 million hectares in 1995 (when the scheme started) to 1096 sites in 80 countries covering 150 million hectares in 2012 (FSC, 2012). The rate of site designation was greatest between 2000 and 2005, with an average of 51 million hectares being designated per year, but declined since, averaging 13 million hectares per year between 2006 and 2008.

The growth in FSC certification has not been equal among forest types; with a far greater area of boreal and temperate forests being certified (78.83 and 54.99 million hectares, respectively) than tropical and subtropical forests (16.02 million hectares: based on data up to February 2012). This difference reflects both the complexities of managing tropical forests and the logistical and financial costs associated with obtaining certification in relation to the resources available to forest owners in different regions. The total area certified also differs substantially among tropical regions, with the largest area of certified forest in the Americas. This reflects both the larger total area of forest and the more rapid development of national certification standards under the FSC scheme in this region. In the Asia-Pacific region, other regional schemes for certifying forest management have been more important than FSC. Globally, the strong concentrations of FSC certified sites in some countries such as the United States, Brazil, South Africa, and European countries, are where the development of national standards for certification under the FSC principles and criteria has been most rapid.

An increase in the area of FSC certified forest represents an increase in the area of commercial forest managed sustainably in respect to biodiversity. However, data are unavailable to calculate a more meaningful indicator: the proportion of production forest that is sustainably managed. Furthermore, the indicator assumes that FSC certification correlates with enhanced biodiversity trends. Future development needs to address these issues, as well as incorporate data from other certification schemes, and ideally include those forests that are sustainably managed but not certified because the certification process demands time and financial resources that many forest owners, especially small ones, are unwilling or unable to commit (such small producers are actively encouraged to join the Rainforest Alliance certification scheme, for example). While the indicator implies a general trend towards improved forest management, this remains to be validated.

Fisheries: the proportion of fisheries being sustainably fished

According to the most recent SOFIA report by the FAO, global production of marine capture fisheries reached a peak of 86.3 million tonnes in 1996 and declined slightly to 79.5 million tonnes in 2008 (FAO, 2010b). The FAO reports the proportion of fish stocks that are fully exploited (and therefore producing catches at or close to their maximum sustainable limits, with no room for further expansion), over-exploited, or depleted (and, thus, yielding less than their maximum potential owing to excess fishing pressure in the past, with no possibilities in the short or medium term of further expansion and with an increased risk of further declines and a need for rebuilding: for more detailed definitions, see FAO, 2009).

In 2008, slightly more than half of the world's fish stocks – 53% – were fully exploited and producing catches at or close to their maximum sustainable limits with no room for further expansion. The proportion of stocks that are over-exploited, depleted, or recovering from depletion has stabilised since the mid-1990s, reaching 32% in 2008 (28% were overexploited, 3% depleted, and 1% recovering from depletion). This combined percentage is the highest since the start of the index in 1974. In 2008, 15% of the stock groups monitored by FAO were estimated to be underexploited (3%) or moderately exploited (12%), and therefore able to produce more than their current catches.

The top 10 species caught worldwide account for approximately 30% of the world's marine capture fisheries production in terms of quantity. In 2008, most stocks of these species were fully exploited, leaving no potential for increased production. Some of the main stocks of some species, such as the anchoveta (*Engraulis ringens*) and the Alaskan pollock (*Theragra chalcogramma*) were found to be over-exploited, and increases in their production could only be possible with rebuilding plans in place. Although there is great uncertainty surrounding these estimates, the report concluded that, 'the increasing trend in the percentage of overexploited, depleted and recovering stocks and the decreasing trend in underexploited and moderately exploited stocks do give cause for concern'.

This indicator is frequently referred to in international policy documents and the media to draw attention to the degree of sustainability of the world's marine capture fisheries. While this information provides a unique global overview, SOFIA 2010 stresses that the stocks included in the analysis represent only a fraction of the exploited stocks around the world. The stocks within the index include the largest single-species stocks and account for almost 80% of the officially declared landings. However, the FAO estimates that only 10% of the fish stocks people exploit are assessed, and these are not always monitored regularly. It is therefore clear that for most of the exploited fisheries stocks there is little or no information about their sustainability. Although ideally, all stocks would be regularly assessed and their sustainability closely monitored, the FAO admits that the collection of such data is expensive, time consuming, and requires specialist expertise. This inevitably leads to a bias in data collection towards developed countries and means that it is lowest in tropical multi-species fisheries exploited by fleets from developing countries or by distant-water fleets. In recognition of the need to identify and develop methods that are less data-demanding, the FAO is preparing a set of guidelines for the assessment of stocks in data-poor situations. These data will be incorporated into assessments in the future and it is hoped that it will enable the production of indicators that include a wider range of fisheries. Other global fisheries datasets, such as those collated by the Sea Around Us Project at the University of British Columbia (UBC, 2009) include catch statistics from both the FAO and other datasets and have also been used as a basis for a range of indicators of the sustainability of marine fisheries (see Pauly, 2007).

Future directions for indicators of exploitation and trade in wild species

The indicators outlined above provide broad overviews of the changes in the status of utilised species and populations, and some measures of the sustainability of the marine fish and timber trades. However, when considering the state of wild populations, the benefits their products provide, and the pressures and impacts of how, where, and how much people are using, each indicator can only tell part of the story.

As already discussed, a key difference between the RLI and LPI is the scale of the information they are based upon. The RLI is primarily a species-level indicator; although the Red List categories and criteria can be applied at the subspecies or population level, and hence RLIs calculated at these scales, only one such example exists to date (Szabo *et al.*, 2012). By contrast, the LPI is more sensitive, as it tracks population-level changes. However, suitable data are available for a much more restricted (and less geographically representative) number of species, and data on drivers of trends of the populations monitored are not available.

Both the LPI and the RLI are treated broadly as 'state' indicators (Butchart *et al.*, 2010; Walpole *et al.*, 2009). However, disaggregations to show trends for utilised species can be considered indicators of benefits (Butchart *et al.*, 2010), because as population abundance or survival probability of utilised species declines, so do the opportunities for people to obtain benefits from them. In the case of the LPI, this interpretation requires caution as populations are included within it if the species as whole is used as data on utilisation of individual populations are not currently available. Obtaining such data would require examination of each of the data sources, and considerable supplementary information. Even if this information could be obtained, the resulting index would show trends in utilised populations, not necessarily the impact of use per se on their abundance.

From source to consumer

Once an organism is harvested, there are many steps before it reaches the end consumer. In an increasingly globalised world, this journey – commonly referred to as the trade chain – may involve tens of people and cross country boundaries, seas, and even continents. In order to provide an overview of the trade and use of a given species, information needs to be collected on both the status of species in the wild (the 'supply' side of the equation) and the volume of products from those species in the market (the 'demand' side). Indicators have the potential to provide an overview of trends and drivers on both sides of this equation. Indicators for the supply side include trends in individual source populations over time or trends in the amount being harvested. On the demand side, indicators can be used to track changes in market value and market size, how much end consumers are willing to pay for products from wild species, and what motivates them to buy them. Within each step of the trade chain, indicators can also summarise information on the effectiveness of the measures taken to limit impacts on wild populations, including trade bans or certification.

Here we use the trade chain as a basis to explore the kinds of data that future indicators of use and wildlife trade could be based on to monitor both the scope and sustainability of wildlife trade and use. We focus on three potential areas of future indicator development: (i) indicators of the sustainability of use and trade in wild animals and plants; (ii) economic indicators; and (iii) indicators that measure responses such as certification and wildlife trade interventions.

Sustainability indicators for marine, terrestrial, and freshwater species

Indicators of sustainable use are essential for assessing how well the conservation of wild species and ecosystems is being balanced with the needs of people, and for incorporating this into policy- and decision-making processes. Determining whether use is sustainable requires information on the off-take (the amount harvested from an area), and the effect that this has on the taxon or population (Milner-Gulland and Akçakaya, 2001). At its most simplistic, if the harvest is causing wildlife populations to decline, then it is unsustainable.

Supply-side indicators of biological sustainability are best developed for marine fisheries. Catch landing data are the foundation of many of these indicators, including those developed by the FAO and the Sea Around Us project (Pauly, 2007) and form the basis for modelling past and future effects of fisheries on the world's oceans (for an example analysis see Tremblay-Boyer *et al.*, 2011). Fisheries catch data have also been used to look at changes in where people are fishing and how by much fishing fleets are expanding. A recent study graphically shows the expansion of the human fisheries 'footprint' in the ocean by using fisheries catch data to calculate the primary productivity needed to supply the fish harvested every year between 1950 and the present day (Swartz *et al.*, 2010; for an animated map showing these changes over time and also the expansion of the EU fishing fleet, see http://www.wwf.eu/fisheries/cfp_reform/external_dimension/). The study indicated a spatial expansion of the industry by 1 million km^2 per annum from the 1950s to the end of the 1970s, resulting in one-third of the world's ocean and two-thirds of continental shelves being exploited by fisheries today. This growth in marine fisheries was made possible through exploitation of new fishing grounds and new fishing technology. Only the unproductive waters of the high seas, and relatively inaccessible waters in the Arctic and Antarctic remain unexploited. Although this study did not directly measure the sustainability of these catches, comparison of the primary productivity needed to provide the harvest with the amount assessed as being produced would provide a spatially explicit measure of the sustainability of fisheries. Other data can also be used to evaluate sustainability indirectly. For example, a reduction in the size of caught predatory fish (Myers and Worm, 2003) or an increase in the distance travelled by fishing fleets to fulfil their fishing quota or to catch fish above a certain size can be used as proxies to indicate whether the harvest of a particular species is sustainable.

Developing similar sustainable use indicators for terrestrial animals and plants has proved to be more problematic and as a consequence, national- and global-level indicators are not yet very well developed or widely used. This is partly due to the effort required to collect data at the scale needed to assess sustainability. Although detailed statistics about the size, number of individuals, and origin are available for some species exploited for fur or game hunting, adequate information is unavailable for the majority of utilised species. The lack of detailed data on specific populations has led to the development of a number of indicators based on highly simplified population models, for example, for bush meat hunting (Robinson and Redford, 1991). These authors developed a simple algorithm to combine data on the density of a given population and predicted rates of population increase (based on life history traits and mortality) to estimate the maximum sustainable level of production. An estimate of whether the harvest is sustainable is obtained by comparing this to the number of individuals harvested from the population.

Milner-Gulland and Akçakaya (2001) reviewed this model and a number of others to test how their assessment of the sustainability of bush meat hunting compared with assessments using other methods. The models ranged from the use of Bayesian statistics to more complex models of population dynamics. Their conclusion was that many of these methods are not precautionary and tend to overestimate the amount that can be sustainably harvested from a population. The simpler the model, the harder it is to incorporate uncertainty within it and the more generalisations have to be made. For example, the Robinson and Redford (1991) method does not include survival rates of individuals or the actual population growth rate. Milner-Gulland and Akçakaya (2001) stress that incorporating such uncertainty is vital for effective indicators.

An additional problem with many of the existing approaches is that they require detailed information, both about the species and the exploited population, which is often lacking. Broad-scale indicators will necessarily be based on data that are widely available. Marine fisheries are an example of a sector where such data are easily and publicly accessible, for example through the Sea Around Us project (http://www.seaaroundus.org/) or FishBase (www.fishbase.org). These sources contain a wealth of information ranging from the biological characteristics of marine fish, to where they are caught, the methods used, and the volume landed. Data on wild animals and plants are much sparser, more disparate, less synthesized, and scattered across multiple institutions and organisations. However, the approaches and methods used to assess the sustainability of fisheries provide a potentially fertile area for ideas and lessons learnt. Milner-Gulland and Akçakaya (2001) framed this as 'the possibility of cross-fertilisation' between disciplines, and such exchange of techniques has the potential to lead to the development of indicators that could both monitor the sustainability of harvests and improve the management of utilised species.

Economic indicators of wild species

As any commodity becomes rarer, the amount it will cost is expected to rise. Collectors in particular value rare objects, and economic modelling of so-called 'market value of rarity' has been carried out with collectors in a range of different fields (Koford and Tschoegl, 1998). The high price placed on rarity applies to biodiversity too (Brook and Sodhi, 2006).

A simple model of the economics of animal and plant collection may make the assumption that harvesting will cease when the cost of finding the species exceeds the amount of money the harvester receives for their work. Although intuitively correct, a study has shown that for some rare species, the high value placed on rarity by collectors may mean that the price will continue to rise as the species becomes harder to find (Courchamp *et al.*, 2006). This means that the incentives and the financial rewards for finding the species remain high and keep pace with the cost of acquiring it. The authors concluded that assuming that consumers are always willing to pay what it takes to obtain a particularly desirable rare species, this model makes both ever increasing prices and extinction by collection inevitable.

Data showing positive relationships between price and rarity are scarce, but Courchamp *et al.* (2006) support their model by providing a number of examples where the collection of rare birds, insects, mammals, and fish has been driven by trophy hunting, by collectors, and by consumers wanting luxury goods and traditional medicine and who are willing to pay or risk substantial amounts to obtain them. For example, the authors carried out an analysis of the average price paid for hunting trophies of 57 species of wild sheep (Caprinae). The authors found that the price of the trophy was correlated with rarity, regardless of size: the rarer the trophy (using IUCN Red List category as a proxy), the more valuable and expensive it was. Publishing the location of populations of rare species may in rare cases leave them vulnerable to collection and to extinction (Stuart *et al.*, 2006). As Brook and Sodhi (2006) discuss, it is not necessary to take all individuals from the wild before a species is 'doomed' to extinction or a population disappears, just enough individuals to push the population below a viable size.

If the price of a species or a product increases as it becomes rarer then it may be possible to use market prices as a very crude indicator of wild population trends. Price data have to be treated with caution as they may vary for a multitude of reasons, ranging from trends in fashion to marketing. There are also a multitude of prices along the trade chain, ranging from the amount of money a hunter or gatherer is paid to find an animal or plant to the final price paid by the consumer. Nevertheless, the models put forward by Courchamp *et al.* (2006) suggest that this may be possible under some circumstances, such as when a species is endangered or when it is targeted by collectors. As circumstances will vary markedly from one species to the next, such indicators may be best suited to national or sub-national levels or for very localized monitoring programmes targeted at specific populations.

Response indicators: certification and wildlife trade interventions

Response indicators track the policies or actions taken to mitigate biodiversity loss. Those used by the CBD to track progress towards the 2010 target included the extent of protected areas (based on the World Database on Protected Areas: IUCN/UNEP-WCMC, 2010) and the number of international and national policy measures related to invasive alien species (McGeoch *et al.*, 2010) (summarised in Walpole *et al.*, 2009 and Butchart *et al.*, 2010). Within the context of utilised species, two potential areas where response indicators could be developed are certification indicators and those that evaluate the effectiveness of wildlife trade interventions.

Certification indicators

Indicators similar to the area of forest under FSC certification are being used to track adoption of certification schemes in other sectors, such as the Marine Stewardship Council (MSC: www.msc.org). In the future, data from multiple certification schemes should be included in such indicators to provide a broader cross-sector picture of the spread of such schemes and to track changes in consumer behaviour.

The effectiveness of wildlife trade interventions

Because the trade in wild animals and plants crosses borders between countries, its regulation requires international cooperation to safeguard species of conservation concern from over-exploitation. One convention in particular, the Convention on International Trade in Endangered Species of Wild Fauna and Flora (CITES), is key to the regulation of the international wildlife trade. This convention came into force in 1975 and aims to ensure that international trade in specimens of wild animals and plants does not threaten their survival (for more details see www.cites.org).

Approximately 5000 species of animals and 28 000 species of plants are currently listed on the CITES appendices. Although some of these species, such as tigers, are endangered, some are not and CITES aims to ensure the sustainability of the trade in order to safeguard these resources for the future.

Existing indicators such as the RLI can be used to explore trends in the conservation status of CITES-listed species; however, caution is needed in interpreting the results. For example, there are 1447 bird species listed on CITES Appendix I and II. Unsurprisingly, the RLI shows that these species are more threatened overall than bird species on average (i.e. their RLI values are lower). In a separate analysis of internationally traded species, those listed on CITES Appendix I or II have been declining at a faster rate between 1988 and 2008 than internationally traded species that are not listed on CITES. This could be interpreted to mean that CITES is not effective in mitigating threats from trade to threatened birds. However, CITES operates only at an international level, and significant unsustainable trade may also take place at a local and national level. Furthermore, CITES-listed species are impacted by multiple threats in addition to trade. Therefore it is not possible to draw any firm conclusions from this analysis about the effectiveness of CITES.

This highlights the need for taking care in interpreting indicator trends, particularly those aiming to tease out the effects of use, as it is only one of a myriad of factors that influence species' status or population abundance. There may also be a substantial illegal trade in wild species in addition to the legal one. For example, although the CITES Trade database (held at UNEP-WCMC) can be used to collate statistics that provide an overview of the scope and scale of CITES permits issued for specific species or taxa at both a national and a global scale, these alone cannot provide a complete overview of the trade in any given species as they reflect the trade permits issued under the CITES convention. As highlighted by TRAFFIC, by its very nature it is almost impossible to obtain reliable figures for the value of illegal wildlife trade, but the figure is likely to be hundreds of millions of dollars. The value of illegal, unreported, and unregulated (IUU) fisheries alone has been estimated to be in the range of US$10–23 billion per year (MRAG and FERR, 2008), while the high prices paid for products such as elephant ivory or tiger bones means that the economic value of the trade in individual species and the impact on wild populations may also be substantial. In order for future indicators to fully reflect the scale of the trade in wild species, measures of the scope of both the legal and the illegal trade should be included where possible.

Applying species exploitation indicators at a national level

The indicators currently available can be applied at multiple scales, including the national (summarised in Table 8.2). These indicators therefore have the potential to provide information that allows countries to both prioritise efforts to address potential threats and incorporate information on use and trade into policy- and decision-making processes. The extent to which existing indicators can be used at a sub-global level will largely be determined by how much data are available. In the case of the RLI, this requires multiple national Red List assessments (Zamin et al., 2010; see also Chapter 2); examples exist for Australia (Szabo et al., 2012) and Denmark (Pihl and Flensted, 2011), with a number of others being developed. The Red List database is currently being developed to incorporate a set of 'livelihoods' fields for collecting additional information on how harvesting a species within a particular area contributes to people's livelihoods, the species' economic value, and methods used to harvest it. For indicators based on abundance data, such as the LPI, a representative national framework for data collection is the ideal scenario under which to develop a national-level indicator. In the meantime, if sufficient data are available then a subset of the global dataset may provide an informative overview of trends in utilised species. The addition of information on the extent to which use is driving changes within the individual populations is essential if the effects of utilisation are to be teased apart from the effects of other drivers such as climate change, pollution, or invasive species. Although potentially time consuming to collect, the addition of such information could greatly increase how informative the indicator is and how easily it can be incorporated into policy- and decision-making processes.

The collation of national-level statistics on the use and trade in wild species as part of assessments such as the CBD National Strategies and Action Plans (NBSAPs) and the FAO Forest Resources Assessment combined with the release of open-source national-level datasets means that data on the use and trade in wild species are becoming much more readily available. Many of these databases are online, making them easy to search by country or region, and of potential use in creating national- and regional-level indicators of wildlife use and trade.

Concluding remarks

Given the importance of wild animals and plants to people's health, wealth, and well-being, there are currently few global- or national-level indicators that summarise trends in the use and trade in wild animals and plants. This chapter has provided an overview of the existing indicators of wildlife use and trade, and explored some opportunities to develop new national- and global-level indicators in the future. We have outlined what specific information each indicator provides within the pressure/state/response/benefits framework.

If indicators of exploitation are to provide information that can be integrated into national-level policy- and decision-making processes, they should provide a measure of not only the status or the supply of wild animals and plants, but also the market for those products and the potential impacts of changes of supply upon the lives and

Table 8.2 Application of existing indicators to a national and regional scale

Indicator	Scale of application			Explanation
	Global	National	Local	
Red List Index (RLI) of utilised species	✓	✓	✓*	Available globally for all utilised species in a number of taxonomic groups, as well as those species traded internationally versus nationally or locally. National RLIs for utilised species can be calculated either by disaggregating the global index, or by repeatedly assessing extinction risk at the national scale. Many countries have compiled national Red Lists, which form the basis of the latter approach (see www.nationalredlist.org), and some have done this twice or more using consistent methods in order to permit calculation of a national RLI (see Szabo et al., 2012, and Pihl and Flensted, 2011, for examples). Local-scale RLIs are more problematic, but an example exists at the provincial level in Canada (Quayle et al., 2007)
Living Planet Index of utilised species	✓	✓*	✓*	The data for this global indicator consist of national- and population-level measures, and can therefore potentially be disaggregated to look at trends at national and regional scales, if sufficient data are available
Forest Stewardship Council (FSC) certification indicator	✓	✓	✓*	The data for this indicator originate from individual FSC sites and as a result this indicator can be produced at the national and regional levels. The FSC Certification Database contains up-to-date information on sites and is available online (www.info.fsc.org). Country certification schemes' websites and databases should also be examined, to ensure a complete picture
UN Food and Agriculture Organisation (FAO) fisheries indicator	✓	✓	?	This indicator is based on assessments of sustainability within FAO major fishing areas (FAO, 2002–2012). There are currently 27 of these, many of which cross the borders of the Exclusive Economic Zones (EEZs) surrounding multiple countries, or include areas of the high seas. National-level catch data are available from other sources such as the Sea Around Us Project database (http://www.seaaroundus.org/) so these could potentially be used to construct national-level indicators

*The indicators can conceptually be applied at these scales, but the results will only be meaningful if sufficient data are available.

livelihoods of people who use and rely upon them. In reality, no single indicator can communicate all the information that is needed to give a 'bird's-eye view' of the trade in a given species and to determine whether its use is sustainable. Exploring a number of different parts of the trade chain highlights some areas where future indicators could be developed, ranging from those that could assess the biological sustainability of the harvest from a particular population, to the economic drivers behind the trade and the effectiveness of responses such trade interventions and certification.

Some sectors such as marine fisheries and forestry already have well-developed indicators and routinely use them as part of national and global assessments of trade and sustainability. It could be argued that the information needed to develop these indicators for marine fish and timber trees is more readily available, partly because of their economic importance and also because of the global nature of their trade. However, the increasing number of open-source national and global datasets on a wide range of wild species is beginning to open the door for the development of indicators that can be applied to a much broader range of terrestrial, freshwater, and marine species. Techniques and approaches developed for creating indicators to assess marine fisheries could potentially be adapted and applied to other species. This 'cross-fertilisation' of techniques together with data from a range of points within the trade chain could help to produce a 'dashboard' of indicators as described here. These indicators have the potential to provide information that allows countries to both prioritise efforts to address threats and incorporate information on use and trade into policy- and decision-making processes. Indicators of exploitation are primarily population- and species-based, and therefore are ideally suited to being scaled to a regional or national level.

Acknowledgements

We would like to thank Matt Walpole and Ben Collen for their invaluable help in editing the document. The development of the Living Planet Index for utilised species brought together knowledge and expertise from a wide range of organisations, including IUCN, UNEP-WCMC, TRAFFIC, the Convention on International Trade in Endangered Species (CITES), the Zoological Society of London (ZSL), WWF, and BirdLife International, and funding from the Global Environment Facility (GEF) and the Shell Research Foundation. LM is supported by WWF International.

References

Angulo, E. and Courchamp, F. (2009) Rare species are valued big time. *PLoS ONE*, 4(4): e5215.
Brook, B.W. and Sodhi, N.S. (2006) Conservation biology: rarity bites. *Nature*, 444, 555–556.

Butchart, S.H.M. (2008) Red List Indices to measure the sustainability of species use and impacts of invasive alien species. *Bird Conservation International*, 18, S245–S262

Butchart, S.H.M., Stattersfield, A.J., Bennun, L.A., et al. (2004) Measuring global trends in the status of biodiversity: Red List Indices for birds. *PLoS Biology*, 2, 2294–2304.

Butchart, S.H., Stattersfield, A.J., Baillie, J., et al. (2005) Using Red List Indices to measure progress towards the 2010 target and beyond. *Philosophical Transactions of the Royal Society of London, Series B*, 360, 255–268.

Butchart, S.H.M., Akcakaya, H.R., Kennedy, E. and Hilton-Taylor, C. (2006) Biodiversity indicators based on trends in conservation status: strengths of the IUCN Red List Index. *Conservation Biology*, 20, 579–581.

Butchart, S.H.M., Akcakaya, H.R., Chanson, J., et al. (2007) Improvements to the Red List Index. *PLoS ONE*, 2(1), e140; doi:10.1371/journal.pone.0000140.

Butchart, S.H., Walpole, M., Collen, B., et al. (2010) Global biodiversity: indicators of recent declines. *Science*, 328, 1164–1168.

CBD (2004) *COP 7 Decision VII/30*. Convention on Biological Diversity, Montreal, Canada. Available at: www.cbd.int/decision/cop/?id=7767.

CBD (2010) *Strategic Plan for Biodiversity 2011–2020, including Aichi Biodiversity Targets (COP 10, Decision X/2)*. Convention on Biological Diversity (CBD), Montreal, Canada.

Collen, B., Loh, J., Whitmee, S., Mcrae, L., Amin, R., and Baillie, J.E.M. (2009) Monitoring change in vertebrate abundance: the Living Planet Index. *Conservation Biology*, 23, 317–327.

Corey, S.J. and Wait, T.A. (2008) Phylogenetic autocorrelation of extinction threat in globally imperilled amphibians. *Diversity and Distributions*, 14, 614–629.

Courchamp, F., Angulo, E., Rivalan, P., et al. (2006) Rarity value and species extinction: the anthropogenic Allee effect. *PLoS Biology*, 4(12), e415.

FAO (2002–2012) CWP Handbook of Fishery Statistical Standards. Section H: Fishing Areas for Statistical Purposes. CWP Data Collection. FAO Fisheries and Aquaculture Department, Rome. Available from: http://www.fao.org/fishery/cwp/handbook/H/en.

FAO (2009) *The State of World Fisheries and Aquaculture 2008 (SOFIA)*. FAO Fisheries and Aquaculture Department, Rome.

FAO (2010a) *Global Forest Resources Assessment, 2010: Key Findings*. FAO, Rome.

FAO (2010b). *The State of World Fisheries and Aquaculture 2010 (SOFIA)* FAO, Rome, Italy.

FSC (2012) *Global FSC Certificates: Type and Distribution (March 2012)*. Forest Stewardship Council, Paris.

Henschel, P., Hunter, L., Breitenmoser, U., et al. (2008) *Panthera pardus*. IUCN Red List of Threatened Species Version 2010.4. IUCN, Gland, Switzerland.

Hilton-Taylor, C, Pollock, C.M., Chanson, J.S., Butchart, S.H.M., Oldfield, T.E.E., and Katariya, V. (2009) State of the world's species. In: Vié, J.C., Hilton-Taylor, C., and Stuart, S.N. (eds), *Wildlife in a Changing World: An Analysis of the 2008 IUCN Red List of Threatened Species*. IUCN, Gland, pp. 15–42.

Issac, N.J.B. and Cowlishaw, G. (2004) How species respond to multiple extinction threats. *Proceedings of the Royal Society of London, Series B*, 271, 1135–1141.

IUCN (2001) *Red List Categories and Criteria version 3.1*. IUCN, Gland, Switzerland.

IUCN (2011) *IUCN Red List of Threatened Species. Version 2011.2*. IUCN, Gland, Switzerland. Available at: http://www.iucnredlist.org.

IUCN/UNEP-WCMC (2010) The World Database on Protected Areas (WDPA). WDPA, IUCN/UNEP-WCMC, Cambridge, UK. Available at: http://www.wdpa.org/.

Koford, K. and Tschoegl, A.E. (1998) The market value of rarity. *Journal of Economic Behaviour and Organisation*, 34, 445–457.

Kotiaho, J.S., Kaitala, V., Komonen, A., and Paivinen, J. (2005) Predicting the risk of extinction from shared ecological characteristics. *Proceedings of the National Academy of Sciences of the USA*, 102, 1963–1967.

Loh, J., Green, R.E., Ricketts, T., *et al.* (2005) The Living Planet Index: using species population time series to track trends in biodiversity. *Philosophical Transactions of the Royal Society B-Biological Sciences*, 360, 289–295.

Lomolino, M.V. and Channell, R. (1995) Splendid isolation: Patterns of geographic range collapse in endangered mammals. *Journal of Mammalogy*, 78, 335–347.

McGeoch, M.A., Butchart, S.H.M., Spear, D., *et al.* (2010) Global indicators of alien species invasion: threats, biodiversity impact and responses. *Diversity and Distributions*, 16, 95–108.

MEA (2005) *Ecosystems and Human Well-Being: Biodiversity Synthesis*. Millennium Ecosystem Assessment, World Resources Institute, Washington, DC.

Milner-Gulland, E.J. and Resit Akçakaya, H. (2001) Sustainability indices for exploited populations. *Trends in Ecology and Evolution*, 16, 686–691.

MRAG and FERR (2008) *The Global Extent of Illegal Fishing*. Marine Resources and Fisheries Consultants (MRAG) and Fisheries Ecosystems Restoration Research (FERR), Fisheries Centre, University of British Columbia.

Myers, R.A. and Worm, B. (2003) Rapid worldwide depletion of predatory fish communities. *Nature*, 423, 280–283.

Owens, I.P.F. and Bennett, P.M. (2000) Ecological basis of extinction risk in birds: Habitat loss versus human persecution and introduced predators. *Proceedings of the National Academy of Sciences of the USA*, 97, 12144–12148.

Pauly, D. (2007) The Sea Around Us Project: Documenting and communicating global fisheries impacts on marine ecosystems. *Ambio*, 36, 290–295.

Pihl, S. and Flensted, K.N. (2011) A Red List Index for breeding birds in Denmark in the period 1991–2009. *Dansk Ornitologisk Forenings Tidsskrift*, 105, 211–218.

Price, S.A. and Gittleman, J.L. (2007) Hunting to extinction: biology and regional economy influence extinction risk and the impact of hunting in artiodactyls. *Proceedings of the Royal Society of London B*, 274, 1845–1851.

Purvis, A., Gittleman, J.L., Cowlishaw, G., and Mace, G.M. (2000) Predicting extinction risk in declining species. *Proceedings of the Royal Society of London B*, 267, 1947–1952.

Quayle, J.F., Ramsay, L.R., and Fraser, D.F. (2007) Trend in the status of breeding bird fauna in British Columbia, Canada, based on the IUCN Red List Index method. *Conservation Biology*, 21, 1241–1247.

Robinson, J.G. and Redford, K.H. (1991) Sustainable harvest of neotropical mammals. In: Robinson, J.G. and Redford, K.H. (eds), *Neo-tropical Wildlife Use and Conservation*. Chicago University Press, Chicago, pp. 415–429.

Rodriguez, J.P. (2002) Range contraction in declining North American bird populations. *Ecological Applications*, 12, 238–248.

Sparks, T.H., Butchart, S.H.M., Balmford, A., *et al.* (2011) Linked indicator sets for addressing biodiversity loss. *Oryx*, 45, 411–419.

Stuart, B.L., Rhodin, A.G., Grismer, L.L., and Hansel, T. (2006) Scientific description can imperil species. *Science*, 312, 1137.

Swartz, W., Sala, E., Tracey, S., Watson, R., and Pauly, D. (2010) The spatial expansion and ecological footprint of fisheries (1950 to present). *PLoS ONE*, 5(12), e15143.

Szabo, J.K., Butchart, S.H.M., Possingham, H.P., and Garnett, S.T. (2012) Adapting global biodiversity indicators to the national scale: A Red List Index for Australian birds. *Biological Conservation*, 148, 61–68.

Thomas, G.H. (2008) Phylogenetic distributions of British birds of conservation concern. *Proceedings of the Royal Society of London B*, 275, 2077–2083.

TRAFFIC (2007) *Opportunity or Threat: The Role of the European Union in Global Wildlife Trade.* TRAFFIC International, Cambridge, UK.

TRAFFIC (2012) Wildlife trade: what is it? TRAFFIC International, Cambridge, UK. Available at: http://www.traffic.org/trade/.

Tremblay-Boyer, L., Gascuel, D., Watson, D.R., Christensen, V., and Pauly, D. (2011) Modelling the effects of fishing on the biomass of the world's oceans from 1950 to 2006. *Marine Ecology – Progress Series*, 442, 169–185.

UBC (2009) Sea Around Us Project. Global database on marine fisheries and ecosystems. University of British Columbia. Available at: www.seaaroundus.org.

Vie, J.C., Hilton-Taylor, C., and Stuart, S.N. (2009) *Wildlife in a Changing World: An Analysis of the 2008 IUCN Red List of Threatened Species.* IUCN, Gland, Switzerland.

Walpole, M., Almond, R.E.A., Besancon, C., et al. (2009) Tracking progress toward the 2010 biodiversity target and beyond. *Science*, 325, 1503–1504.

Zamin, T.J., Baillie, J.E.M., Miller, R.M., Rodríguez, J.P., Ardid, A., and Collen, B. (2010) National red listing beyond the 2010 target. *Conservation Biology*, 24, 1012–1020.

9

Personalized Measures of Consumption and Development in the Context of Biodiversity Conservation: Connecting the Ecological Footprint Calculation with the Human Footprint Map

Eric W. Sanderson

Global Conservation Programs, Wildlife Conservation Society,
New York, USA

Consumption is the sole end and purpose of all production.

Adam Smith, *The Wealth of Nations*, 1734

Introduction

The most important component of biodiversity to monitor may be us, the human species. Many if not all of the current threats to biodiversity can be traced back to the large and expanding, voracious, and technologically advanced human population, currently standing at over 7 billion individuals and expected to reach somewhere between 8 and 10 billion by the close of the twenty-first century (United Nations, 2009). Anthropogenically induced climate change, habitat fragmentation and destruction, biological homogenization, and disease dynamics related to humans and human commensal organisms are all linked to human social and economic activities in profound and inconvenient ways (Lockwood and McKinney, 2001; Millennium Ecosystem Assessment, 2005; Intergovernmental Panel on Climate Change, 2007; Jones et al., 2008).

Two key measures of human impact on the biosphere are denoted metaphorically as footprints: the human footprint map (Sanderson et al., 2002) and the ecological footprint calculation (Wackernagel and Rees, 1996). Both the human footprint map and the ecological footprint calculation have had important impacts on conservation policy and public education efforts, but these two similar sounding conservation tools actually represent two quite different approaches to measuring development and consumption, starting from different premises and yielding different, but complementary, measures.

The human footprint map is a map of the collective human influence on the planet, derived by summing spatial measures (i.e. maps) of human influence on the land; specifically maps of population density; land use; access from roads, rivers and rails; and power use (Sanderson et al., 2002). The human footprint shows that humanity influences 83% of the Earth's land surface. Compared with maps of where it is possible to grow wheat, corn and rice, the human footprint indicates that 98% of these arable lands have already been accessed by people. This kind of geographical comparison is made possible because the human footprint shows the spatial distribution of human influence, allowing the human footprint to be compared to other maps, like crop productivity or maps of priorities for biodiversity conservation, using technologies such as a geographic information system (GIS). The trade made for spatial explicitness is that the global pattern of the human footprint cannot be traced back to any one individual – the human footprint is a spatial, but generalized estimate of human impact globally.

The ecological footprint, in contrast, is a calculation at the individual level (Wackernagel and Rees, 1996). It begins with the consumption patterns of an individual, evaluated through a set of questions that begin with the nation where the respondent lives. From a small set of answers about individual consumption, a calculation is made about how much 'productive area' a person needs. From the individual level, the ecological footprint can be extrapolated to a given human population, for example, for a country, to suggest the ecological 'overshoots', where ecological consumption exceeds ecological production. Like the human footprint map, ecological footprint calculations suggest the extent to which humanity is overusing the sustainable biological capacity of the planet. The trade-off the ecological footprint makes for its individualized focus is that no one knows where the places of production are that satisfy consumption. In the globalized twenty-first century economy most consumption in one place by one person is supplied by numerous places, in many countries and connected through articulating, variably efficient, opaque, economic networks.

Thus we have one footprint that is spatially explicit but generalized to the world population; and we have another footprint that is individualized but geographically vague; the natural thought is to connect the two in some way. This chapter suggests a tentative method to make that connection.

Connecting human ecological footprints large and small requires weaving economic geography into the calculations supporting both footprints, so that we can measure national-level consumption patterns not only against their domestic sources, but also versus their various international supplies. A spatially explicit, personal human footprint could provide a baseline against which gains in sustainable development and improved environmental choices could be monitored, while demonstrating how individual, business, and national consumption patterns are connected to wild nature at home and over the horizon, and thus form a basis for conservation policy and action. First, however, we need to delve into the details of each footprint in more detail.

Ecological footprint calculation

The ecological footprint concept and supporting method for its calculation was first suggested by Matthias Wackernagel and William Rees (Wackernagel, 1994; Wackernagel and Rees, 1996) and has subsequently been revised several times (Wackernagel *et al.*, 1999; Loh and Wackernagel, 2004; Monfreda *et al.*, 2004; Wackernagel *et al.*, 2006; Ewing *et al.*, 2008). Other authors have offered refinements to the methodology (e.g. Venetoulis and Talberth 2006), or related analyses like Material Energy Flow Accounting (Haberl *et al.*, 2004). Recently some divergence in methods underlies the several ecological footprint calculators available on the internet (Best Foot Forward, 2009; Global Footprint Network, 2009a; Ecological Footprint, 2009). Here I focus on the ecological footprint calculation methodology suggested by Kitzes (Kitzes *et al.*, 2008) and Ewing (Ewing *et al.*, 2008).

In their landmark work, Wackernagel and Rees (1996) defined ecological footprint analysis as 'an accounting tool that enables us to estimate the resource consumption and waste assimilation requirements of a defined human population or economy in terms of a corresponding land area'. Later this definition was modified to include land and sea areas (see Monfreda *et al.*, 2004). As suggested by the definition, an important frame for thinking about ecological footprints is accounting: ecological footprints are conceptually about tracking transactions, comparing inputs (resources) from the biosphere to the human economy and outputs in terms of consumption and waste assimilation. As an accounting process, ecological footprint analysis requires a standard currency and assumes that outputs must balance total inputs (plus or minus any accounting errors).

The standardized currency of the ecological footprint is not money, but area, though not 'area' in the conventional sense. Ecological footprints depend on the abstracted notion of the 'biocapacity', where biocapacity is defined as 'a measure of the amount of biologically productive land and sea area available to provide the ecosystem services that humanity consumes' (Ewing *et al.*, 2008). Because land and sea vary in the amounts and kinds of resources that can be produced given different technologies and inputs, some system of conversion factors is required to make the accounting work. In the words of Ewing and colleagues (2008), 'By weighting each area in proportion to its bioproductivity, different types of areas can be converted into the common unit of *global hectares* [my emphasis], hectares with world average bioproductivity'. A global hectare is a one-hectare area of land or sea with average bioproductivity; this measure is the standardized unit of ecological footprint analysis; it is often abbreviated as gha.

The 'global hectare' (gha) makes ecological footprint accounting possible. Ecological footprint calculators estimate how many gha are needed to supply one person's (or by extrapolating, one population's) consumption. Commonly these calculators compare consumption to estimates of production, specifically, how much 'bioproductive area' or biocapacity (measured in global hectares) exists in a country or on Earth in total. The difference between ecological footprint and biocapacity is a measure of how much we are using, or overusing, the Earth's ecological potential. In 2005 the world overshoot was estimated as 4 082 670 718 gha, or approximately 1.31 times the Earth's biocapacity (Global Footprint Network, 2009b). Because the ecological footprint is calculable for different populations, startling comparisons are possible: for example, it has been said that 'if every human being consumed at the rate of the average American, then we would need four Earths' to provide enough commodities (Wilson, 2003). Such claims are used to bolster the argument for changing human behaviour in service of biodiversity conservation, while pointing a finger at the worst offenders (Loh and Wackernagel, 2004).

Interestingly, although ecological footprints are often made at the individual level, in fact most of the statistical data on consumption and production to support these calculations are available only nationally, through a system of 'national accounts' (Wackernagel *et al.*, 1999, 2006), which are fundamentally based on national economic accounting. These ecological footprint national accounts measure the amounts of different goods and services that different national economies produce, import, and export, based on published statistics; the ecological footprint methods provide conversion factors for each of these flows, to convert them to the standard unit of global hectares, using methods detailed below. National accounts are 'individualized' by comparing a person's self-reported consumption habits to a statistically generated 'average citizen' of that nation. Thus the first question of all individual ecological footprint calculators is 'where does the user live?' because a person's country of residence points to a particular national account.

It is easy to demonstrate this effect by playing with the online ecological footprint calculators. For example, I live in a modest, one-thousand square foot house in the Bronx, New York, in the United States; my family has three persons; we have one car, but fairly often take public transportation (subway or bus) or walk to shopping or school; we eat meat six to ten times per week, generally chicken, but sometimes beef or pork; and so on. When I put in these answers to a typical internet calculator (Ecological Footprint, 2009), the estimate of my ecological footprint is 200.93 global acres (81.31 gha; Figure 9.1a). Furthermore I am told that it would require 5.18 Earths if everyone in the world consumed the way I do. If, as an experiment, I answer all the same questions identically, except tell the calculator that I live in Bolivia, then the same calculator estimates that I use 35.86 global acres per year (14.51 gha) and it would take only 0.92 Earths if everyone consumed like me (Figure 9.1b). From these statistics, I can conclude that I am somewhat less consumptive than the average American, somewhat more consumptive than the average Bolivian, and (probably more importantly) that Americans consume much, much more than Bolivians on a per capita basis. Indeed one might draw the incorrect conclusion that the best way for

Your results ↓		HIDE WINDOW X X
(in global acres)	Footprint	Country Average
Carbon footprint	59.48	91.43
Food footprint	99.31	65.74
Housing footprint	21.77	31.58
Goods and services footprint	20.37	57.66
My total footprint	200.93	246.41
(in global acres)	Footprint	Country Average
Cropland footprint	23.95	29.61
Pastureland footprint	56.61	68.02
Marine fisheries footprint	40.72	49.33
Forestland footprint	79.64	99.45
My total footprint	200.93	246.41
Number of Earths	5.18	6.35

Monday, June 15, 2009

Figure 9.1 **Results from a representative ecological footprint calculator (Ecological Footprint, 2009) for the author's consumption, given residence claimed in (a) the United States, and (b) Bolivia. Ecological footprints are usually measured in a standardized unit, the global hectare (gha), but in this case are shown in global acres. Ecological Footprint, 2009.**

(in global acres)	Footprint	Country Average
Carbon footprint	20.37	3.90
Food footprint	11.10	8.75
Housing footprint	1.10	1.58
Goods and services footprint	3.29	3.17
My total footprint	35.86	17.40
(in global acres)	Footprint	Country Average
Cropland footprint	4.35	2.39
Pastureland footprint	13.84	8.04
Marine fisheries footprint	5.87	2.23
Forestland footprint	11.80	4.73
My total footprint	35.86	17.40
Number of Earths	0.92	0.45

Monday, June 15, 2009

Figure 9.1 *(continued)*

me to help to the Earth's biodiversity is to move to Bolivia, thus dramatically lowering my ecological footprint.

More importantly, this experiment also indicates the importance of the national accounts in the ecological footprint calculation. If we could express the human footprint at a national scale, then perhaps an equivalency can be drawn.

The human footprint map

The human footprint map as conceptualized by Sanderson and colleagues (2002, 2006) starts from an entirely different basis. Its intellectual history lies in the tradition of wilderness mapping (McCloskey and Spalding, 1989; Hannah *et al.*, 1994; Leslie and Malsen 1995; Aplet *et al.*, 2000; Nelleman *et al.*, 2001; Carver *et al.*, 2002; Mittermeier *et al.*, 2003), which is a type of negative space mapping. Measures of human activity and infrastructure are compiled, and then the places with least levels of human influence, infrastructure, or activity are classified as wilderness areas or, less formally, as wild places. Common measures of activity and infrastructure include human population density, land use, roads, urbanized areas, and in the oceans, fishing, pollution, maritime traffic, and global warming. The human footprint methodology has been applied and modified at a variety of different scales and using different measures (Levin *et al.*, 2007; Halpern *et al.*, 2008; Leu *et al.*, 2008; Woolmer *et al.*, 2008), including a study of forward projection based on demographic and infrastructure trends (Trombulak *et al.*, 2008).

The assumption that underlies this kind of mapping is that areas with less human influence – all other things being equal – are more likely to have a full range of biodiversity and fully functioning ecological processes than places with more influence; hence wild places are vital conservation targets. They are also likely to be less expensive to conserve, because by definition they experience lower levels of threat, which enables conservationists to save more with less resources and ideally take a more proactive approach to dealing with threats that have not yet materialized. This line of thinking, like the ecological footprint, masks a lot of underlying complexity and assumptions, which have been evaluated by a variety of critics, including those who note that people are necessary for conservation and are not always 'threats' (e.g. Colchester, 1994); those who believe that conservation should tackle immediate problems, where threats are greatest, not conserve places furthest from the fray (e.g. Brooks *et al.*, 2006); and those who believe that the very concept of wilderness or a wild place is a human construct, which ranges from unhelpful to destructive depending on the context (Cronon, 1995; Nelson and Callicott, 2008; see also Chapter 12). Supporting the argument for wild place conservation are studies that observe that protected areas, which explicitly limit human influence through regulation, are key tools in conserving biodiversity (Bhagwat *et al.*, 2001; Bruner *et al.*, 2001; Gaston *et al.*, 2008), and observations, like those of the human footprint, that suggest much of the world is already under threat, and hence there isn't really any place on Earth where conservation activity is not a necessity.

In this chapter, I am concerned not with the value of wilderness per se, but the human footprint map as a measure of human activity. The recognition of the importance of the map of human influence itself as an interesting and valuable product, apart from its use in mapping wilderness, was a key insight emerging from the human footprint work (Sanderson *et al.*, 2002). By summing up data on population density (Center for International Earth Science Information Network, Columbia University *et al.*, 2000), land use (Loveland *et al.*, 2000), access from roads, rivers, and rail (derived from the National Imagery and Mapping Agency, 1997), and an energy-use measure (e.g. night-time lights visible to a low-light detecting satellite; Elvidge *et al.*, 1997) at 1 km^2 resolution, a visible gradient of human influence was apparent. At local scales this gradient is recognizable as the familiar pattern of roads, towns, and cities, which we

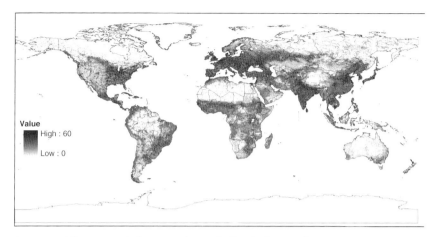

Figure 9.2 The Human Influence Index underlies the human footprint map. It is a standardized index combining human population density, land use, transportation access and infrastructure, and electrical power usage (based on satellite-based night-time light maps). These different data are mapped at 1 km² resolution worldwide. Scale: 1:270 000 000. Center for International Earth Science Information Network (CIESIN).

think of as the political and economic landscape. Globally the pattern indicates the comprehensive extent and differing degrees of influence marking humanity's current habitation on the planet.

The human footprint, like the ecological footprint, requires combining a set of diverse factors measured in different ways, into a common, standardized metric. The human footprint has its own standardization, in this case an invented measure called the Human Influence Index (HII; see Figure 9.2). My colleagues and I defined weighting schemes for each of the layers, which translated, for example, human population density values or distance from roads, into index values scaled 0–10 (Sanderson et al., 2002). The input maps are translated into these index values through the weighting schemes and then added together to produce the HII. These weighting schemes are hypotheses and specific to the scale of the mapping exercise. They are derived through literature review, expert opinion, and sensitivity testing. In other contexts, other authors have modified these weighting schemes to match their own purposes and geographies (e.g. Levin et al., 2007; Woolmer et al., 2008).

The Human Influence Index has been interpreted in some analyses as a generic measure of threat, following the wilderness mapping tradition; conservation biologists have used the index to test hypotheses about the causes of mammalian range collapse (Laliberte and Ripple, 2004), patterns of African biodiversity (Burgess et al., 2007), and tiger anti-habitat (Dinerstein et al., 2007). Others have used it to ask what role conservation organizations can play in poverty alleviation (Redford et al., 2008).

This last usage suggests another way to see the human footprint – not as a map of threats, but as a map of development. Many groups, including most governments, actively promote extending the human footprint of population and infrastructure as a means of economic development. In fact, it is generally assumed that the infrastructure inputs to the human footprint (i.e. roads, towns, cities, factories, power plants, etc.) are essential to national economies, as they are the means by which natural resources are extracted, transformed into goods, and transported to market. Moreover, the act of extending the human footprint (e.g. building roads, converting lands to agriculture, building power infrastructure) translates into jobs for the citizenry; therefore investment in infrastructure is often used as an economic stimulus (111th US Congress, 2009).

This potential for a dual reading of the human footprint – economic panacea and environmental conundrum – provides an opportunity to connect the ecological footprint calculation to the human footprint map through the system of national accounts. Because of the relatively fine resolution (1 km^2) of the human footprint mapping, it is possible to summarize the human influence index over larger geographies, whether those are ecosystems, biomes, or nation state boundaries (e.g. the sum Human Influence Index of the United States is larger than the sum Human Influence Index of Bolivia). For nation states, national accounts about consumption and production lie at the heart of the ecological footprint calculations. Since both the human footprint and the ecological footprint are measured in abstracted, invented units, representing either side of the economic equation of production and consumption, one can posit an equivalency between them. This equivalency will be the trunk on which we connect both feet to the same body and see the combined footprint as a measure of human society, striding across the continents and around the world, and connected back to each individual – you, me, and everyone else we know.

Connecting the ecological footprint and human footprint

The ecological footprint calculation depends on a system of national accounts measured in global hectares. Theoretically, national consumption is measured as a sum across all the goods and services used within a country and the amount of land or sea required to assimilate the wastes resulting from that consumption. In practice, it is not possible to measure all goods and services, but rather a subset. Recent iterations of the methodology have striven to make these subsets as comprehensive as possible, and now typically include summaries for carbon footprint, food footprint, housing footprint, and a catch-all bin for other goods and services (Ewing *et al.*, 2008; Monfreda *et al.*, 2004).

These national accounts are driven by analysis of available international databases of annual production and trade in commodities. Using these sources (which have their own problems, which will not be dealt with here – UN Statistics Division, 2009), a given national account measures national consumption for a given product, in terms of ecological footprint (EF), as the sum of domestic production, plus imports minus exports (Ewing et al., 2008):

$$EF = EF_{domestic\ production} - EF_{exports} + EF_{imports} \quad (9.1)$$

This calculation in turn requires estimating the ecological footprint of each product in each nation each year. These estimates are made for a set of standardized land- and sea-use categories: Cropland, Grazing Land, Forest, Fishing Ground, Carbon Uptake Areas, and Built-up Lands, yielding a Cropland Footprint, Grazing Land Footprint, Forest Footprint, etc. For each product produced in each land-use type, these footprints are estimated using the following formulation:

$$EF = (P/YN) \times YF \times EQF \quad (9.2)$$

Where:
  EF = ecological footprint associated with product or waste (in gha)
  P = amount of product extracted or waste generated (tons year^{-1})
  YN = national-average yield for product extraction or waste absorption (tons (national ha)$^{-1}$ year^{-1}
  YF = yield factor of a given land type within a country (world ha × national ha^{-1})
  EQF = equivalence factor for given land type (gha × world ha^{-1})

In other words, one measures the amount of commodity production (P) and divides that production by the national yield (YN) to estimate the amount of area (land or sea) used in that nation, measured in hectares within a given nation (the 'national ha'). To convert the area used in national ha to gha requires two conversion factors. The yield factor (YF) for a country estimates how productive its land use is relative to the global average. Nations with more productive lands and technology (e.g. for a crop, with an amenable climate and soils, perhaps enhanced by technology) will produce more per hectare than a less fortunate nation, and so YF calibrates their production to the global mean for that land-use type. The YF is calculated as the ratio of the national yield to the global yield for that product (abbreviated Y_W). Finally the equivalence factor (EQF) converts from a global mean for a land-use type to the single, generalized gha measure.

Note that in Equation 9.2 above, because the Yield Factor is defined as the national yield divided by world yield, so the two national-average yields can be cancelled out. Although it seems counterintuitive that one can calculate the ecological footprint of production of any nation using only the worldwide value, mathematically it follows from how the ecological footprint measure is defined (in fact, one could say it is the defining characteristic of the ecological footprint's 'gha' unit). Taking this step provides a significant short-cut in the calculations, avoiding compilation of national production yields on an annual basis. Taking this short-cut means that the Ecological Footprint formula can be written in this simplified form:

$$\text{EF} = (P/Y_W) \times \text{EQF} \tag{9.3}$$

Where:
Y_W = world-average yield for product extraction or waste absorption (tons × world ha^{-1} × year^{-1})

Additional calculations are made for some secondary products, reflecting inputs from diverse sources and the industrial process change (see Ewing *et al.*, 2008), but those extended methods will not be summarized here. For study and education purposes, the Global Footprint Network makes a sample dataset freely available; the sample provides global estimates and one national account – for Hungary in 2005 (full national accounts are available on a schedule of licensing fees).

It is helpful to work through an example for a single commodity to see how this works – for purposes of illustration, let us consider apple consumption in Hungary. Hungarian domestic production in 2005 was 510 361 tons of apples according to its national account (Global Footprint Network, 2009a). The average global yield for apples (Y_W) that year was estimated as 12.84 (tons per world hectares of cropland, calculated from the total world production of apples divided by the total world acreage of apples). This average global yield is assumed to be the same for apples produced anywhere in the world; moreover, apples are assumed to be produced only in cropland (if apples were produced in part from other land-use types, we would need to add in those contributions as well). The equivalence factor (EQF) for cropland to gha is 2.64, so the estimated contribution to Hungary's ecological footprint from domestic apple production is 105 129 gha. Hungary exported 39 000 tons of apples in 2005, so in effect, it is subsidizing the nations that imported those apples with an estimated 8034 global hectares of its national biocapacity. Hungary also imported 11 000 tons of apples in 2005, and so received, in effect, a subsidy from the biocapacity of other nations, amounting to 2266 gha, for its imported apples. The total contribution of apples to Hungary's global footprint therefore is 99 361 gha, of which 98% was supplied domestically; the remainder was imported.

Taken as a whole, adding up similar calculations across 177 other crops, 7 kinds of livestock (divided into 59 products), 33 wood products, 1171 fish products, 32 energy sectors, infrastructure and areas flooded for hydroelectric dams (Global Footprint Network, 2009b), Hungary's ecological footprint in 2005 was 35 839 284 gha. Remarkably 29 779 559 gha (or 83% of the total consumption) was imported; only 17% was produced domestically. Another 34 070 342 gha worth of commodities were exported, representing 85% of domestic production. In other words, Hungary is not only a productive nation, but highly involved in world trade in commodities, importing 'biocapacity' and exporting its own. In other contexts, one would refer to the difference as its foreign trade balance of a nation participating in the advantages of a globalized economy.

To connect Hungary's ecological footprint to Hungary's influence on the global human footprint map requires sourcing the areas of production that support consumption in Hungary. Hungary is able to produce, export, and import its commodities (read: its ecological footprint) because of the distribution and abundance of its population and infrastructure (read: its human footprint). The ecological footprint is a measure of consumption; the human footprint is a measure of production. They are two sides of the same coin.

Equivalency between the ecological footprint and the human footprint

My thought is to make an equivalency between the ecological footprint and the human footprint map at the national scale, matching up consumption to production, ecological footprint's gha to the human footprint's HII. This match is made at the national scale, without breaking down each production chain into its components (population, land use, transportation, and power infrastructure) or breaking down consumption into its components (urban land, agricultural productivity, carbon use, etc.). Rather we match a measure of overall production (HII) to a measure of overall consumption (gha). For example, summing the Human Influence Index across Hungary yields a total of 4 419 060 HII units. Hungary's domestic production from its biocapacity is 40 105 988 gha. In other words, Hungary harvests 9.07 gha of biocapacity for every unit of HII. This ratio, which we can call the production to infrastructure yield (PIY), provides a way to allocate ecological footprints geographically.

If one allows for a moment this equivalency (discussed further below), then to allocate Hungary's ecological footprint out into the world, we need to know (i) the domestically produced ecological footprint (production minus exports), (ii) the source nations and their contributions to the imported component of the ecological footprint, and (iii) the PIY of all the nations involved. I will use statistics on imports and exports to allocate the international portion of Hungary's ecological footprint. Part (i) we already have in the ecological footprint calculations; part (ii) can be derived from a reanalysis of the ecological footprint data, simply by tracking the source countries for all imports, rather than lumping them together; and part (iii) can be calculated by comparing the production ecological footprints to the human influence index on a national basis. To show Hungary's ecological footprint on the human footprint map, we would need to highlight 17% of HII units within Hungary (the portion of consumption sourced domestically) and HII units representing the 83% of Hungary's consumption (the portion supplied through imports.)

Ideally the import and export statistics we use would be expressed in gha and would allow us to calculate PIY values for all nations. For the purposes of this analysis, however, unfortunately the funding was not available to actually obtain the ecological footprint national accounts from the holding agencies. Fortunately, a weaker substitute is available in the public domain, taking advantage of yet another invented unit of standardization: money.

Money is the premier example of a fictional unit created to standardize value between diverse entities. When a buyer agrees to pay $3.00 for a bag of apples, the seller and the buyer establish an exchange ratio for an abstraction, money, for a real commodity, apples. The seller can then use those $3.00 to buy something else – a bag of bananas, a pair of shoelaces, or anything else, where the new seller is willing to trade a good or service for that amount of money. One great discovery behind the adoption of the monetary economy is that the exchange rates are not fixed between money and any given good or service, but are allowed to float in all transactions depending on the relative interests of the buyer and the seller through the mechanism of prices. Prices have further attractive qualities: they are scalable and traceable through conventional accounting, from individuals to nations to the world economy. For our purposes, most importantly, because trade balance is of great interest, monetized data on imports and exports are widely available. The equivalency is weaker for this purpose because monetary value does not translate as readily into the population and infrastructure metrics of the human footprint. Some lucrative economic activities (i.e. investment banking) may have relatively small physical footprints per unit of monetary value, while other activities (i.e. industrial agriculture) may have relatively large physical footprints per unit of value, though assessing exactly how much environmental impact any economic activity has is difficult and revealing as shown by life cycle assessments.

For purposes of this chapter, however, using monetary (as opposed to ecological footprint) trade accounts provides a path of connection between the ecological footprint and the human footprint, which eventually could lead to a personalized, spatialized ecological human footprint (see Figure 9.4). Using US importation statistics from 2002 (US Census Bureau 2009), and comparing them to gross domestic product (GDP) statistics from the United Nations (UN Statistics Division, 2009), allows me to estimate the fraction of GDP imported into the United States from 201 other nations that year. I assume for the purposes of illustration that this fraction of GDP imported is equivalent to the fraction of population and infrastructure dedicated to that production in the exporting country (its HII). Fractions of domestic production exported to the United States varied widely in 2002, from 48.0% of GDP from Lesotho to 0.0007% from the People's Republic of Korea (North Korea). US domestic production used to satisfy domestic consumption represents 93.35% of its GDP, making the complementary adjustment for US production exported to other nations

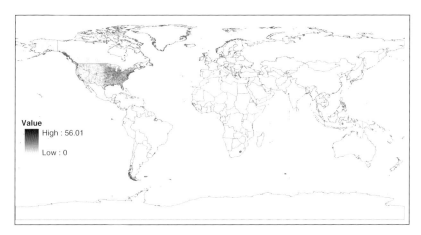

Figure 9.3 A first attempt at calculation of a globally distributed human ecological footprint for the United States using monetary balances of overall trade and domestic production (see text for details). The author's consumption represents slightly less than 1 (equivalent to 201/246ths) in 300 million of the pattern shown on this map. Similar calculations are possible for other nations using the methods discussed in the chapter. Scale: 1:270 000 000.

(e.g. GDP – exports). These percentages are then multiplied by the HII values for their respective nations (including zeros for nations that did not export any GDP to the United States in 2002) to yield the maps in Figure 9.3 (details in Figure 9.4).

At a personal scale, my consumption represents a fractional amount (approximately 0.00000000272) of the human influence shown in Figure 9.3. My influence is estimated using the same personal apportionment methods of the ecological footprint (in my case, my consumption is equivalent to 201/246ths or 81.7% of an average American) in the context of an overall population of 300 million people (the approximate US population in 2002.) If I moved to Bolivia, then not only would my ecological footprint decrease, but its spatial pattern would change dramatically.

Discussion

Scepticism is warranted as we consider the pile of assumptions, invented measures, and intricate calculations discussed above, to connect the ecological and human footprints. What is proposed is much more a research programme in human ecological footprint monitoring than an established method – further refinements and much testing are required before we can find firm footing. Many people may question the premise that lies at the heart of this chapter: that we can use the human footprint as a surrogate for

PERSONALIZED MEASURES OF CONSUMPTION AND DEVELOPMENT 203

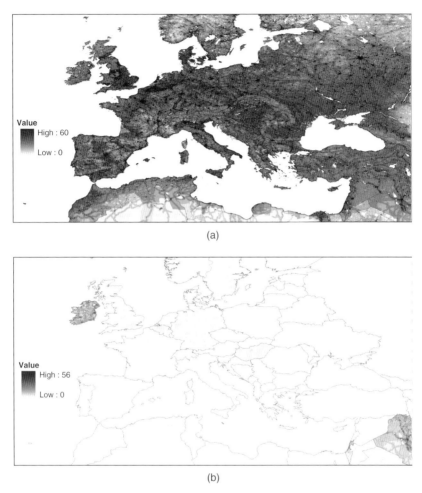

Figure 9.4 Detail maps comparing the Human Influence Index and the United States' human ecological footprint for Europe and North America. (a) The Human Influence Index in Europe, scale 1:48 000 000. (b) The United States' human ecological footprint in Europe, scale 1:48 000 000. (c) The Human Influence Index in portions of North America, scale 1:65 000 000. (d) The United States' human ecological footprint in portions of North America, scale 1:65 000 000. Center for International Earth Science Information Network (CIESIN).

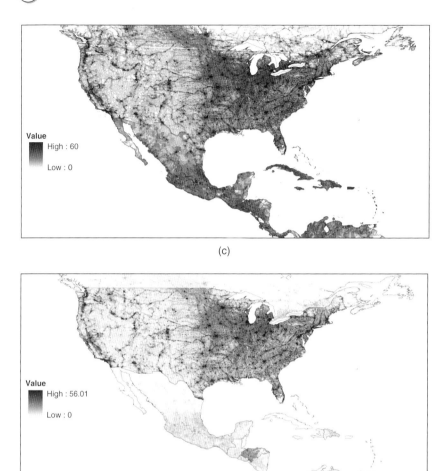

Figure 9.4 (*continued*)

production, which is equivalent in some sense to the ecological footprint as a surrogate for consumption. Surrogacy is an issue and equivalency is an issue.

Taking equivalency first, it is common in economics to see the balance between production and consumption. The equivalency between these two economic forces is the basis for accounting methods for national economies as well as households and companies. Households, companies, and nations can be out of balance; they can consume more than they produce, resulting in a deficit, or they can produce more than they consume, resulting in a surplus; these surpluses or deficits can be measured for individual commodities, groups of commodities, or for economies as wholes. In the globalized economy of our world, national deficits and surpluses are levelled through trade, which is why it is reasonable, and in fact necessary, that any personalized ecological human footprint take into account trade networks. Finally, it is worth noting that the idea of equivalence between consumption and production underlies not only economic relationships, but also the ecological footprint's notion of 'overshoot' and in general mass balance equations for natural resources.

Surrogacy is a more difficult issue for both the human footprint map and the ecological footprint calculation. The ecological footprint posits a measure of relative consumption, which as we have seen, is based on a complicated rendering of national economic accounts into inventories of global hectares, customized to the individual through comparison to a statistically average national citizen. Although the developers of the ecological footprint calculation have taken considerable effort to extend the kinds of consumption the ecological footprint covers, it is in no way complete (water consumption, e.g., is not considered).

Similarly the human footprint cannot claim to be a complete rendering of all the sources of production. What it can claim is to have some surrogacy for land, labour, and capital: land, to the extent that some nations are larger than others, and in the sense that human population and infrastructure tends to locate in the most productive parts of a country; labour, as measured by the population density; and capital, as expressed through the built infrastructure of a country – its transportation and power networks and its land use. One can argue that each of these layers could be improved, to be more indicative of economic production (agricultural or grazing productivity could be incorporated, the skills and ages of the population could be mapped, etc.). Nevertheless the human footprint seems, at least to me, to stand up as a crude spatial measure of economic production.

Finally, although both the human footprint map and the ecological footprint calculation rest on globally validated statistical sources (whether economic statistics or maps of population and infrastructure), these underlying data sources are also studded with gaps, interpolations, and inaccuracies, which potentially weaken the overall connection being attempted here. It would take another chapter to detail these, so I won't describe them here, but they are worth keeping in mind.

What could improve the personal ecological human footprint? A commodity-specific equivalence would help. Such equivalence would require that we relate the production of individual commodities to the specific population and infrastructure components of the human footprint required for its production. For example, to return to apples in Hungary, we could specify how many people are involved in apple production, how much agricultural land, and how much transportation infrastructure (e.g. roads and railways) and power is involved in the production and movement of apples within Hungary, including to border crossings for export. These specifications would require additional analysis of agricultural and industrial processes, employment, transportation, and power consumption. If we could equate the production of a given commodity directly to its population and infrastructure requirements, however, we might be able to make an estimate of the human footprint required while not delving as deeply into the details of the ecological footprint accounting. Such an approach would be even stronger if we could work directly from the national economic accounts that exist in the public domain and are routinely updated.

It would also be grand if we could identify the locations of production and consumption more exactly. The user of an ecological footprint calculator knows where he or she is (and that information could be obtained by postal code or some other non-exact, but more specific spatial measure). A New Yorker, for example, might have a different consumption pattern than a Californian. Such intra-national comparisons might be more meaningful for some users than international ones. Similarly, production is not evenly spread across a country's human footprint, as implied here; rather production tends to centre in particularly economically potent areas, which could be taken into account by combining sub-national datasets. Both of these kinds of improvements would take massive, additional data collection and synthesis to be operable, but are at least conceptually, if not practically, possible.

While this work remains to be completed, this analysis suggests how a personalized ecological human footprint might be constructed. Such a map is valuable not only for what it reveals about our individual accounts of production and consumption, but also how our individual consumption may impact biodiversity. The value of a map, in today's world of geographic overlays and computer synthesis, is finding new intersections: for example, intersecting the map of my consumption with a map of the habitat of tigers, the distribution of globally significant ecoregions, or the distribution of biodiversity hotspots. Such an overlap can help us monitor our choices against our priorities for biodiversity conservation, and perhaps, eventually lead to a different set of decisions that would allow human production, consumption, and biodiversity to coexist peacefully on Earth.

References

111th US Congress (2009) American Recovery and Reinvestment Act of 2009. Pub.L. 111–5. US Congress.

Aplet, G., Thomson, J., and Wilbert, M. (2000) Indicators of wildness: using attributes of the land to assess the context of wilderness. In: McCool, S.F., Cole, D.N., Borrie, W.T., O'Loughlin, J. (eds), *Wilderness Science in a Time of Change, Vol. 2: Wilderness within the Context of Larger Systems*. Conference proceedings, 23–27 May 1999, Missoula, MT. USDA, US Forest Service, Rocky Mountain Research Station, Ogden, Utah, Proceedings RMRS-P-15, pp. 89–98.

Best Foot Forward (2009) Ecological Footprint Calculator. Available at: http://www.ecological footprint.com/.
Bhagwat, S., Brown, N., Evans, T., Jennings, S., and Savill, P. (2001) Parks and factors in their success. *Science*, 293, 1045–1046.
Brooks, T.M., Mittermeier, R.A., da Fonseca, G.A.B, *et al.* (2006) Global biodiversity conservation priorities. *Science*, 313, 58–61.
Bruner, A.G., Gullison, R.E., Rice, R.E., and da Fonseca, G.A.B. (2001) Effectiveness of parks in protecting tropical biodiversity. *Science*, 291, 125–128.
Burgess, N.D., Balmford, A., Cordeiro, N.J., *et al.* (2007) Correlations among species distributions, human density and human infrastructure across the high biodiversity tropical mountains of Africa. *Biological Conservation*, 134, 164–177.
Carver, S., Evans, A., and Fritz, S. (2002) Wilderness attribute mapping in the United Kingdom. *International Journal of Wilderness*, 8, 24–29.
Center for International Earth Science Information Network, Columbia University, *et al.* (2000) Gridded Population of the World (GPW). Version 2. Center for International Earth Science Information Network (CIESIN).
Colchester, M. (1994) *Salvaging Nature: Indigenous Peoples, Protected Areas and Biodiversity Conservation*. Diane Publishing Company, Darby, PA.
Cronon, W. (1995) The trouble with wilderness; or, getting back to the wrong nature. In: *Uncommon Ground: Rethinking the Human Place in Nature*. W.W. Norton and Co., New York, pp. 69–90.
Dinerstein, E., Loucks, C., Wickramanayake, E., *et al.* (2007) The fate of wild tigers. *BioScience*, 57, 508–514.
Ecological Footprint (2009) Ecological Footprint Quiz. Available at: http://myfootprint.org/.
Elvidge, C., Baugh, K., Kihn, E.A., Kroehl, H.W., and Davis, E.R. (1997) Mapping city lights with nighttime lights data from the DMSP Operational Linescan System. *Photogrammetric Engineering and Remote Sensing*, 63, 772–734.
Ewing, B., Reed, A., Rizk, S.M., Galli, A., Wackernagel, M., and Kitzes, J. (2008) *Calculation Methodology for the National Footprint Accounts, 2008 Edition*. Global Footprint Network, Oakland, CA.
Gaston, K.J., Jackson, S.F., Cantú-Salazar, L., and Cruz-Piñón, G. (2008) The ecological performance of protected areas. *Annual Review of Ecology, Evolution and Systematics*, 39, 93–113.
Global Footprint Network (2009a) Footprint Calculator. Available at: http://www.footprint network.org/en/index.php/GFN/page/calculators/.
Global Footprint Network (2009b) *National Footprint Accounts, 2008 Edition – Data Year 2005 (Academic Edition)*. Global Footprint Network, Oakland CA. Available at: www.footprintnetwork.org.
Haberl, H., Fischer-Kowalski, M., *et al.* (2004) Progress towards sustainability? What the conceptual framework of material and energy flow accounting (MEFA) can offer. *Land Use Policy*, 21, 199–213.
Halpern, B.S., Walbridge, S., Selkoe, A.K., *et al.* (2008) A global map of human impact on marine ecosystems. *Science*, 319, 948–952.
Hannah, L., Lohse, D., Hutchinson, C., Carr, J.L., and Lankerani, A. (1994) A preliminary inventory of human disturbance of world ecosystems. *Ambio*, 23, 246–250.

Intergovernmental Panel on Climate Change (2007) *Climate Change 2007: Synthesis Report. Contribution of Working Groups I, II and III to the Fourth Assessment*. Report of the Intergovernmental Panel on Climate Change. IPCC, Geneva, Switzerland.

Jones, K.E., Patel, N.G., Levy, M.A., et al. (2008) Global trends in emerging infectious diseases. *Nature*, 451, 990–993.

Kitzes, J., Galli, A., Rizk, S.M., Reed A., and Wackernagel, M. (2008) *Guidebook to the National Footprint Accounts: 2008 Edition*. Global Footprint Network, Oakland, CA.

Laliberte, A.S. and Ripple, W.J. (2004) Range contractions of North American carnivores and ungulates. *BioScience*, 54, 123–128.

Leslie, R. and Malsen, M. (1995) *National Wilderness Inventory Handbook of Procedures, Content and Usage*. Australian Government Publishing Service, .Canberra.

Leu, M., Hanser, S., and Knick, S.T. (2008) The human footprint in the west: A large-scale analysis of anthropogenic impacts. *Ecological Applications*, 18, 1119–1139.

Levin, N., Lahav, H., Ramon, U., et al. (2007) Landscape continuity analysis: a new approach to conservation planning in Israel. *Landscape and Urban Planning*, 79, 53–64.

Lockwood, J.L. and McKinney, M.L. (eds) (2001) *Biotic Homogenization*. Springer, New York.

Loh, J. and Wackernagel, M. (2004) *Living Planet Report 2004*. World-Wide Fund for Nature International, Gland, Switzerland, Global Footprint Network, Oakland, CA; UNEP World Conservation Monitoring Centre, Cambridge, UK.

Loveland, T., Reed, B., Brown, J.F., et al. (2000) Development of a global land cover characteristics database and IGBP DISCover from 1-km AVHRR data. *International Journal of Remote Sensing*, 21, 1303–1330.

McCloskey, M. and Spalding, H. (1989) A reconnaissance-level inventory of the amount of wilderness remaining in the world. *Ambio*, 8, 221–227.

Millennium Ecosystem Assessment (2005) *Ecosystems and Human Well-being: Biodiversity Synthesis*. World Resources Institute, Washington, DC.

Mittermeier, R.A., Mittermeier, C.G., Brooks, T.M., et al. (2003) Wilderness and biodiversity conservation. *Proceedings of the National Academy of Sciences of the USA*, 100, 10309–10313.

Monfreda, C., Wackernagel, M., and Deumling, E. (2004) Establishing national natural capital accounts based on detailed ecological footprint and biological capacity accounts. *Land Use Policy*, 21, 231–246.

National Imagery and Mapping Agency (1997) *Vector Map Level 0 (VMAP0)*, ed. 003. NIMA.

Nellemann, C., Kullerud, L., Vistnes, I., et al. (2001) *GLOBIO: Global Methodology for Mapping Human Impacts on the Biosphere*. UNEP/DEWA/TR.01-3. United Nations Environment Programme, 47 pp.

Nelson, M.P. and Callicott, J.B. (eds) (2008) *The Wilderness Debate Rages on: Continuing the Great New Wilderness Debate*. University of Georgia Press, Athens, GA.

Redford, K.H., Levy, M.A., Sanderson, E.W., and de Sherbinin, A. (2008) What is the role of conservation organizations in poverty alleviation in the world's wild places? *Oryx*, 42, 516–528.

Sanderson, E.W., Jaiteh, M., Levy, M.A., Redford, K.H., Wannebo, A.V., and Woolmer, G. (2002) The human footprint and the last of the wild. *BioScience*, 52, 891–904.

Sanderson, E.W., Robles Gil, P., Mittermeier, C.G., Martin, V.G., and Kormos, C.F. (eds) (2006) *The Human Footprint: Challenges for Wilderness and Biodiversity*. CEMEX, WILD Foundation, Agrupacion Sierra Madre, Wildlife Conservation Society.

Trombulak, S.C., Anderson, M.G., Baldwin, R.F., *et al.* (2008) *2040 Future Human Footprint – Current Trends Scenario. The Northern Appalachian/Acadian Ecoregion: Priority Locations for Conservation Action*. Two Countries, One Forest Special Report No. 1.

UN Environment Programme, Nelleman, C., Kullerud, L., *et al.* (2001) *GLOBIO: Global Methodology for Mapping Human Impacts on the Biosphere*. UNEP Division of Early Warning and Assessment, Nairobi.

United Nations (2009) *World Population Prospects: The 2008 Revision*. Population Division of the Department of Economic and Social Affairs of the United Nations Secretariat.

UN Statistics Division (2009) National Accounts Main Aggregates Database. Available at: http://unstats.un.org/unsd/snaama/Introduction.asp.

US Census Bureau (2009) USA Trade Online. Economics and Statistics Administration, US Department of Commerce.

Venetoulis, J. and Talberth, J. (2006) *Refining the Ecological Footprint*. Redefining Progress, Oakland, CA.

Wackernagel, M. (1994) Ecological footprint and appropriated carrying capacity: a tool for planning toward sustainability. PhD thesis, School of Community and Regional Planning. The University of British Columbia, Vancouver, BC.

Wackernagel, M. and Rees, W. (1996) *Our Ecological Footprint: Reducing Human Impact on the Earth*. New Society Publishers, Gabriola Island, BC, Canada.

Wackernagel, M., Onisto, L., Bello, P., *et al.* (1999) Natural capital accounting with the Ecological Footprint concept. *Ecological Economics*, 29, 375–390.

Wackernagel, M., Monfreda, C., Moran, D., *et al.* (2006) *National Footprint and Biocapacity Accounts 2005: The underlying calculation method*. Global Footprint Network, Oakland, CA.

Wilson, E.O. (2003) *The Future of Life*. Abacus, New York.

Woolmer, G., Trombulak, S.C., Ray, J.C., *et al.* (2008) Rescaling the human footprint: a tool for conservation planning at an ecoregional scale. *Landscape and Urban Planning*, 87, 42–53.

Part III
The Next Generation of Biodiversity Indicators

10

Indicator Bats Program: A System for the Global Acoustic Monitoring of Bats

Kate E. Jones[1,2], Jon A. Russ[2], Andriy-Taras Bashta[3,4], Zoltán Bilhari[5], Colin Catto[2], István Csősz[6], Alexander Gorbachev[7,8], Péter Győrfi[5], Alice Hughes[9], Igor Ivashkiv[3,4], Natalia Koryagina[7], Anikó Kurali[5], Steve Langton[2], Alanna Collen[1], Georgiana Margiean[6], Ivan Pandourski[10], Stuart Parsons[11], Igor Prokofev[7,8], Abigel Szodoray-Paradi[6], Farkas Szodoray-Paradi[6], Elena Tilova[12], Charlotte L. Walters[1,2], Aidan Weatherill[1] and Oleg Zavarzin[7]

[1]Institute of Zoology, Zoological Society of London, London, UK
[2]Bat Conservation Trust, London, UK
[3]Animals Research and Protection Association "Fauna", Lviv, Ukraine
[4]Institute of Ecology of the Carpathians, National Academy of Sciences of Ukraine, Lviv, Ukraine
[5]Nature Foundation, Nyíregyháza, Hungary
[6]Romanian Bat Protection Association, Satu Mare, Romania
[7]PERESVET, Bryansk, Russia
[8]University of Bryansk, Bryansk, Russia
[9]School of Biological Sciences, University of Bristol, Bristol, UK
[10]Institute of Zoology, Bulgaria Academy of Sciences, Sofia, Bulgaria
[11]School of Biological Sciences, University of Auckland, Auckland, New Zealand
[12]Green Balkans – Stara Zagora, Stara Zagora, Bulgaria

Biodiversity Monitoring and Conservation: Bridging the Gap between Global Commitment and Local Action, First Edition. Edited by Ben Collen, Nathalie Pettorelli, Jonathan E.M. Baillie and Sarah M. Durant.
© 2013 John Wiley & Sons, Ltd. Published 2013 by John Wiley & Sons, Ltd.

Introduction

Technological development, globalization, and increases in human population and consumption have resulted in an unprecedented and precipitous loss of biodiversity and ecosystem services over the last 500 years (Mace *et al.*, 2005). Concern about the global decline in biodiversity, the degradation of ecosystem services, and the combined impact on people gave rise in 1992 to the Convention on Biological Diversity (CBD) and a target 'to achieve by 2010 a significant reduction of the current rate of biodiversity loss at the global, regional and national levels as a contribution to poverty alleviation and to the benefit of all life on Earth'. The global community has failed to meet the 2010 target (Butchart *et al.*, 2010), unsurprising as the dominant direct and indirect drivers of biodiversity loss are increasing or are at best stable (Mace *et al.*, 2005; Mace and Baillie, 2007; Collen *et al.*, 2009; Butchart *et al.*, 2010). Part of this failure is due to lack of knowledge about the effects of global change on ecosystems and the services they provide and how to mitigate this. There are few measurable indicators of global change at global, regional, national, and local levels (the Living Planet Index being one of the notable exceptions; Loh *et al.*, 2008; see also Chapter 4). Looking forward from the 2010 target, it is now essential for both the scientific and development communities to develop joint targets that are specific, time-limited, and achievable and that can be quantifiably measured (Sachs *et al.*, 2009). Currently, quantifiable biodiversity monitoring data result from programmes that vary in geographical coverage, methodology, length, and species coverage (Collen *et al.*, 2009; Butchart *et al.*, 2010), and previous efforts that involve standardized approaches operating at large scales have been focused on monitoring populations of a few taxa (e.g., birds and butterflies), due in part to the availability of data and lack of suitable monitoring tools. This is unfortunate as national biodiversity monitoring is a fundamental obligation of all nations under the Convention on Biological Diversity (2003).

Monitoring ultrasonic biodiversity

The use of sound is beginning to be recognized as a potentially useful tool in biodiversity surveillance and monitoring programmes. For example, the North American Amphibian Monitoring Program uses volunteer networks to collect frog and toad abundances through 'calling surveys' where volunteers identify species by their unique breeding vocalizations along transects (Weir *et al.*, 2009). Conservation International's Tropical Ecology Assessment and Monitoring Network has adopted audio monitoring as part of its bird monitoring protocol due to this sampling method's increased efficiency compared to traditional field observational approaches and because all vocalizations can potentially be identified, even without experienced field observers (Lacher, 2008). Recent studies have demonstrated that acoustic recordings of birds can perform as well as or even better in sampling communities than using field observers (Celis-Murillo *et al.*, 2009). Cornell University's Ornithology Laboratory has been the first to apply these acoustic monitoring approaches to a broad range of species, from night migrant birds, to endangered songbirds, to the infrasonic calls of elephants and ultrasonic sounds made by bowhead whales (www.birds.cornell.edu). With the exception of this example from bowhead whales, the widespread use of ultrasonic sound as a monitoring tool remains largely under-explored.

Many species produce ultrasound including many bats and cetaceans, some rodents, insectivores, birds, and insects (Sales and Pye, 1974; Thomas and Jalili, 2004; Thomas *et al.*, 2004b). However, the complexity, sophistication, and diversity of designs of bat ultrasonic calls are unparalleled (Maltby *et al.*, 2009). Bats tend to use ultrasonic frequencies (>20 kHz) for their calls, although some species do use calls that are audible to humans, for example, *Euderma maculatum*, which calls with most energy at 9 kHz (Fullard and Dawson, 1997). Bats use their calls and the resulting echoes to detect, localize and classify objects (echolocation or biosonar) as well as to communicate (Fenton, 1985; Jones and Siemers, 2011). The potential for using the information that bats 'leak' about themselves whilst foraging and orientating to monitor population densities and population trends across space and time has been recognized for some time (Walsh *et al.*, 2001, 2004; Battersby, 2010). However, a number of valid concerns have been raised over this approach (e.g. Hayes *et al.*, 2009) as bat calls can vary intraspecifically as well as interspecifically, which makes species identification challenging (Neuweiler, 1989). For example, there is some evidence that call characteristics vary with sex, age, and body size (Jones *et al.*, 1992), geographical location (Russo *et al.*, 2007), presence of conspecifics (Obrist, 1995), habitat (Neuweiler, 1989; Barclay *et al.*, 1999), and with the type of foraging activity (Parsons *et al.*, 1997).

Calls can also vary in their detectability. For example, some species produce high-amplitude, easily detectable calls that can be heard up to 100 m away (e.g. the noctule, *Nyctalus noctula*) while others have low-amplitude calls that can only be detected within a few metres (e.g. the brown long-eared bat *Plecotus auritus*), and some do not echolocate at all (e.g. most Old World fruit bats, family Pteropodidae) (Jones and Teeling, 2006; Maltby *et al.*, 2009). The appearance of the structure of bat calls can also vary depending on the type of equipment used to record them (see later), and the relationship between the number of calls heard and bat population abundance is not well understood (Hayes *et al.*, 2009). Despite these considerations, species identification of bats from echolocation calls has been used extensively in the studies of habitat use, reviewed in, for example, Vaughan *et al.* (1997a), Russo and Jones (2003), and Wickramasinghe *et al.* (2003). We show here that when used appropriately, surveillance of and monitoring changes in bat ultrasonic biodiversity can provide a fast, efficient method to generate a global bat biodiversity indicator.

Bats are an important component of mammalian biodiversity (one-fifth of all mammal species) (Simmons, 2005) and fill such a wide array of ecological niches that they may offer an important multisensory bioindicator role in assessing ecosystem health (Jones *et al.*, 2009). Bats are sensitive to two key environmental stressors – habitat conversion and climate change. For example, they are sensitive to water quality (Vaughan *et al.*, 1996), noise and light pollution (Stone *et al.*, 2009), agricultural intensification (Wickramasinghe *et al.*, 2003), urbanization (Loeb *et al.*, 2009), and deforestation and fragmentation (Fenton *et al.*, 1992). Bat hibernation cycles and migration routes are dependent on environmental temperatures and may provide important climate change indicators (Newson *et al.*, 2009), and migration routes in particular are also sensitive to renewable energy technologies (wind turbines) (Baerwald *et al.*, 2008).

Interest in bats as indicators is also focused on their role in provision of ecosystem services, that is, those services from ecosystems that are of benefit to humans (Sukhdev, 2008). Bats have been shown to play a role in pollination and seed dispersal of a number of ecologically and economically important plants (Fujita and Tuttle, 1991; Hodgkison *et al.*, 2003) and in regulating pests of economically important crops (Federico *et al.*, 2008). Bats are reservoirs of a wide range of infectious diseases (Dobson, 2005) whose epidemiology may reflect environmental stress and therefore may provide another indicator of impact on ecosystems (Keesing *et al.*, 2010). Indeed, the importance of bats as biodiversity indicators has been recognized nationally by the UK government, establishing them along with birds and butterflies as UK2010 Headline Indicators (www.bipnational.net), using data from the National Bat Monitoring Program (Walsh *et al.*, 2001). Work continues to develop bats as an indicator on a wider pan-European level through EUROBATS (www.eurobats.org). There is also a need to monitor population trends of bats for their own sake because many populations face numerous environmental threats related to climate change, habitat loss, fragmentation, hunting, and emerging diseases (Schipper *et al.*, 2008). A recent worrying illustration of the effect of emerging diseases is provided by the catastrophic declines in little brown bats, *Myotis lucifugus*, in North America from white-nose syndrome (Frick *et al.*, 2010).

Developing an ultrasonic indicator

To be able to establish bat ultrasonic biodiversity trends as a reliable indicator, it is important to standardize monitoring protocols, data management, and analyses. We discuss the main issues to be considered in developing a bat ultrasonic indicator below.

Monitoring methodology

It is critical to develop statistically rigorous methodologies for monitoring. Most existing bat monitoring programmes undertake visual counts of individuals within or emerging from a roost (generally summer or winter hibernation sites) (reviewed in Battersby, 2010, for European studies). Such counts may be prone to bias as occupancy of bat roosts may be flexible and a decline of individuals at a roost may indicate a genuine local population change, or reflect that individuals from the colony have switched to using another site. Some species in the temperate regions switch roosts regularly during the summer, for example, *Pipistrellus pipistrellus* (Feyerabend and Simon, 2000); *Barbastella barbastellus* (Russo *et al.*, 2005); and *Plecotus auritus* (Entwistle *et al.*, 1997). For species that are known to switch roosts frequently, bat abundance activity data gathered along random habitat transects is predicted to be more statistically robust (e.g. Bat Conservation Trust, 2008). Individuals are typically counted at fixed points along transects or continuously recorded using direct observation, capture (e.g. using mist nets or harp traps), or ultrasonic techniques (Hayes *et al.*, 2009; Battersby 2010). Knowing the variability of the data along these transects allows statistical techniques such as power analyses to determine how many, how often, and for how long transects should be monitored to detect statistically robust population trends (see Box 10.1 for an example) (Gibbs *et al.*, 1998; Sims *et al.*, 2006; Meyer *et al.*, 2010) (refer to chapter 15). Statistical power is a measure of the confidence with which a statistical test can detect a particular effect when such an effect does indeed exist. It is influenced by many factors, including the magnitude of population change over time, between-year population variation, the number of years of data, frequency of surveillance, the number of sites surveyed, proportion of samples in which the species is present, and sampling error (Battersby, 2010; Meyer *et al.*, 2010).

Box 10.1 **Power analyses of monitoring data**

Power analyses can be used to estimate how much data (given the variability in the data) are needed to generate statistically robust estimates of population trends over time. There are two basic approaches to power analyses: either the variance of the trend can be calculated based on standard statistical theory, or data are simulated with estimates of the variance present in these data. The simulation approach is often more powerful as it does not assume data normality nor a simple linear trend in the data. One of the challenges facing this type of analysis is the uncertainty in the variance estimators (Table 10.1), which can lead to misleading power estimates (Sims *et al.*, 2006). Table 10.1 uses data from routes monitored in July and August between 2005 and 2008 to estimate the sources of variation for three species of bats in the Bats and Roadside Mammals Programme (now the iBats UK project). Although the mean abundance per event for both pipistrelle species differs (common pipistrelle, *Pipistrellus pipistrellus*, = 44.3 calls per event, and soprano pipistrelle,

P. pygmaeus, = 10.9 calls per event), the key sources of variation are evenly split between the differences in abundance between different routes and differences between the replicates done of each route. The low value for the differences between routes within years for all species indicates that replicate events within the same year are only marginally more similar than events of routes in different years. The less commonly encountered noctule (*Nyctalus noctula*) (1.8 calls per event) has proportionally more random variation between replicates.

Table 10.1 **Estimates (standard error, SE) of the variance components (scaling them using normal scores) of total counts per monitoring event carried out in July and August in the UK for three bat species for (i) routes (differences between different routes); (ii) routes within years (differences between different routes within each year); and (iii) replicates (differences between replicates of the same route)**

Source of variation	*P. pipistrellus*	*P. pygmaeus*	*Nyctalus noctula*
Routes	0.488 (0.19)	0.358 (0.15)	0.216 (0.13)
Routes within years	0.068 (0.14)	0.091 (0.13)	0.007 (0.13)
Replicates	0.446 (0.14)	0.385 (0.11)	0.468 (0.14)

Using simulations based on these estimates of variance it is possible to calculate the numbers of years taken to achieve 80% power with different numbers of routes and replicates per year (Figure 10.1). Figure 10.1a-c shows the number of years to achieve 80% power with between 20 and 100 routes, each monitored twice a year for all three species for Amber (25% decline over 25 years) and Red (50% decline over 25 years) Alerts. Amber and Red Alerts are a common approach originally established for birds (Gregory *et al.*, 2002). For the common pipistrelle, a Red Alert will be statistically significant after only around 7 years, even with just 20 routes replicated twice each year. Amber Alerts take longer, but can still be detected in around 10 years with 50 routes repeated twice each year. For the less abundant species, Amber Alerts are very challenging, requiring around 20 years of data even with 100 routes repeated twice. Red Alerts can, however, be detected with a reasonable number of years and routes. Figure 10.1d,e shows the impact of increasing the number of replicates of each route per year. Figures are based on 20 routes for the common pipistrelle and 100 routes for the other two species. It is apparent that moving from one to two replicates generally yields a good reduction in the numbers of years, thereafter the gain is more limited. These figures also suggest that a similar number of years are required if a number of routes are monitored twice compared to twice that number monitored once (e.g. Amber Alert for the soprano pipistrelle is 23.4 years with 50 routes twice per year, or 25.1 years with 100 once a year).

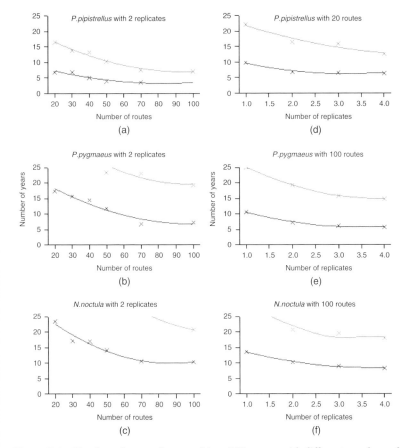

Figure 10.1 Number of years taken to achieve 80% power with different numbers of routes and replicates for three different species for Amber (upper line) and Red (lower line) Alerts.

Sustainable monitoring

To obtain statistically robust trends, it is often necessary to collect data for decades rather than years (Bat Conservation Trust, 2008; Meyer *et al.*, 2010). This means that monitoring programmes must be sustainable and have long-term funding (Greenwood, 2007). Many organizations have adopted a citizen science approach to monitoring, where they use unpaid non-professional volunteers to collect monitoring data (e.g. the British Trust for Ornithology (BTO), the North American Breeding Bird Survey, the Bat Conservation Trust). This means that monitoring data can potentially be collected at a large scale for many years and cost-effectively. However, citizen science brings its own challenges (Greenwood, 2007). For example, there has to be a culture of volunteering in the monitoring region (more difficult in developing countries), and methodologies have to be appropriate to the available skills base and communicated effectively to achieve consistency among surveyors. Volunteers may also require coordination, training, and feedback on their activities, and data need to be managed efficiently (Battersby and Greenwood, 2004; Battersby, 2005). However, this type of monitoring programme is gaining rapidly in popularity and its value is being seen globally (Lowman *et al.*, 2009; Devictor *et al.*, 2010).

Ultrasonic detection

Currently, ultrasound is detected by several different methods, all of which have strengths and weaknesses (see Box 10.2; also reviewed in Parsons and Szewczak, 2009). Some methods that convert the ultrasound into audible sound do so by reducing the information content of the call (i.e. heterodyne and frequency division detection). These methods have the advantage that detectors are inexpensive and the resulting files are smaller and easier to store. The disadvantages are that species identification is harder as information is lost. In other detection methods, such as the more expensive time-expansion and direct recording ultrasonic detectors, the ultrasound is recorded intact. Within time-expansion detectors, a number of settings can be used. For example, snapshots of ultrasound can be recorded for different lengths of time (e.g. 1.28 s, 320 ms, and 40 ms) and then played back at either $\times 10$ or $\times 32$ slower than the original, and the detectors can be set to trigger on hearing ultrasound (at a specified amplitude threshold) or can be set to cycle continuously. Selection of ultrasonic equipment should broadly depend on the importance of species identification versus file size and cost although even with the best recording equipment there is no guarantee of capturing the true signal because of factors like signal attenuation (Pye, 1993).

Box 10.2 **Comparisons of ultrasonic detection methods**

Heterodyne detectors mix ultrasound with a signal produced by the detector and produce the audible difference between the two sounds. This method does not record information about the call such as the frequency structure, duration, or absolute frequency (Parsons *et al.*, 2000; Parsons and Szewczak, 2009). This means that species identification can be difficult as it relies on an observer to recognize rhythm, repetition rate, pitch, and tonal differences. For example, the closer the detector is tuned to the peak frequency of the ultrasonic call, the deeper the tonal quality of the call. Also as the detector can be tuned to only a narrow frequency range at once, bats calling outside of the tuned frequency range can be missed (Parsons and Szewczak, 2009).

Frequency division detectors convert ultrasound to audible sound by dividing the frequency of the incoming signal by a predetermined factor (most often 10), so reducing the frequency into audible ranges (Andersen and Miller, 1977). This method does not record information about the amplitude-time structure of the call, and much of the frequency-time structure is lost (Parsons *et al.*, 2000). As only the loudest part of the call is recorded, any harmonics are ignored. Again species identification can be difficult because many important parameters are lost, but this method has the advantage that the detector does not need to be tuned to a particular frequency and so picks up a wide range of frequencies (broadband).

Time expansion detectors convert ultrasound to audible sound by slowing down the ultrasound and playing it back through the detector (Parsons *et al.*, 2000). Frequency is inversely related to time, so slowing down a sound will reduce the frequency linearly. No information is lost from the ultrasound so all necessary parameters can be collected from the sound (Parsons *et al.*, 2000) and these can be used to identify species more accurately. Although these detectors are broadband, so do not miss calls, they cannot record ultrasound continuously as they cannot record ultrasound whilst playing back a recording. For example, if the time expansion factor is set to 10 and the recording time is set to 1.28 s, then the detector will record for 1.28s, and playback for 12.8 s, meaning there is a 12.8 s gap between each recording. The gap between recordings can of course be reduced by decreasing the time the detector records sound for (e.g. 320 ms or 40 ms are common settings). Long recording sequences may mean that some bats may be missed while the detector is not recording. On the other hand, short recording sequences may only record partial call sequences making species identification from the recordings more difficult. **Direct sampling** of ultrasound is possible by attaching an ultrasonic microphone to a laptop with a high-speed sampling card, and a few high-specification detectors are specifically designed to do this.

A lot of debate surrounds the use of frequency division detectors in scientific research (Ahlén and Baagøe, 1999, Fenton, 2000, Corben and Fellers, 2001). Time expansion detectors have been shown to be more sensitive to low-amplitude ultrasound and therefore record more calls than frequency division detectors (Ahlén and Baagøe 1999, Fenton, 2000). Species identification was also shown to be easier with time expansion detectors because all the sound parameters are collected (Ahlén and Baagøe, 1999). However, frequency division detectors create much smaller files and therefore require much less data storage space and have faster processing times (Corben and Fellers, 2001).

Acoustic recording

If species identification is not done manually in the field, then the converted audible sound must be digitally recorded in an uncompressed format (such as *.wav format) that can be transferred to a computer for later acoustic analysis. Useful devices include some MP3 players, high-end mini-disc recorders, hard disk recorders (including laptops and other notebook computers), iPods, and some mobile phones. Price, ease of use, and reliability are all issues to consider before deciding on a particular model/make. Of great importance is the recording level (gain), which determines the sensitivity of the device to sound. For example, too high a recording level can result in distorted or 'clipped' sound, whereas too low a recording level may result in a high level of background noise. Some recording devices are equipped with an optional automatic gain control, which adjusts the gain to an appropriate level for the recording signal. Although non-digital sound recorders can be used, it takes extra time to convert the sound into a digital format for sonogram analysis.

Geo-referencing data

Although geo-referencing records of bats encountered is not essential for long-term monitoring, spatially referenced data enable research into habitat selection of bats, as records can be plotted onto Geographical Information Systems (GIS) after the survey. A Global Positioning System (GPS) device links the sounds recorded to their position.

GPS devices can be stand-alone units or are more commonly incorporated into other communication devices such as personal digital assistants (PDAs), mobile phones, or computers.

Analysis and species identification

Species identification from heterodyne detectors is often carried out in the field with no post-processing (Limpens, 2004). Matching these calls to a particular species often requires considerable training in recognition of pitch, repetition rate, and rhythm of the call; hence identification of heterodyne calls can be rather subjective. However, heterodyne detectors have been successfully deployed in a number of programmes where the species targeted were easy to identify with well-documented echolocation calls (e.g., Walsh *et al.*, 2001). Calls recorded using frequency division detectors also lose a lot of information and so identification of these calls can also be problematic. However, in contrast to heterodyne, calls can be identified with post-processing, allowing identifications to be verified and making the process less subjective. Frequency division detector monitoring may be useful in well-described acoustic bat communities where identification is relatively simple. Full-spectrum recordings of calls (such as those recorded with real-time or time-expansion detectors) preserve the original characteristics, and a number of call parameters are typically measured with post-processing, such as frequency of the call containing maximum energy, minimum and maximum call frequency, and call duration.

Whether the parameters have been extracted from frequency division or full-spectrum calls they can then be compared to existing call libraries, for example, frequency division calls within the 'Bat Call Library' hosted at the University of New Mexico, USA (www.msb.unm.edu/mammals/batcall/html/calllibrary.html) and the South-eastern Australian bat call library (http://batcall.csu.edu.au/batcall1.html). However, these libraries currently tend to be only regional (Waters and Gannon, 2004) and often do not cover the full plasticity present in the structure of calls of some species (Parsons and Szewczak, 2009). There has been some discussion of the need for more globally comprehensive call libraries (Korine and Kalko, 2001; Waters and Gannon, 2004) retaining any regional characteristics and work is progressing on this (Collen, 2012). Although qualitative assessment of calls to identify species from reference library calls has been used for many years, more quantitative approaches are now available and are growing in sophistication and efficacy. For example, these approaches include multivariate techniques such as discriminant function analysis and decision trees (Vaughan *et al.*, 1997b; Russo and Jones, 2003), as well as machine learning algorithms such as artificial neural networks (Parsons and Jones, 2000; Skowronski and Harris, 2006; Parsons and Szewczak, 2009; Redgwell *et al.*, 2009).

Data storage and management

Data storage requirements from an acoustic monitoring programme are large owing to the size of the acoustic files. It is therefore important to allow for adequate storage of the data recorded in an accessible format. Additionally, for a citizen science programme, it is important that data submission by volunteers is simple and feedback from their efforts is provided to them as soon as possible (Battersby, 2005; Devictor *et al.*, 2010). Online data portals can provide an integrated solution to this problem. The web interface allows volunteers to enter and download data, and the data are stored within an established data model that is accessible to all programme participants. Programme managers may also manage the programme and the volunteers remotely via the website (Battersby, 2010). Such web tools can be expensive to develop, requiring input from professional computer programmers, but are extremely useful and are gaining in popularity in biodiversity conservation and other citizen science programmes (Devictor *et al.*, 2010). The BTO and Royal Society for the Protection of Birds (RSPB)'s BirdTrack (www.bto.org/birdtrack) and the eBird program from Cornell Lab of Ornithology (ebird.org/content/ebird) are recent examples.

Comparisons with existing indicators

There are currently no regional, continental, or global indicators of bat biodiversity (Battersby, 2010; Meyer *et al.*, 2010). However, at national scales there are a large number of monitoring programmes whose data are beginning to be developed into indicators, mostly based in northern temperate regions. The longest running monitoring programmes are those in Western and Eastern Europe, which count bat abundances at roost sites using volunteers, usually on an annual cycle (Haysom, 2008; Battersby, 2010). This is despite the fact that most bat biodiversity is situated in the tropics (Schipper *et al.*, 2008). Some European programmes started in the 1940s but most date back to 1995–2000 (Haysom, 2008). These data have often been analysed to produce population abundance indicator trends (Walsh *et al.*, 2001; Bat Conservation Trust 2008; Meschede and Rudolph, 2010) using statistical techniques such as Generalized Linear Modelling (GLMs) and or Generalized Additive Modelling (GAMs) following standardized approaches in other taxa, especially birds (Fewster *et al.*, 2000; Gregory *et al.*, 2005).

Other programmes monitor bat occurrence or abundance using randomized transects with bat detectors across different habitats (e.g., Ahlén and Baagøe, 1999). The longest running multi-species programme incorporating detector surveys is the Bat Conservation Trust's (BCT's) National Bat Monitoring Programme (NBMP), which has used volunteers since 1997 to collect records of bats both visually and ultrasonically (using heterodyne detectors) to detect trends in populations of five species – *Pipistrellus pipistrellus, Pipistrellus pygmaeus, Nyctalus noctula, Eptesicus serotinus*, and *Myotis daubentonii* (Walsh *et al.*, 2001; Bat Conservation Trust, 2008). The BCT has developed a fully integrated programme comprising surveys tailored for 11 out of the 18 resident UK species. These data are used to produce UK population trends and a national indicator of the status of widespread species, which has now been adopted by the UK government. These data also contribute to the UK biodiversity strategy, the National Biodiversity Network (NBN), and a global biodiversity assessment programme, the Living Planet Index (Loh *et al.*, 2008), which monitors how effective efforts have been to halt biodiversity loss for the Convention on Biological Diversity (2003).

The NBMP relies on a large trained volunteer network to carry out its monitoring (1018 volunteers in the latest report) drawing on a long tradition of enthusiastic amateur naturalists in the UK (Bat Conservation Trust, 2008). However, most countries do not have such a volunteering culture and during development of new national and regional monitoring projects for Ireland, the BCT developed an approach that reduced the number of volunteers required to monitor bats. By recording the ultrasound from bats from time-expansion detectors attached to cars, it is possible for a small number of people to monitor large areas. This approach was piloted in 2003 (Catto *et al.*, 2003) and has been implemented in Ireland since 2004, administered by Bat Conservation Ireland (Roche *et al.*, 2011). The programme involves volunteers driving set transects at a speed of 24 km/hr for 58 miles (93 km) after sunset, recording for 1 mile (1.6 km) stretches every 3 miles (4.8 km), with 20 transects in total (Roche *et al.*, 2011). The Irish government has adopted this method as its official monitoring technique for three species of bats (*Pipistrellus pipistrellus, Pipistrellus pygmaeus*, and *Nyctalus leisleri*), and since 2006 the programme has included Northern Ireland with funding from the Northern Ireland Environment Agency. The programme has recently started to produce its first population abundance trends for these three species although it is too early to produce reliable estimates of long-term trends (Roche *et al.*, 2011).

A similar project based on this approach was started in France in 2006, which included both driven and walked transects as part of the Vigie Nature biodiversity programme (http://www2.mnhn.fr/vigie-nature/spip.php?rubrique6). Fixed 30-km transects were driven at a speed of 25 km/h with recording occurring for 2 km every 3 km (with 10 2-km transects being monitored in total). With the increased availability of low-cost Global Positioning System equipment, BCT modified the Irish protocol to record the transect's acoustic and GPS data simultaneously, recording 40 km in one transect at a driven speed of 25 km/h. This new protocol was implemented in the UK in 2005 (as the Bats and Roadside Mammals Programme, a joint initiative with Mammals Trust UK). The project aimed to supplement the NBMP, providing data from regions with few volunteer surveyors in order to increase overall programme coverage (Russ *et al.*, 2008). These data are now starting to produce trend lines, which can be compared with those collected with BCT's other methods (Pagan, 2010).

Indicator Bats Program

Following on from the success of the car survey monitoring programmes in Ireland, France, and the UK, we recognized the potential to use these new protocols to develop similar monitoring programmes in other countries. The Indicator Bats Program (iBats) was established in 2006 as a partnership project between the Zoological Society of London (ZSL) and BCT with the aim of establishing regional bat monitoring programmes that together could internationally monitor the impact of global change on bat populations. With funding from the Darwin Initiative (UK's Department of Environment, Food and Rural Affairs), the project started in Transylvania (Romania) in May 2006 in partnership with volunteers from the Romanian Bat Protection Association (RBPA). This project was established in Bulgaria in 2007 (in partnership with Green Balkans and Bulgarian Academy of Sciences) and also in eastern Hungary and Croatia. In 2009 the programme expanded in Hungary (with the NGO Nature Foundation), Ukraine (in collaboration with Animals Research and Protection Association and the National Academy of Sciences of Ukraine), and to western Russia (with Grassroots Alliance PERESVET). The region of focus is one of the highest areas of bat species richness on the continent of Europe (Figure 10.2) and includes a number of different ecoregions (Olson *et al.*, 2001), such as Pannonian mixed forests (the dominant habitat in Hungary), the Carpathian and Balkan mixed forests of Romania and Bulgaria, and the central European mixed forests and east European forest steppe of Ukraine and Russia (Figure 10.3). The iBats Program also incorporated the UK's Bats and Roadside Mammals Programme in 2008, and we are currently piloting projects in New York, Mongolia, Mexico, Madagascar, Japan, New Zealand, Thailand, and Zambia (see www.ibats.org.uk). In this chapter, we focus on the results from the program in Eastern Europe.

Monitoring methodology

Monitoring was carried out using 40-km road transect routes driven at 25 km/h and starting 30–45 minutes after sunset in the period of peak seasonal activity for most northern European species (July to August) when young of the year are also foraging (Russ *et al.*, 2003, 2005). Bats emerge at different times in relation to sunset (Jones and Rydell, 1994). However, data from the UK Bats and Roadside Mammals Survey demonstrated that by starting surveys 30–45 minutes after sunset the likelihood of encountering detectable bat species is maximized (Russ *et al.*, 2005) and we assumed this to be also appropriate for Eastern Europe. Encounter rates differ depending on speed of travel along a transect (Roche *et al.*, 2007). A speed was chosen to maximize encounter rates but to minimize the chances of a bat being encountered more than once on the transect (Russ *et al.*, 2003); for example, the average flight speed for

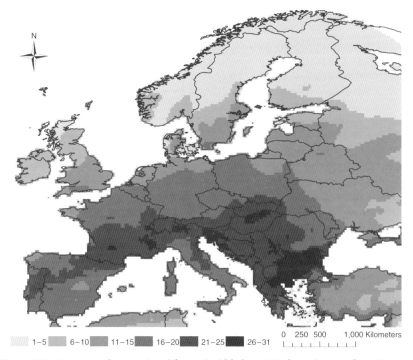

Figure 10.2 **European bat species richness (gridded at 0.25 degree). Data from Grenyer *et al.*, 2006.**

N. leisleri is 21 km/h (min. 14 km/h, max. 34 km/h; Shiel *et al.*, 1999), and that for *P. pipistrellus* is 16 km/h (min. 12 km/h, max. 19 km/h; Baagøe, 1987). As the car travels at a constant speed of 25 km/h, which is above the average speed of bats, recorded bat passes were assumed to represent individual bats. This speed was also chosen because higher speeds increase background noise and the effect of the Doppler shifts on recorded calls (Catto *et al.*, 2003).

Statistical power analyses of the UK Bats and Roadside Mammals Survey data (see Box 10.1) suggest that monitoring a number of transects once yearly was statistically equivalent to monitoring half the number twice yearly. However, as missing data is a problem for citizen science programmes such as this one, it was felt that volunteers were more likely to remember to monitor transects that were repeated twice each year. Monitoring routes were selected by the volunteers but with respect to safety of travelling at a slow speed along roads, and so avoided large roads and motorways. The geographical coverage of the monitoring routes depended on the distribution of the volunteer network, but efforts were made to cover the important habitats and ecoregions of each region (Olson *et al.*, 2001) (Figure 10.2). In addition, the Program also encouraged survey transects to be made throughout the entire period of bat activity (May to October in Europe) to gather bat distributional and abundance information (Figure 10.3). The exact monitoring strategy for each region was a balance between the funds available for fuel, storage and analysis of the generated data, equipment, and the distribution of the volunteer network. Between June 2006 and October 2009, a total of 226 routes were monitored or surveyed in 529 events (i.e. an event is an occasion where the route was monitored or surveyed in its entirety) (Table 10.2), with 83 of these routes monitored in July and August in 335 events. Power analysis suggests regionally 3 to 10 years (based on data for *Pipistrellus pipistrellus* and *Nyctalus noctula*, respectively in the UK) of monitoring using the current number of sites would be needed to detect a statistically significant Red Alert in population change, or 8 to 25 years for an Amber Alert (see Box 10.1). This assumes that the abundances and sources of variance in Eastern Europe are the same as within the UK. More precise power analyses will be obtained over the next few years as regional monitoring data accumulate.

Sustainable monitoring

We made strong links with national NGOs and academic institutions across the region, developing a large network of in-country volunteers. In Romania, the partnership

Table 10.2 **Number of events (number of routes) for all transects (monitoring and survey) and only monitoring transects conducted as part of the iBats Program from 2006 to 2009 across countries of Eastern Europe. From Vaughan *et al.* 1997b, Russ 1999, Parsons and Jones 2000, Holderied 2001, Russo and Jones 2002, Skiba 2003.**

	Country						
Year	Romania	Bulgaria	Hungary	Croatia	Ukraine	Russia	Regional
All transects (monitoring and survey)							
2006	16(15)						16(15)
2007	52(40)	74(46)	8(6)	1(1)			135(93)
2008	58(40)	87(53)	7(5)				152(98)
2009	84(48)	73(43)	26(16)		23(15)	20(11)	226(133)
Total	210(102)	234(77)	41(20)	1(1)	23(15)	20(11)	529(226)
Monitoring transects only							
2006	2(2)						2(2)
2007	22(13)	53(27)	4(2)				79(42)
2008	29(19)	64(31)	4(2)				97(52)
2009	51(28)	60(31)	20(10)		16(8)	10(5)	157(82)
Total	104(29)	177(31)	28(10)		16(8)	10(5)	335(83)

was built on an existing relationship between BCT and the Romanian Bat Protection Association (RBPA). The RBPA then helped the Program to establish links within neighbouring countries; the Green Balkans and the Bulgarian Academy of Sciences in Bulgaria, the Nature Foundation in Hungary, the Animals Research and Protection Association and Ukraine Academy of Sciences in the Ukraine, and the Grassroots Alliance PERSEVET in Russia. We focused on partnerships with established national NGOs that already had an existing network of volunteers interested in biodiversity and conservation. Knowledge of bats was less important than a stable, volunteer-based organization. For example, the Green Balkans was founded in 1988 with a strong volunteer ethos (modelled on the British Trust for Conservation Volunteers) and has a focus on bird conservation. We employed a project manager in each country and directly recruited 74 volunteers across the region. These volunteers ranged from professional biologists to interested citizens. We delivered training in bat biology and conservation, monitoring techniques, acoustic detection, and analysis at 15 workshops or meetings between 2006 and 2009. These volunteers went on to train and recruit a further 115 volunteers, bringing the total of people involved in the Program to 189 by the end of 2009. We also donated sets of monitoring equipment to each national project. Each national project manager loaned equipment to the volunteers on the basis that if the equipment was not used to collect data it would be returned to the national equipment pool. The Program has also supported a number of students within each country, including Masters and PhD studentships and international conservation internships.

This Program currently needs three types of expertise: volunteer management (to ensure the data are collected and analysed correctly), surveyors (to collect the acoustic data), and sound analysers (to analyse the resulting acoustic data). One of the strengths of this Program was to recognize these different roles and that not everyone is interested in all aspects. For example, volunteers interested in analysing sound could become involved in the Program without collecting sound data. We created an online data portal that enabled each project manager to manage the data submitted online by the volunteers (www.ibats.org.uk). The web portal enabled results and the project's progress to be fed directly back to the volunteers. This is a critical step when using volunteers as it is a form of 'payment' for their services and helps with their continued motivation and Program involvement (Battersby, 2005).

Ultrasound detection

We used time-expansion detectors equipped with a wideband capacitance microphone within the frequency range 10–160 kHz and a sample rate of 409.6 kHz to record sound (Tranquility Transect, Courtpan Design Ltd, UK). Time-expansion was selected in preference to heterodyne or frequency division because the transformation method retains all the information from the original signal and is thus suitable for detailed digital signal analysis and species identification (Ahlén and Baagøe, 1999; Parsons and Szewczak, 2009). Time expansion was chosen in preference to real-time recording because of the prohibitively high cost of the equipment required for high speed analogue-to-digital conversion for such a large-scale programme as this. We used the Tranquility Transect detector as it was the cheapest, has a higher quality capacitance microphone that is less sensitive to low frequencies (given off by the car) and more sensitive to higher frequencies, and has a wider range of detectable frequencies than similar detector makes.

We used a time-expansion factor of 10 and a sampling time of 320 ms. The 'sensitivity' of the detector was set to maximum, which ensured the detector was continuously triggered. Thus the detector recorded sound for 320 ms, played back the 3200 ms time-expanded signal (during which time the detector was unable to record sounds), then recorded again in a continuous loop through the transect. We chose 320 ms over the other settings of 40 ms and 1.28 s as it gave enough time for calls of longer duration to be recorded and did not unnecessarily limit the time the detector was unable to acquire sounds. We selected continuous triggering as this standardized the recording effort along the transect. The detector was attached to the front or back passenger window of the vehicle with an adapted camera window clamp, with the microphone pointing up and back at a 45° angle towards the curb of the road.

Acoustic recording

We used either Sony MiniDiscs (Sony-MZ-NH600 Silver-Hi-MiniDisc Walkman) or a Zoom H2 to record time-expanded sound from the Tranquillity Transect. For the minidiscs the sound was recorded in HiSP format (256 kbit/s) and we transferred the minidisc ATRAC3plus files across to the computer using Sonic Stage 3.4 (Sony Corporation, Japan), which transformed the files into wav format (see www.ibats.org.uk for detailed protocols). For Zoom H2 devices, the audio was recorded in wav format (44.1 kHz, 16 bit) and transferred from the internal microSD card to the computer.

Geo-referencing data

We used PDAs (Mio A201, Mio 350, Mio 360) with a built in Global Positioning System (GPS) taking a reading every 5 seconds via a mapping software program (GPSTuner v.4 and v.5) to record GPX tracks of the transect route. We also used a basic GPS (Garmin Etrex Vista, Garmin Etrex Legend HCx), and the transect tracks were also recorded as GPX files. Recording was started and stopped simultaneously with the audio recording. For volunteers using the PDAs, maps were uploaded onto the devices allowing simultaneous recording, navigation, and speed checking (see www.ibats.org.uk for detailed protocols). Volunteers were provided with standardized recording sheets to record additional survey details such as temperature, humidity, rainfall, cloud cover, and which additional volunteers were present on the event.

Analysis and species identification

Calls were analysed by trained volunteers using Batsound v3.31 (Pettersson Elektronik AB, Sweden). In all instances a fast Fourier transform (FFT) size of 512 and a temporal overlap of 99% was used, giving a frequency resolution of 121 Hz. Although we record the presence of social calls in the recordings, we focus parameterization on echolocation calls only. To enable volunteers to identify isolated echolocation calls to species, a simple key was constructed based on a qualitative assessment of 'call shape' and the frequency containing maximum energy (*fmaxE*) for European bat species (following the taxonomy of Simmons, 2005) (Table 10.3). Call shape was determined from the sonograms (frequency vs time) from the calls of known species flying in varying degrees of clutter (Figure 10.4). Some species have more than one type of echolocation call. For example, the noctule (*Nyctalus noctula*) often produces two discrete calls, one with a call shape B with a *fmaxE* of 18–21 and the second with a call shape C with a *fmaxE* of 23–26. The key captures this variation in call shape within the same species (Table 10.3).

To ensure that call shapes were standardized, computer monitors with a 4:3 width-to-height screen ratio were used and the software window was opened to fill the monitor screen. A time-expansion software setting of ×10 was selected, the maximum frequency range was set to 113 500 kHz, and the milliseconds per plot was set to 352 ms. Once an echolocation call had been assigned to a call shape category the *fmaxE* was obtained from the power spectrum (frequency vs amplitude). Calls were then assigned to species (or multiple species groups if call parameters overlapped) (Table 10.3). Bats of the genus *Myotis* and *Plecotus* were grouped due to the difficulty in separating these species by their echolocation calls. Therefore measurements of *fmaxE* were not recorded for the call shapes for these species. The positions in time in the sequence and the parameters of the call (or the clearest call where there were multiple calls from a call sequence from the same individual) were recorded. Detailed protocols and training were given to sonogram analysis volunteers to ensure standardization of the analysis.

Data storage and management

We designed an innovative web data portal (www.ibats.org.uk) written in VB.NET and ASP.Net v.3.5 with a Microsoft SQL Server (2005) database and served online through

Table 10.3 Range of frequency values containing maximum energy (*fmaxE*) reported for echolocation calls of European bat species for different call shape categories (see Figure 10.4 for call categories). *fmaxE* values derived from (Vaughan *et al.*, 1997b, Russ, 1999, Parsons and Jones, 2000, Holderied, 2001, Russo and Jones, 2002, Skiba, 2003)

Call shape	Description	Species	Range (*fmaxE*) Min.	Max.
A	FM/CF/FM calls typical of Rhinolophid bats	*Rhinolophus blasii*	91	97
		Rhinolophus euryale	101	104
		Rhinolophus ferrumequinum	78	<87
		Rhinolophus hipposideros	<106	118
		Rhinolophus mehelyi	<106	112
B	QCF calls typical of *Eptesicus* spp., *Hypsugo* spp., *Miniopterus* spp., *Nyctalus* spp., *Pipistrellus* spp. *Tadarida* spp., and *Vespertilio* spp. flying away from clutter	*Eptesicus nilssonii*	27	29
		Eptesicus serotinus	24	26
		Hypsugo savii	30	33
		Miniopterus schreibersii	50	<52
		Nyctalus lasiopterus	14	17
		Nyctalus leisleri	23	25
		Nyctalus noctula	18	21
		Pipistrellus kuhlii	34	38
		Pipistrellus nathusii	35	38
		Pipistrellus pipistrellus	41	45
		Pipistrellus pygmaeus	50	65
		Tadarida teniotis	9	12
		Vespertilio murinus	22	24
C	FM-QCF calls typical of *Eptesicus* spp., *Hypsugo* spp., *Miniopterus* spp., *Nyctalus* spp., *Pipistrellus* spp. *Tadarida* spp., and *Vespertilio* spp. flying within or near to clutter	*Eptesicus nilssonii*	29	<32
		Eptesicus serotinus	25	<30
		Hypsugo savii	32	38
		Miniopterus schreibersii	50	<55
		Nyctalus lasiopterus	18	<23
		Nyctalus leisleri	25	<29
		Nyctalus noctula	23	26
		Pipistrellus kuhlii	36	<42
		Pipistrellus nathusii	36	41
		Pipistrellus pipistrellus	43	<51
		Pipistrellus pygmaeus	52	65
		Tadarida teniotis	11	<18
		Vespertilio murinus	22	26

Table 10.3 (*continued*)

Call shape	Description	Species	Range (*fmaxE*) Min.	Max.
D	FM calls typical of *Myotis* spp. and *Plecotus* spp.	*Myotis alcathoe*	N/A*	N/A*
		Myotis aurescens	N/A	N/A
		Myotis bechsteinii	N/A	N/A
		Myotis blythii	N/A	N/A
		Myotis brandtii	N/A	N/A
		Myotis capaccinii	N/A	N/A
		Myotis dasycneme	N/A	N/A
		Myotis daubentonii	N/A	N/A
		Myotis emarginatus	N/A	N/A
		Myotis lucifugus	N/A	N/A
		Myotis myotis	N/A	N/A
		Myotis mystacinus	N/A	N/A
		Myotis nattereri	N/A	N/A
		Plecotus auritus	N/A	N/A
		Plecotus austriacus	N/A	N/A
E	FM or QCF-FM calls typical of those produced by *Barbastella* spp.	*Barbastella barbastellus*	29	45

FM, steep frequency modulated; CF, constant frequency; QCF, quasi-constant frequency.
*$fmaxE$ values for *Myotis* spp. are not presented as they were not used in the key.

an IIS6 web server. The web portal database allows user registration and for users to be assigned different 'roles'. These roles grant different rights to each user such as the ability to set up new projects, upload data to projects, or analyse data from a particular project. Through the portal, acoustic and GPS files were uploaded for each survey, along with meteorological and other metadata. Following upload, the audio files could be downloaded in 5-minute chunks suitable for analysis, and the call parameters entered through the portal, maintaining the link between the audio file and the extracted parameters. Different reports about these data can be generated automatically, such as the number of transects uploaded and analysed, parameterizations of the call data, species identities and location of the calls (by reference to the GPS file), and the number of volunteers involved in each project. For example, each identified call can be instantly plotted on Google Maps from within the portal, as well as downloaded for further analysis offline. The web portal allows automatic categorization of the call parameters into species (based on Table 10.3), returning these results to the user. The portal also automatically calculates the length of the uploaded GPS file (using a haversine

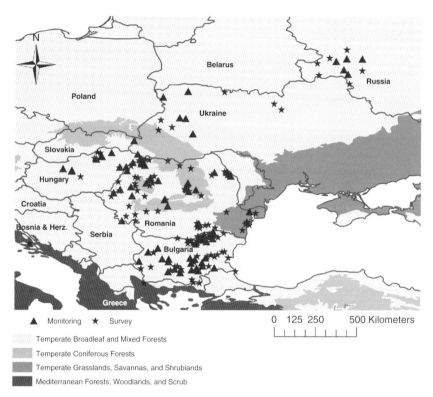

Figure 10.3 Locations of survey (stars) and monitoring transects (triangles) across the region within ecoregions (Olson *et al.*, 2001).

formula) and reports this in the database after the transect has been uploaded. The web data portal allows extremely efficient project management (as national project leaders can manage the data uploaded and analysed by volunteers in their project) and is a very secure and professional data storage and management system. The web portal code is available from the senior author on request.

Project outcomes

The project has gathered data from 303 transects (11 169 km) across Romania, Bulgaria, and parts of Hungary and Croatia between 2006 and 2008. The total area covered will likely increase to over 21 160 km once the 2009 data analysis is completed. From the

2006–2008 transects, 37 could not be used for analysis as they were either lost due to errors in recording or in transferring them into our portal, or were recorded with different settings by mistake. Many volunteers found the transfer of the audio files from minidisc to wav files difficult and confusing and that caused the majority of the errors. We subsequently switched to Zoom H2s, which are simpler and allow direct storage and transfer of sound data in standard wav format. The remaining 272 transects covered 10 118 km, from which we identified 18 602 call sequences (consisting of one or more calls from the same individuals). Of these calls, 6656 could be identified by the key to species or species groups, while for 11 946 calls the identity was uncertain (i.e. split between one or more species groups or genera) (Table 10.4). We generated data for 14 protected or threatened bat species or species groups. All European bat species are protected under the Agreement on the Conservation of Populations of European Bats (1994) through legislation, education, and conservation measures and international cooperation between the Agreement members. Additionally, two species for which this Program can produce data are classified as 'Near Threatened' (*Miniopterus schreibersii*) or 'Vulnerable' (*Barbastella barbastellus*) on the IUCN Red List (IUCN, 2010).

Table 10.4 **Number of encounters (encounters per km) for 14 species or species groups in four countries recorded in 272 transects from 2006 to 2008**

Family	Species	Romania $n = 106/126$	Bulgaria $n = 152/161$	Hungary $n = 13/15$	Croatia $n = 1/1$
Miniopteridae	*Miniopterus schreibersii*	40 (0.009)	52 (0.010)	4 (0.008)	2 (0.06)
Molossidae	*Tadarida teniotis*	10 (0.003)	18 (0.003)	–	2 (0.06)
Rhinolophidae	*Rhinolophus euryale*	–	1 (0.0002)	–	–
Vespertilionidae	*Barbastella barbastellus*	3 (0.001)	5 (0.001)	–	–
Vespertilionidae	*Eptesicus nilssonii*	226 (0.059)	113 (0.020)	23 (0.046)	1 (0.032)
Vespertilionidae	*Eptesicus serotinus*	85 (0.022)	228 (0.040)	21 (0.042)	1 (0.032)
Vespertilionidae	*Hypsugo savii*	97 (0.025)	314 (0.054)	8 (0.016)	14 (0.44)
Vespertilionidae	*Myotis* or *Plecotus spp.*	331 (0.086)	151 (0.026)	31 (0.063)	8 (0.253)
Vespertilionidae	*Nyctalus lasioterus*	583 (0.152)	169 (0.029)	28 (0.057)	–
Vespertilionidae	*Nyctalus noctula*	520 (0.136)	734 (0.128)	34 (0.069)	2 (0.063)
Vespertilionidae	*Pipistrellus kuhlii*	52 (0.014)	246 (0.043)	–	10 (0.316)
Vespertilionidae	*Pipistrellus pipistrellus*	439 (0.114)	1177 (0.205)	2 (0.004)	18 (0.569)
Vespertilionidae	*Pipistrellus pygmaeus*	182 (0.047)	70 (0.012)	11 (0.022)	8 (0.253)
Vespertilionidae	*Vespertilio murinus*	234 (0.061)	322 (0.056)	23 (0.046)	3 (0.095)
Uncertain		5759 (1.501)	5644 (0.981)	448 (0.906)	95 (3.001)
Total		8561 (2.231)	9244 (1.606)	633 (1.280)	164 (5.181)

n, number of transects analysed out of total number surveyed.

Nyctalus noctula (noctule) and *Pipistrellus pipistrellus* (common pipistrelle) have consistently the highest abundance along roads across this region, with the abundance of calls per km broadly consistent across countries (Croatia's higher abundance is probably confounded by small sample sizes) (Table 10.4). Comparing the data collected using the same techniques in the UK, we find similar call encounter rates per km (*Barbastella barbastellus*, 0.001; *Eptesicus serotinus*, 0.026; *Myotis* or *Plecotus* spp., 0.032) (Russ *et al.*, 2008). However, noctules were encountered more frequently than in the UK (*Nyctalus noctula*, 0.048), and the UK densities of common pipistrelles (1.009) and soprano pipistrelles (0.337) (*P. pipistrellus and P. pygmaeus*, respectively) are much higher in comparison to this region. This is the first time to our knowledge that comparisons in bat abundances have been possible on such a continental scale, demonstrating the power of the Program.

The number of monitoring transects (routes repeated in July and August) has increased steadily per national project from 2006 to 2009, and for established projects like those in Romania and Bulgaria volunteers can cover around 30 routes twice per annum (Table 10.2). Potentially this survey effort may detect trends in approximately 5–15 years for a Red or Amber Alert, respectively. If the data are considered regionally with 60 transects surveyed twice then the figure is 5–10 years (Box 10.1, based on the UK power analysis). As well as monitoring trends, these data generated by the Program may also be used to investigate habitat associations of different species and model likely occurrences in areas that have not been surveyed (Guisan and Zimmermann, 2000). Such distribution models may also be extended to understand changes in abundance under different models of global changes in climate and land use (Thomas *et al.*, 2004a; Harrison *et al.*, 2006).

As an example of this approach, we plotted the geo-referenced call data obtained from surveys carried out in 2006–2008 for all species. We used Maximum Entropy Modelling (MaxEnt) (Phillips *et al.*, 2006) to predict habitat suitability over a given area by generating the habitat preferences of each species and finding the dependence on 15 ecological and climatic parameters. We used data such as land cover, altitude, and a number of different variables derived from precipitation and temperature (data sources: WorldClim – Global Climate Data (www.worldclim.org) and http://ionial.esrin.esa.int). We calculated the relative contribution of each variable to each species model through a jack-knifing process, and intercorrelated variables were later removed from the final model. For each species MaxEnt produces a habitat suitability map (Figure 10.5a). The model is split into three threshold suitability types, where black is suitable habitat, dark grey is marginal, and light grey is unsuitable. The thresholds used were based on the balance training omission value, where values under this value were defined as unsuitable, values between the lower and upper thresholds (10 percentile training threshold) were defined as marginal, and values above the upper threshold defined as suitable. By combining the predictions of suitable habitats

INDICATOR BATS PROGRAM

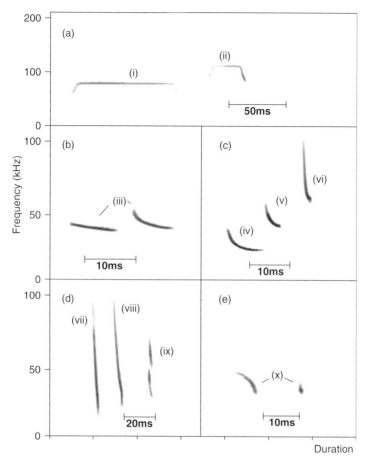

Figure 10.4 Echolocation call-shape categories of European species. (i) *Rhinolophus ferrumequinum*, (ii) *Rhinolophus hipposideros*, (iii) *Pipistrellus pipistrellus*, (iv) *Nyctalus leisleri*, (v) *Pipistrellus nathusii*, (vi) *Pipistrellus pygmaeus*, (vii) *Myotis nattereri*, (viii) *Myotis daubentonii*, (ix) *Plecotus auritus*, and (x) *Barbastella barbastellus*.

for each species and grouping marginal and unsuitable ones together it is possible to create hotspot maps, which show the number of species that find a particular area suitable (Figure 10.5b). Using predictions of future climatic conditions and land cover, we can use the generated models to predict possible responses of these species to global change and understand their changing distributions.

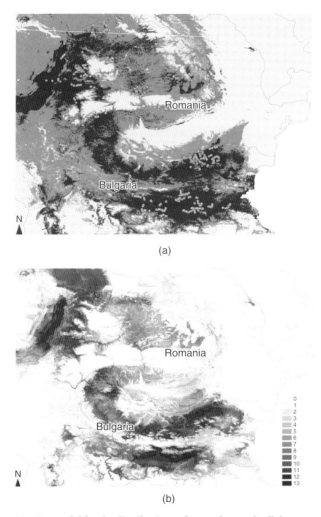

Figure 10.5 MaxEnt model for the distributions of geo-referenced call data generated across Eastern Europe from 2006 to 2008, for (a) *Pipistrellus pipistrellus* and (b) all recognized species or species groups ($n = 13$). Colours in (a) represent level of habitat suitability where black is suitable, dark grey is marginal and light grey is unsuitable. Dots represent locations of *Pipistrellus pipistrellus*. Colours in (b) scale from black to white where black represents the areas where most species' suitable habitats overlap and white represents the areas where no species' habitats overlap.

The future

Looking forward, we see that the iBats Program has great potential to deliver global bat monitoring sustainably using networks of volunteers. We review the strengths and weaknesses of the Program and consider the opportunities and threats that it may face in the future.

Strengths

The iBats Program has developed a standardized and innovative method of monitoring ultrasonic biodiversity and is delivering regional and international sustainable monitoring programmes to track the occurrences and abundances of bats – an important indicator group. This method combines novel methodologies in recording geo-referenced ultrasonic calls with a cutting-edge web portal management system to manage, store, and analyse data within existing volunteer networks. The programme enhances existing volunteer networks and strengthens in-country capacity in ecological research by providing equipment and training in bat ecology, ultrasonic recording technology, analysis software, and conservation. We have for the first time provided host countries with a valuable means of generating bat biodiversity data rapidly and sustainably across large regions cost-effectively, generating over 18 360 records to date.

Weaknesses

Biased sampling

Population sampling in the iBats Program takes place along roads and therefore may not be representative of bat communities in other habitats. Recent studies in the UK have suggested that the relative abundances of some species along road habitat are very similar to those found in other landscapes, in particular agricultural habitats (e.g. *Pipistrellus pygmaeus*, *Nyctalus*, and *Eptesicus* species) but others were over- or under-represented in road habitats (Linton, 2009). In particular, a higher percentage of calls recorded along roads were produced by the common pipistrelles (*Pipistrellus pipistrellus*) compared to calls recorded on agricultural land (64.4% vs 42.4%), and *Myotis* species were under-represented along roads (3.8% vs 21.7% on agricultural land) (Linton, 2009). Some species have been shown to avoid street lights (Stone *et al.*, 2009), and a large-scale street-lighting project may alter the species composition along the road but not indicate population decline. Trends may also differ between habitats; for example, a decline might have a big impact on populations in suboptimal habitats before it becomes evident in optimal ones. Hence, if the habitats monitored differ from those in the wider environment there is a risk of bias, even if there is no initial difference in abundance between the habitats (Buckland *et al.*, 2005).

It is important to understand the biases of any monitoring programme, and to some extent all bat surveillance methods do have different inherent biases (Hayes *et al.*, 2009). For example, acoustic sampling in any habitat will not be representative of quiet or non-echolocating species. However, the key element for the iBats Program is estimating relative changes in abundance along the same habitats to develop an indicator of change, not estimating the absolute population abundances. Additionally, once volunteer networks are trained and in place, then other more volunteer-intensive monitoring methods can be implemented to understand population changes in other species.

Equipment

Our equipment is complex and presents training challenges for volunteers, and we have experienced some loss of data from a number of transects as a result. To address this, we are developing applications for the iPhone and phones running the Android operating system (e.g. Google G1 or HTC Desire phones), which will enable some volunteers to use their own mobile devices to record time-expanded audio and GPS data from the transects. Phones will be directly attached to the detector and the geo-referenced sound files can then be automatically uploaded onto the web portal after the survey. This software development will reduce the amount of equipment that the iBats Program has to supply and also decrease the complexity of the survey, opening it up to other less technical volunteer groups, for example, school children.

Acoustic analysis

The most expensive and unsustainable part of the project is the time-consuming nature of the analyses of the resulting acoustic data by volunteers (e.g. a 90-minute transect may take up to 6–8 hours to process). In addition, analysis can be subjective and the data extracted may vary depending on analyser experience. We are addressing this by working on automatic methods of extracting and identifying calls from long sequences. These algorithms can then be built into the web portal so that calls are found and their parameters extracted automatically once uploaded.

Species identification

Over half the calls recorded could not be identified to species using our simple identification key. We are addressing this in several ways. Firstly we are building a global call library from a network of collaborators (EchoBank) (Collen, 2012) to provide an online reference set for comparisons. Secondly, we are building a neural classification network to identify species more accurately (Walters *et al.*, 2012). We hope to include these algorithms into the mobile phone applications in the future to give real-time analysis of surveys to the volunteers in the field.

Opportunities and threats

The iBats Program for the first time offers an opportunity to monitor bats consistently, cost-effectively, and efficiently on regional, international, and global scales. Developing such a programme has been recently proposed to be of high importance because of the potential role of bats as bioindicators (Jones *et al*., 2009). This method has the potential to be appropriate not only for temperate species (where there is a considerable bias in monitoring programmes in general) but also in tropical areas where most bat biodiversity is present (MacSwiney *et al*., 2008). Indeed, a recent analysis showed that acoustic monitoring of aerial insectivores in the Neotropics and Palaeotropics gave the highest statistical power and chance of detecting a trend when compared to traditional capture techniques (Meyer *et al*., 2010). Potential threats to the Program are that monitoring is not sustained because of lack of funds, resulting in a loss of power to detect statistical trends. Finding long-term funding for a programme such as this is notoriously difficult. However, we hope that our approach of using volunteers, building in-country capacity, technological advances, and bringing the Program to the attention of national and international policy-makers will mean that ongoing costs remain low and long-term costs are funded through government agencies.

Acknowledgements

The Indicator Bats Program would be impossible without the help of its amazing volunteers (over 300 in 2009) and we are enormously grateful for their efforts and enthusiasm. This work was supported financially through the Darwin Initiative (Awards 15003, 161333, EIDPR075), the Leverhulme Trust (Philip Leverhulme Prize for KEJ), the People's Trust for Endangered Species, the British Ecological Society, the Rufford Maurice Laing Foundation, Bat Conservation International, Conservation International, and the Zoological Society of London. We also thank Paul Fletcher for assistance and Karen Haysom, Gareth Jones, and Kimberly Pollard for comments on previous versions of the manuscript.

References

Ahlén, I. and Baagøe, H.J. (1999) Use of ultrasound detectors for bat studies in Europe: experiences from field identification, surveys and monitoring. *Acta Chiropterologica*, 1, 137–150.

Andersen, B.B. and Miller, L.A. (1977) A portable ultrasonic detection system for recording bat cries in the field. *Journal of Mammalogy*, 58, 226–229.

Baagøe, H.J. (1987) The Scandinavian bat fauna – adaptive wing morphology and free flight in the field. In: Fenton, M.B., Racey, P.A., and R. J. M. V.(eds), *Recent Advances in the Study of Bats*. Cambridge University Press, Cambridge, pp. 57–74.

Baerwald, E.F., D'Amours, G.H., Klug, B.J., and Barclay, R.M.R. (2008) Barotrauma is a significant cause of bat fatalities at wind turbines. *Current Biology*, 18, 695–696.

Barclay, R.M.R., Fullard, J.H., and Jacobs, D.S. (1999) Variation in the echolocation calls of the hoary bat (*Lasiurus cinereus*): influence of body size, habitat structure, and geographic location. *Candian Journal of Zoology*, 77, 530–534.

Bat Conservation Trust (2008) *The National Bat Monitoring Programme: Annual Report 2008*. Bat Conservation Trust, London.

Battersby, J. (2005) *Engaging with Volunteers. Setting up and Managing Volunteer Networks*. Tracking Mammals Partnership and National Biodiversity Network Trust, London.

Battersby, J. (2010) Guidelines for surveillance and monitoring of European bats. *EUROBATS Publication No. 5*. UNEP/EUROBATS Secretariat, Bonn.

Battersby, J. and Greenwood, J.D. (2004) Monitoring terrestrial mammals in the UK: past, present and future, using lessons from the bird world. *Mammal Review*, 34, 3–29.

Buckland, S.T., Magurran, A.E., Green, R.E., and Fewster, R.M. (2005) Monitoring change in biodiversity through composite indices. *Philosophical Transactions of the Royal Society of London, Series B*, 360, 243–254.

Butchart, S.H.M., Walpole, M., Collen, B., *et al*. (2010) Global biodiversity: indicators of recent declines. *Science*, 328, 1164–1168.

Catto, C., Russ, J., and Langton, S. (2003) *Development of a Car Survey Monitoring Protocol for the Republic of Ireland*. Prepared on behalf of the Heritage Council by the Bat Conservation Trust, UK, The Heritage Council, Ireland.

Celis-Murillo, A., Deppe, J., and Allen, M.F. (2009) Using soundscape recordings to estimate bird species abundance, richness, and composition. *Journal of Field Ornithology*, 80, 64–78.

Collen, A. (2012) Evolution of echolocation in bats. *PhD thesis*. University College London, London.

Collen, B., Loh, J., Whitmee, S., McRae, L., Amin, R., and Baillie, J.E.M. (2009) Monitoring change in vertebrate abundance: the Living Planet Index. *Conservation Biology*, 23, 317–327.

Convention on Biological Diversity (2003) *Handbook of the Convention on Biological Diversity*. Earthscan, London.

Corben, C. and Fellers, G.M. (2001) Choosing the 'correct' bat detector – a reply. *Acta Chiropterologica*, 3, 253–256.

Devictor, V., Whittaker, R.J., and Beltrame, C. (2010) Beyond scarity: citizen science programmes as useful tools for conservation biogeography. *Diversity and Distributions*, 16, 354–362.

Dobson, A.P. (2005) What links bats to emerging infectious disease? *Science*, 310, 628–629.

Entwistle, A.C., Racey, P.A., and Speakman, J. (1997) Roost selection by the brown long-eared bat *Plecotus auritus*. *Journal of Applied Ecology*, 34, 399–408.

Federico, P., Hallam, T.G., McCracken, G.F., *et al*. (2008) Brazilian free-tailed bats (*Tadarida brasiliensis*) as insect predators in transgenic and conventional cotton crops. *Ecological Applications*, 18, 826–837.

Fenton, M.B. (1985) *Communication in the Chiroptera*. Indiana University Press, Bloomington.

Fenton, M.B. (2000) Choosing the 'correct' bat detector. *Acta Chiropterologica*, 2, 215–224.

Fenton, M.B., Acharya, L., Audet, D., *et al*. (1992) Phyllostomid bats (Chiroptera: Phyllostomidae) as indicators of habitat disruption in the Neotropics. *Biotropica*, 24, 440–446.

Fewster, R.M., Buckland, S.T., Siriwardena, G.M., Baillie, S.R., and Wilson, J.D. (2000) Analysis of population trends for farmland birds using generalized additive models. *Ecology*, 81, 1970–1984.

Feyerabend, F. and Simon, M. (2000) Use of roosts and roost switching in a summer colony of 45 kHz phonic type pipistrelle bats (*Pipistrellus pipistrellus* Schreber, 1774). *Myotis*, 38, 51–59.

Frick, W.F., Pollock, J.F., Hicks, A.C., *et al.* (2010) An emerging disease causes regional population collapse of a common North American bat species. *Science*, 329:679–682.

Fujita, M.S. and Tuttle, M.D. (1991) Flying foxes (Chiroptera: Pteropodidae): threatened animals of key economic importance. *Conservation Biology*, 5, 455–463.

Fullard, J.H. and Dawson, J.W. (1997) The echolocation calls of the spotted bat *Euderma maculatum* are relatively inaudible to moths. *Journal of Experimental Biology*, 200, 129–137.

Gibbs, J.P., Droege, S., and Eagle, P. (1998) Monitoring populations of plants and animals. *BioScience*, 48, 935–940.

Greenwood, J.J.D. (2007) Citizens, science and bird conservation. *Journal of Ornithology*, 148, 77–124.

Gregory, R.D., Wilkinson, N.I., Noble, D.G., *et al.* (2002) The population status of birds in the United Kingdom, Channel Islands and Isle of Man: an analysis of conservation concern 2002–2007. *British Birds*, 95, 410–450.

Gregory, R.D., von Strein, A., Vorisek, P., *et al.* (2005) Developing indicators for European birds. *Proceedings of the Royal Society, Biological Sciences Series B*, 360, 269–288.

Grenyer, R., Orme, C.D.L., Jackson, S.F., *et al.* (2006) The global distribution and conservation of rare and threatened vertebrates. *Nature*, 444, 93–96.

Guisan, A. and Zimmermann, N.E. (2000) Predictive habitat distribution models in ecology. *Ecological Modelling*, 135, 147–186.

Harrison, P., Berry, P., Butt, N., and New, M. (2006) Modelling climate change impacts on species' distributions at the European scale: implications for conservation policy. *Environmental Science and Policy*, 9, 116–128.

Hayes, J.P., Ober, H.K., and Sherwin, R.E. (2009) Survey and monitoring of bats. In: Kunz, T.H. and Parsons, S. (eds), *Ecological and Behavioral Methods for the Study of Bats*. Johns Hopkins University Press, Baltimore, pp. 112–129.

Haysom, K. (2008) *Streamlining European 2010 Biodiversity Indicators (SEBI 2010): Developing a methodology for using bats as indicator species and testing the usability of GBIF data for use in 2010 biodiversity indicators*. Bat Conservation Trust, London.

Hodgkison, R., Balding, S.T., Zubaid, A., and Kunz, T.H. (2003) Fruit bats (Chiroptera: Pteropodidae) as seed dispersers and pollinators in a lowland Malaysian rain forest. *Biotropica*, 35, 491–502.

Holderied, M.W. (2001) *Akustische Flugbahnverfolgung von Fledermäusen: Artvergleich des Verhaltens beim Suchflug und Richtcharakteristik der Schallabstrahlung*. Friedrich-Alexander-Universität, Erlangen-Nürnberg.

IUCN (2010) *IUCN Red List of Threatened Species*. IUCN, Gland, Switzerland.

Jones, G. and Rydell, J. (1994) Foraging strategy and predation risk as factors influencing emergence time in echolocating bats. *Philosophical Transactions of the Royal Society of London, Series B*, 346, 445–455.

Jones, G. and Siemers, B.M. (2011) The communicative potential of bat echolocation pulses. *Journal of Comparative Physiology A*, 197, 447–457.

Jones, G. and Teeling, E.C. (2006) The evolution of echolocation in bats. *Trends in Ecology and Evolution*, 21, 149–156.

Jones, G., Gordon, T., and Nightingale, J. (1992) Sex and age differences in the echolocation calls of the lesser horseshoe bat, *Rhinolophus hipposideros*. *Mammalia*, 56, 189.

Jones, G., Jacobs, D.S., Kunz, T.H., Willig, M.R., and Racey, P.A. (2009) Carpe noctem: the importance of bats as bioindicators. *Endangered Species Research*, 8, 93–115.

Keesing, F., Belden, L.K., Daszak, P.A., et al. (2010) Impacts of biodiversity on the emergence and transmission of infectious diseases. *Nature*, 468, 647–652.

Korine, C. and Kalko, E.K.V. (2001) Towards a global bat-signal database. *IEEE Engineering in Medicine and Biology*, 20, 81–85.

Lacher, T. (2008) *Tropical Ecology, Assessment and Monitoring (TEAM) Network. Avian Monitoring Protocol Version 3.1*. Conservation International, Washington, DC.

Limpens, H.J.G.A. (2004) Field identification: using bat detectors to identify species. In: Brigham, M., Kalko, E.K.V., Jones, G., Parsons, S., and Limpens, H.J.G.A. (eds), *Bat Echolocation Research*. Bat Conservation International, Austin, TX, pp. 46–57.

Linton, D.M. (2009) *Bat Ecology and Conservation in Lowland Farmland*. University of Oxford, Oxford.

Loeb, S.C., Post, C.J., and Hall, S.T. (2009) Relationship between urbanization and bat community structure in national parks of the south-eastern U.S. *Urban Ecosystems*, 12, 197–214.

Loh, J., Collen, B., McRae, L.G., et al. (2008) Living Planet Index. In: Loh, J. (ed.) *2010 and Beyond: Rising to the Biodiversity Challenge*. WWF, Gland, Switzerland.

Lowman, M., D'Avanze, C., and Brewer, C. (2009) A national ecological network for research and education. *Science*, 323, 1172–1173.

Mace, G.M. and Baillie, J.E.M. (2007) The 2010 biodiversity indicators: challenges for science and policy. *Conservation Biology*, 21, 1406–1413.

Mace, G.M., Masundire, H., and Baillie, J.E.M. (2005) Chapter 4: Biodiversity. In: *Millennium Ecosystem Assessment. Current State and Trends: Findings of the Condition and Trends Working Group. Ecosystems and Human Well-being*. Island Press, Washington, DC.

MacSwiney, G., Cristina, M., Clarke, F.M., and Racey, P.A. (2008) What you see is not what you get: the role of ultrasonic detectors in increasing inventory completeness in Neotropical bat assemblages. *Journal of Applied Ecology*, 45, 1364–1371.

Maltby, A., Jones, K.E., and Jones, G. (2009) Understanding the origin and diversification of bat echolocation calls. In: Brudzynski, S.M. (ed.), *Handbook of Mammalian Vocalization. An Integrative Neuroscience Approach*. Elsevier, London, pp. 37–48.

Meschede, A. and Rudolph, B.-U. (2010) *1985–2009: 25 Jahre Fledermausmonitoring in Bayern – UmweltSpezial*. Bavarian Environment Agency, Augsburg.

Meyer, C.F.J., Aguiar, L.M.S., Aguirre, L.F., et al. (2010) Long-term monitoring of tropical bats for anthropogenic impact assessment: Gauging the statistical power to detect population change. *Biological Conservation*, 143, 2797–2807.

Neuweiler, G. (1989) Foraging ecology and audition in echolocating bats. *Trends in Ecology and Evolution*, 4, 160–166.

Newson, S.E., Mendes, S., Crick, H.Q.P., et al. (2009) Indicators of the impact of climate change on migratory species. *Endangered Species Research*, 7, 101–113.

Obrist, M.K. (1995) Flexible bat echolocation: the influence of individual, habitat and conspecifics on sonar signal design. *Behavioral Ecology and Sociobiology*, 36, 207–219.

Olson, D.M., Dinerstein, E., Wikramanayake, E.D., et al. (2001) Terrestrial ecoregions of the world: A new map of life on earth. *BioScience*, 51, 933–938.

Pagan, A. (2010) An analysis of iBats UK (Indicator Bats Program UK): How effective is iBats at generating an indicator of bat abundance in comparison to existing programs? MSc thesis, Imperial College, London.

Parsons, S. and Jones, G. (2000) Acoustic identification of 12 species of echolocating bat by discriminant function analysis and artificial neural networks. *Journal of Experimental Biology*, 203, 2641–2656.

Parsons, S. and Szewczak, J.M. (2009) Detecting, recording, and analyzing the vocalizations of bats. In: Kunz, T.H. and Parsons, S. (eds), *Ecological and Behavioral Methods for the Study of Bats*. Johns Hopkins University Press, Baltimore, pp. 91–111.

Parsons, S., Thorpe, C.W., and Dawson, S.M. (1997) The echolocation calls of the long-tailed bat (*Chalinolobus tuberculatus*): A quantitative description and analysis of call phase. *Journal of Mammalogy*, 78, 964–976.

Parsons, S., Boonman, A.M., and Obrist, M.K. (2000) Advantages and disadvantages of techniques for transforming and analyzing Chiropteran echolocation calls. *Journal of Mammalogy*, 81, 927–938.

Partnership, U.B. (2010) *UK Biodiversity Indicators in Your Pocket 2010. Measuring Progress Towards Halting Biodiversity Loss*. Department for Environment, Food and Rural Affairs, London.

Phillips, S., Anderson, R.P., and Schapire, R. (2006) Maximum entropy modelling of species geographic distributions. *Ecological Modelling*, 190, 231–259.

Pye, D. (1993) Is fidelity futile? The "true" signal is illusory, especially with ultrasound. *Bioacoustics*, 4, 271–286.

Redgwell, R.D., Szewczak, J.M., Jones, G., and Parsons, S. (2009) Classification of echolocation calls from 14 species of bat by support vector machines and ensembles of neural networks. *Algorithms*, 2, 907–924.

Roche, N., Langton, S., and Aughney, T. (2007) *The Car-Based Bat Monitoring Scheme for Ireland: Report for 2006*. National Parks and Wildlife Service, Department of Environment, Heritage and Local Government, Dublin.

Roche, N., Langton, S., Aughney, T., et al. (2011) A car-based monitoring method reveals new information on bat populations and distributions in Ireland. *Animal Conservation*, 14, 642–651.

Russ, J.M. (1999) *The Bats of Britain and Ireland*. Alana Ecology Ltd.

Russ, J.M., Briffa, M., and Montgomery, W.I. (2003) Seasonal patterns in activity and habitat use by bats (*Pipistrellus* spp. and *Nyctalus leisleri*) in Northern Ireland determined using a driven transect. *Journal of Zoology*, London, 259, 289–299.

Russ, J., Catto, C., and Wembridge, D. (2005) *The Bats and Roadside Mammals Survey 2005. Final Report on First Year of Study*. The Bat Conservation Trust and People's Trust for Endangered Species, London.

Russ, J., Briggs, P., and Wembridge, D. (2008) *The Bats and Roadside Mammals Survey 2008. Final Report on Fourth Year of Study*. The Bat Conservation Trust and People's Trust for Endangered Species, London.

Walters, C.L., Freeman, R., Collen, A., *et al.* (2012) A continental-scale tool for acoustic identification of European bats. *Journal of Applied Ecology*, 49, 1064–1074.

Russo, D. and Jones, G. (2002) Identification of twenty-two bat species (Mammalia: Chiroptera) from Italy by analysis of time-expanded recordings of echolocation calls. *Journal of Zoology, London*, 258, 91–103.

Russo, D. and Jones, G. (2003) Use of foraging habitats by bats (Mammalia: Chiroptera) in a Mediterranean area determined by acoustic surveys: conservation implications. *Ecography*, 26, 197–209.

Russo, D., Cistrone, L., and Jones, G. (2005) Spatial and temporal patterns of roost use by tree-dwelling barbastelle bats, *Barbastella barbastellus*. *Ecography*, 28, 769–776.

Russo, D., Mucedda, M., Bello, M., Biscardi, S., Pidinchedda, E., and Jones. G. (2007) Divergent echolocation call frequencies in insular rhinolopids (Chiroptera): a case of character displacement? *Journal of Biogeography*, 34, 2129–2138.

Sachs, J.D., Baillie, J., Sutherland, W.J., *et al.* (2009) Biodiversity conservation and achievement of the Millennium Development Goals. *Science*, 325, 1502–1503.

Sales, G. and Pye, D. (1974) *Ultrasonic Communication by Animals*. Chapman & Hall, London.

Schipper, J., Chanson, J., Chiozza, F., *et al.* (2008) The biogeography of diversity, threat, and knowledge in the world's terrestrial and aquatic mammals. *Science*, 322, 225–230.

Shiel, C.B., Shiel, R.E., and Fairley, J.S. (1999) Seasonal changes in the foraging behaviour of Leisler's bats (*Nyctalus leisleri*) in Ireland as revealed by radio-telemetry. *Journal of Zoology, London*, 249, 347–358.

Simmons, N.B. (2005) Order Chiroptera. In: Wilson, D.E. and Reeder, D.M. (eds), *Mammal Species of the World: A Taxonomic and Geographic Reference*. Smithsonian Institution Press, Washington, DC, pp. 312–529.

Sims, M., Wanless, S., Harris, M.P., Mitchell, P.I., and Elston, D.A. (2006) Evaluating the power of monitoring plot designs for detecting long-term trends in the numbers of common guillemots. *Journal of Applied Ecology*, 43, 537–546.

Skiba, R. (2003) *Europäische Fledermäuse*. Westarp Wissenschaften, Hohenwarsleben.

Skowronski, M.D. and Harris, J.G. (2006) Acoustic detection and classifcation of microchiroptera using machine learning: lessons learned from automatic speech recognition. *Journal of the Acoustical Society of America*, 119, 1817–1833.

Stone, E.L., Jones, G., and Harris, S. (2009) Street lighting disturbs commuting bats. *Current Biology*, 19, 1123–1127.

Sukhdev, P. (2008) *The Economics of Ecosystems and Biodiversity. An Interim Report*. European Community, Brussels.

Thomas, C.D., Cameron, A., Green, R.E., *et al.* (2004a) Extinction risk from climate change. *Nature*, 427, 145–148.

Thomas, J.A. and Jalili, M.S. (2004) Echolocation in insectivores and rodents. In: Thomas, J.A., Moss, C.F., and Vater, M. (eds), *Echolocation in Bats and Dolphins*. University of Chicago Press, Chicago and London, pp. 547–564.

Thomas, J.A., Moss, C.F., and Vater, M. (2004b) *Echolocation in Bats and Dolphins*. University of Chcago Press, Chicago and London.

Vaughan, N., Jones, G., and Harris, S. (1996) Effects of sewage effluent on the activity of bats (Chiroptera: Vespertilionidae) foraging along rivers. *Biological Conservation*, 78, 337–343.

Vaughan, N., Jones, G., and Harris, S. (1997a) Habitat use by bats (Chiroptera) assessed by means of a broad-band acoustic method. *Journal of Applied Ecology*, 34, 716–730.

Vaughan, N., Jones, G., and Harris, S. (1997b) Identification of British bat species by multivariate analysis of echolocation parameters. *Bioacoustics*, 7, 189–207.

Walsh, A.L., Catto, C.M.C., Hutson, A.M., Racey, P.A., Richardson, P., and Langton, S. (2001) *The UK's National Bat Monitoring Programme. Final Report*. Department for Environment, Food and Rural Affairs, London.

Walsh, A.L., Barclay, R.M.R., and McCracken, G.F. (2004) Designing bat-activity surveys for inventory and monitoring studies at local and regional scales. In: Brigham, M., Kalko, E.K.V., Jones, G., Parsons, S., and Limpens, H.J.G.A. (eds), *Bat Echolocation Research: Tools, Techniques and Analysis*. Bat Conservation International, Austin, TX.

Waters, D.A. and Gannon, M.L. (2004) Bat call libraries: management and potential use. In: Brigham, M., Kalko, E.K.V., Jones, G., Parsons, S., and Limpens, H.J.G.A. (eds), *Bat Echolocation Research: Tools, Techniques and Analysis*. Bat Conservation International, Austin, TX, pp. 150–157.

Weir, L., Fiske, I., and Royle, J.A. (2009) Trends in anuran occupancy from northeastern states of the North American Amphibian Monitoring Program. *Herpetological Conservation and Biology*, 4, 389–402.

Wickramasinghe, L.P., Harris, S., Jones, G., and Vaughan, N. (2003) Bat activity and species richness on organic and conventional farms: impact of agricultural intensification. *Journal of Applied Ecology*, 40, 984–993.

Occupancy Methods for Conservation Management

Darryl I. MacKenzie[1] and James T. Reardon[2,3]

[1]Proteus Wildlife Research Consultants, Dunedin, New Zealand
[2]Conservations Programmes, Zoological Society of London, London, UK
[3]Southland Conservancy, Department of Conservation, Invercargill, New Zealand

Introduction

Managers of conservation programmes require reliable information about the status of their system in order to make sensible management decisions, or learn about the impact of previous decisions. Without such information, inappropriate decisions may be made leading to undesirable conservation outcomes (Nichols and Williams, 2006; MacKenzie, 2009). However, many conservation programmes have limited resources and may be operating in areas where field crews have little technical training. It is therefore necessary that this reliable information can be collected in a practical, robust, and cost-efficient manner. The example we will use to illustrate the utility of these techniques is the recently established Slender Loris Conservation Programme in Sri Lanka. The red slender loris (*Loris tardigradus tardigradus* and the montane subspecies *Loris tardigradus nycticeboides*) are currently maintained as Endangered according to the IUCN Red List (Endangered C2a(i); Nekaris *et al.*, 2008). This rareness is borne out by the inclusion of *L. t. nycticeboides* in the list of the 25 most endangered primates in the world for 2006–2008 (Mittermeier *et al.*, 2007), in which the account for *L. t. nycticeboides* suggests the range of the species to be less than 40 000 ha (Nekaris, 2003). Continued forest fragmentation due to agricultural and urban development, dieback of forest due to presumed climatic events in montane habitats (Werner, 1984), and more insidiously, the degradation of forest due to fuel-wood collection, may put the area of suitable habitat at a worryingly lower figure

---

Biodiversity Monitoring and Conservation: Bridging the Gap between Global Commitment and Local Action, First Edition. Edited by Ben Collen, Nathalie Pettorelli, Jonathan E.M. Baillie and Sarah M. Durant.
© 2013 John Wiley & Sons, Ltd. Published 2013 by John Wiley & Sons, Ltd.

for the long-term persistence of the species. The anthropogenic land use pressures in Sri Lanka, which is in the top three biodiversity hotspots threatened by population pressure (Cincotta *et al.*, 2000), suggest that strategic conservation management decisions need to be taken as soon as possible to optimize chances of success for security and persistence of such species.

A traditional focus for quantifying the status of systems in many conservation programmes is animal abundance or population size. 'How many?' is a very easy and natural question to ask. However, in only a minority of cases will it be possible to identify every individual within the population of interest, or even within sampled sub-regions of the system. Often, an unknown fraction of the population will be observed by the employed field methods due to the imperfect detection of individuals (i.e. not all individuals within a survey area will be observed). In order to obtain a reliable estimate of population size, it is therefore necessary to account for imperfect detection by using techniques such as capture-recapture or distance sampling (Otis *et al.*, 1978; Buckland *et al.*, 1993; Williams *et al.*, 2002). While these methods can offer detailed information about the population, they do require a certain level of technical competence and can involve substantial resources particularly if they are to be applied at large spatial scales. They may also necessitate the capture, handling, and marking of individuals, which may be undesirable for endangered species of special concern. Capture and handling might also influence subsequent recapture probabilities, adding further uncertainty to estimates. Even if interest is primarily directed towards relative abundance, it is important that detection probability is accounted for otherwise estimates will likely be biased in an unknown direction (e.g. Skalski *et al.*, 1983; Thompson *et al.*, 1998; Nichols *et al.*, 2000; Anderson 2001; Yoccoz *et al.*, 2001; MacKenzie and Kendall, 2002).

Many of the impediments to successful conservation management faced in countries such as Sri Lanka stem from a fundamental lack of data and analysis skills for interpreting biodiversity trends in relation to land use, management practices, and ecological or environmental factors such as invasive species and changing climates. Within the biodiverse wet zone of Sri Lanka, which is regarded as one of the global biodiversity hotspots, less than 10% of the former forest cover remains in a natural state (Mittermeir *et al.*, 1999; Myers *et al.*, 2000). With at least half of Sri Lanka's energy demand provided by uncultivated timber extraction in a region already extensively cleared during the colonial and post-colonial period for tea, rubber, and coffee plantations, it is no surprise that the remaining lowland and montane rainforest is extremely fragmented and under pressures of encroachment (Gunawardene *et al.*, 2007).

Forest fragmentation has an added impact for animals such as loris, which rely on closed and complex vegetation in which to move and roost during the day. It is possible that even when seral vegetation growing in cleared areas reaches some level of maturity and canopy connectivity, the vastly simplified forest structure will remain largely unsuitable for the species. Due to the low detection probability of these discreet nocturnal primates it is unlikely that critical conservation management questions such as forest type preferences can be answered robustly by simple survey observations that do not account accurately for detection probability.

The history of investment in biodiversity research, management, and protection in Sri Lanka has been dominated by taxonomic description and biodiversity baseline type surveys of fauna and flora within protected areas (Green et al., 2009). Whilst these approaches are essential first steps in generating the knowledge required to enable strategic management, so far there is no recognized and practical process in place to test assumptions that protected areas do actually protect the biodiversity they contain. As international conventions such as the Convention on Biological Diversity (CBD) become increasingly recognized as the framework for ensuring the persistence of national biodiversity and ecosystem services, there is a growing understanding of the need for adaptive and informed management. The role of the international conservation and resource management community must therefore now be to assist, wherever possible, the transfer of skills to enable more accountable and adaptive management. However, the act of monitoring fauna and flora can easily become a form of intellectual displacement behaviour when challenging management decisions need to be made (Nichols and Williams, 2006). It is essential in such circumstances that any surveying or monitoring activities be clearly justified by pertinent management or conservation questions.

Thus, in the case of assessing the status, ecotype preference, and relative management efficacy for loris in Sri Lanka, one of the first questions to be addressed is what metric should be recorded within the monitoring programme? Alternative metrics that can be useful in many situations where abundance is an infeasible metric include proportion of area occupied (PAO) and species richness (SR). In essence, the question is changing from 'How many?' to 'Where are they?' or 'Who are they?' PAO is essentially a measure of species distribution, the fraction of places within the region of interest where the species is present. A number of locations are surveyed with the intent of establishing the presence or absence of the target species at each location. Estimating PAO has a number of practical advantages but also is arguably the most reliable metric for landscape-level management decision-making as it is likely to be more robust to local effects and stochasticity than local estimates of abundance or density. SR is a community-based measure, estimating the number of species present at a location or within a defined region. SR may focus upon only certain taxonomic groups, or different species may be assigned different values. As for estimation of abundance, issues of detection probabilities are relevant for PAO and SR as well (although at a different scale). Few species are likely to be so conspicuous that they will always be detected when present at a location, hence the possibility of a false absence may be non-negligible. Unaccounted for, imperfect detection will cause PAO and SR to be underestimated.

PAO and SR are not new ideas, but methods that account for imperfect detection are, relatively, more recent (PAO: MacKenzie et al., 2002, 2003, 2006; Tyre et al., 2003; Wintle et al., 2004; SR: Burnham and Overton, 1979; Bunge and Fitzpatrick, 1993;

Nichols and Conroy, 1996; Dorazio *et al.*, 2006). In this chapter we briefly outline recently developed 'occupancy models' that could be used in either situation. We illustrate some of the key ideas for application of PAO by considering a conservation issue from Sri Lanka, that of the red slender loris (*Loris tardigradus*).

First step: defining conservation management objectives

Any conservation programme needs to have clear, quantifiable management objectives (Yoccoz *et al.*, 2001; Williams *et al.*, 2002; Nichols and Williams, 2006; MacKenzie, 2009). Without such objectives it is impossible to conclude whether the desired conservation outcome is being achieved. The ability to demonstrate success is extremely important when seeking continuing support from host institutions, government agencies, and other funding sources. Worldwide, conservation resources are becoming ever more limited as the need for investment grows, economies contract, and perceived competition with other environmental processes such as climate change are identified by some as diverting investment away from conservation management. Hence, programmes with clear objectives are likely to be more acceptable to donor agencies than programmes that only have vague goals or mission statements, or those not appropriately aligned to applied management. The latter types of programmes are likely just wasting valuable resources.

Note that we do not regard statements that involve the detection of a trend of a specified magnitude over a certain timeframe (e.g. a 5% decline in population size over a 10-year period) as an entirely useful or optimal objective. The intent of such statements often implies that any management intervention of the system will only occur once a trend has been identified with some degree of confidence. By definition, it will take a relatively long timeframe (often 10–20 years for larger vertebrates) before sufficient data have been collected to determine whether there is any statistical evidence of a trend. Indeed, the very definitions used to classify species as critically endangered (population sizes and rates of decline under IUCN Red List criteria) remove the luxury of waiting for those sorts of timeframes before managers decide whether or not they should take some action. Managers may argue that they need some assurance of a problem before embarking on a costly management programme; however, how much more costly is it going to be to restore a population from the brink of extinction compared to intervening early? Ultimately, it is likely to be much more cost-efficient in the long term to intervene as soon as possible, therefore also avoiding the additional cost of rescuing genetic diversity from populations allowed to bottleneck (Allendorf and Luikart, 2006; Mills, 2006). Arguments about a lack of information are also often untrue. While there may be limited 'hard' data available upon which to base management decisions, there will often be a degree of anecdotal information from local communities or historic data that could be useful. We can also rely on basic ecological principles to suggest that certain activities are likely to be impacting upon systems and therefore should be included immediately in comparative management strategies. There will always be various sources of uncertainties associated with management decisions that no amount of monitoring will ever fully resolve. Management should be striving to make smart decisions in the face of these uncertainties, with a well-defined objective being the first key step.

Rather than focus on trends, we suggest that more useful objectives attempt to maintain systems above some level, or maximize returns of some nature. Exactly what level is deemed appropriate should include input from all relevant stakeholders, and while it involves a certain degree of arbitrariness, it is a choice that would be faced anyway once a trend has been detected in order to decide what level the systems should be restored to. We are simply stating that it is more useful to have that discussion sooner rather than later if successful and efficient conservation management is the goal. When working in environments such as Sri Lanka where the field of dynamic and successful conservation management is still in its infancy these discussions can be challenging, yet they are no less essential.

Useful conservation management objectives may also include utilization of natural resources. In many situations involving exploitation of natural resources there will be economic benefits to nations and local communities that simply cannot be ignored. Therefore when objectives are being set, there must be a way of balancing conservation objectives against other, often economic, objectives. One approach is to assign a value to the system to put it onto the same scale as the economic objectives (Pearce and Moran, 1995). Another approach is that one objective has primary focus, with the second objective being used as a constraint; for example,, maximize yield of timber harvesting for next 200 years while maintaining the distribution of a key species across 60% of the forest (Martin *et al.*, 2009).

It is important to realize that useful conservation management objectives for systems can only be made through consultation with all relevant parties (e.g. conservation groups, government and donor agencies, industries, and local communities). The purpose of such discussions is to have all parties share their views on the system, and reach agreement on some long-term goals. We do not expect such discussions to be trivial; however, without agreement on fundamental aims and outcomes the conservation programme will likely fail. Through the discussions it may also become apparent that the combined resources of all parties could be more efficiently used when working towards a common goal. Disagreements about how a system might work, or the impact of natural resource exploitation, are irrelevant at this stage of the process and should only be considered once an objective has been agreed upon. Once it has, then we believe decision-theoretic management frameworks are likely to be most useful, and adaptive resource management in particular as it provides a natural framework for formally incorporating for uncertainties in how systems work or any exploitation impact upon them. In addressing these issues we suggest that PAO and SR could both be useful descriptors of the system.

Proportion of area occupied (PAO)

Overview

The proportion of area occupied is defined as the fraction of landscape units within a region that are occupied by the target species (i.e. units where the species is present). Landscape units may be naturally (e.g. ponds or habitat fragments) or arbitrarily (e.g. grid cells or quadrats) defined. PAO is essentially a measure of species range or distribution within the studied region. Imperfect detection of the species will result in false absences, hence a naive estimate of PAO that does not account for imperfect detection will systematically underestimate the true PAO. To combat imperfect detection, appropriate data need to be collected in the field enabling imperfect detection to be estimated. One data source is to survey each unit multiple times within a relatively short time interval during which it is assumed the species is either always present or always absent from each unit. From those units where the species is detected at least once, the frequency with which the species is not detected provides the essential information enabling the number of units where the species was present but never detected (i.e. the number of false absences) to be estimated. This issue of imperfect detection has long been recognized by field biologists, who have conducted repeat surveys to minimize the likelihood of false absences, but generally the outcomes of the surveys were simply pooled into a single value to indicate overall 'presence' or 'absence', which is better described as detection or non-detection. These recently developed methods are simply exploiting the information contained in those repeat surveys more fully to obtain a value for PAO that is likely closer to the true value.

Estimated change in PAO over time can also be influenced by changes in detection probability in the same manner that relative abundance can, as outlined above. Furthermore, if interest is in local extinction and colonization probabilities, then imperfect detection will also bias those rates as well (Moilenan, 2002; MacKenzie *et al.*, 2006). Naive extinction rates will overestimate true extinction rates as some of the apparent extinctions will actually be caused by non-detection of the species in the second year rather than the species truly being absent. Colonization rates could be biased in either direction as now imperfect detection can be an issue in both the first and second years. Methods have also been developed to explicitly account for imperfect detection in these situations to estimate change in PAO, with a similar data requirement as above where each year (say) landscape units are surveyed repeatedly within a relatively short timeframe (MacKenzie *et al.*, 2003, 2006; Royle and Kéry, 2007).

One advantage of moving to a presence/absence-based measure is that individual animals do not necessarily have to be observed in order to register a 'detection' of the species. Hearing calls or vocalizations may be sufficient evidence to confirm that the target species is present at a location. Similarly, scat, tracks, or other animal signs may constitute the required piece of evidence provided that sufficient knowledge exists regarding the persistence of those signs or scat so as not to violate the appropriate closed population assumptions being applied to the surveyed patches. Some types of information can be collected remotely such as acoustic recording devices (see, e.g., Chapter 10) or camera traps (see, e.g., Chapter 3). Because there is a wide variety of potentially useful techniques for detecting the target species, there is a great deal of flexibility for field implementation of these methods.

Basic methods

To illustrate how these methods account for imperfect detection, here we briefly review the approach of MacKenzie *et al.* (2002; see also MacKenzie *et al.*, 2006), for estimating PAO at a single time point. Essentially both the biological (presence or absence) and sampling (detection or non-detection) processes are considered in order to develop an expression for the probability of observing a particular set of data. For example, suppose a landscape unit is surveyed three times and the following sequence of detections (1) and non-detections (0) is observed: 101. This detection history (h) indicates that:

1. the species must be present at the unit (as it was detected at least once);
2. it was detected in the first survey;
3. it was not detected in the second survey; and
4. it was detected in the third survey.

To develop an expression for the probability of observing the detection history 101, we need to define some quantities that represent the probabilities of presence and detection. Let ψ represent the probability of the species being present at a unit (i.e. the unit is occupied) and p_j be the probability of detecting the species in the jth survey of a unit if the species is present there (hence $1 - p$ is the probability of non-detection). If we substitute in the relevant quantity for each of the above four statements we obtain:

$$\Pr(h = 101) = \psi p_1 (1 - p_2) p_3$$

As it is known that each of those even must have happened, then the quantities are multiplied together.

Now consider the situation where a unit might be surveyed three times and the species never detected (i.e. $h = 000$). Admitting imperfect detection there are now two possible explanations for the data. The first is that this is in fact a false absence; the species was really present, but never detected, which implies the following statements:

1. the species was present at the unit;
2. it was not detected in the first survey;
3. it was not detected in the second survey; and
4. it was not detected in the third survey.

The resulting expression for this possibility is:

$$\Pr\left(h = 000 \text{ given species present}\right) = \psi\left(1 - p_1\right)\left(1 - p_2\right)\left(1 - p_3\right)$$

The second explanation is that the species is actually absent (i.e. not present). In this case, the expression is simply:

$$\Pr\left(h = 000 \text{ given species absent}\right) = (1 - \psi)$$

As the species is absent from the unit, there is no chance of detecting it hence this expression does not involve any detection probabilities. Because we cannot differentiate which of these two possibilities may be correct from the available data (i.e. we never saw the species either because it is a false or a true absence), the overall expression combines them by adding the two possibilities together. Hence:

$$\Pr(h = 000) = \psi\left(1 - p_1\right)\left(1 - p_2\right)\left(1 - p_3\right) + (1 - \psi)$$

Such an expression is developed for all units from which data were collected, which in combination can then be used to obtain estimates of these quantities. We can also include covariates or predictor variables for occupancy and detection probabilities to investigate whether they vary with habitat, elevation, or weather conditions, for example. Finally, there is no requirement that the same number of surveys are required for all units; there could be unequal sampling effort although it is methodologically more robust to set out to survey all units with equal effort until differences in relative detection probabilities between, for example, forest types, is established. For more details see MacKenzie et al. (2002, 2003, 2006). These methods have also been implemented in the free software program PRESENCE, which can be downloaded from http://www.mbr-pwrc.usgs.gov/software/presence.html.

Repeat surveys

The repeat surveys are basically required to provide a measure of search effort and frequency of detection at locations where the species is present. Therefore, given the level of search effort at a location where the species was never detected, what are the chances there was really something there? This repeat survey information could be collected in a number of different ways. Units could be visited multiple times over a relatively short timeframe with a single survey on each visit (e.g. surveyed nightly over a 1-week period). Multiple surveys could be conducted on a single visit either with a single observer (e.g. 3 × 5-minute point counts separated by 5-minute rest periods), or multiple observers (e.g., two observers independently survey for the target species). Spatial replicates could also be used with a single survey of each selected subunit (e.g. survey three 20 × 20 m plots within a 4 ha block). Here, each survey of a subunit is an opportunity to detect the species given the species is present within the larger unit.

Once one is convinced that detectability is a potential issue that needs to be addressed, a common question is "How many repeat surveys need to be conducted?" MacKenzie and Royle (2005) assessed this issue from the perspective of designing a study that either minimized the overall level of effort to achieve a specific standard error on PAO, or minimized the standard error for a fixed level of effort, and found that in either case there was an optimal number of repeat surveys that should be conducted dependent upon the true level of PAO and detectability (Table 11.1). Importantly, designs may actually be less efficient if fewer than the recommended number of surveys are conducted in order to have increased spatial replication. For example, suppose there was sufficient funding for 400 surveys in total and that PAO ≈ 0.4 and $p \approx 0.3$. From Table 11.1, the optimal number of surveys is 5, hence 80 sites could be surveyed and the expected standard error on PAO (from equation 1 of MacKenzie and Royle, 2005) would be 0.07. However, if the researchers decided to maximize spatial replication by going to 200 sites only twice, the expected standard error would be 0.11, approximately 60% greater than the 'optimal' design. To achieve a standard error of 0.07 with only two surveys per site, then total effort would have to be increased by 250%.

Interested readers are directed to MacKenzie *et al.* (2006) for in-depth discussion of study design issues such as the nature of repeat surveys, limiting potential observer biases, and other factors.

Key assumptions

There are four key assumptions required by the above methods, which if not met may result in biased results:

1. no species misidentification;
2. units are closed to changes in occupancy during repeated surveying;
3. no unmodelled additional variation in detection probabilities; and
4. observations are independent.

Table 11.1 Optimal number of repeat surveys per sampling unit for differing levels of proportion of area occupied (PAO; ψ) and detectability (p). Reproduced from MacKenzie and Royle (2005), with permission from John Wiley & Sons.

		\multicolumn{9}{c}{ψ}
		0·1	0·2	0·3	0·4	0·5	0·6	0·7	0·8	0·9
p	0·1	14	15	16	17	18	20	23	26	34
	0·2	7	7	8	8	9	10	11	13	16
	0·3	5	5	5	5	6	6	7	8	10
	0·4	3	4	4	4	4	5	5	6	7
	0·5	3	3	3	3	3	3	4	4	5
	0·6	2	2	2	2	3	3	3	3	4
	0·7	2	2	2	2	2	2	2	3	3
	0·8	2	2	2	2	2	2	2	2	2
	0·9	2	2	2	2	2	2	2	2	2

These assumptions are discussed at length by MacKenzie (2005) and MacKenzie et al. (2006), but are briefly covered here. An important point is that other methods that use similar data, but do not account for imperfect detection, are actually making the same set of assumptions, they are just rarely stated explicitly.

Species misidentification will result in PAO being overestimated. However, regardless of the estimation technique being applied (e.g. occupancy models vs simple logistic regression), species misidentification will create the same problem. Good field protocols and rigorous training of observers will obviously reduce the risk of misidentification, and if there is any uncertainty about which species has been detected in a survey, this should be recorded such that appropriate steps can be taken at the time of data analysis. From an estimation perspective, it is better to consider a dubious detection as a non-detection as we have methods that can account for false absences only, but not false absences and false presences.

It is presumed that the species is either always present, or always absent from a unit during the repeated surveys. This is required such that at those units where the target species has been detected at least once, any non-detection is known to be a genuine non-detection rather than a temporary absence of the species from the unit. However, provided that any changes in the true presence/absence of the species between surveys is random (i.e. the probability of the species being present at the time of the current survey is the same regardless of whether the species was present or absent at the time of the previous survey), the estimate of PAO should be interpreted as the proportion of area used by the species during the surveying period (where use is interpreted as the species being at that unit at some stage; MacKenzie, 2005).

Other non-random changes may cause bias in the estimate of PAO. There are three salient points that need to be considered with respect to 'closure'. First, closure is at the species level, not (necessarily) the individual level. Movement of individual animals does not violate the closure assumption provided that at least one individual of the species is always present. Second, if the species is being detected by sign surveys (tracks, scratching, scat, etc.), then closure applies to the sign rather than the actual organisms. Third, the vast majority of presence/absence studies implicitly make this closure assumption, regardless of whether they believe there is a detection problem or not. When a species is detected in a survey, in most applications it would be assumed that the detection is indicative of the species being present at that unit for some time period that exceeds that of the actual survey. For example, if a bird species is detected during a 5-minute point count, few would seriously interpret that as an indication of the species only being present at that point count for only that 5-minute period. As soon as one wants to interpret that a detection is indicative of the species being present outside of the actual survey, one is using both the concept of a sampling season and closure.

Unaccounted for variation in detection probabilities (detection heterogeneity) will cause PAO to be underestimated. Some forms of variation may be well explained by covariates, random effects, or with specific model structures to account for certain sources of variation (e.g. abundance), but additional variation can be problematic. There is likely never going to be a perfect analytic solution to the problem of detection heterogeneity; the best solution is to account for as many possible sources of variation through good study design and data collection protocols. An important point is that heterogeneity is going to be more problematic when the average overall probability of detecting the species (i.e. probability of detecting the species at least once during the repeat surveys of the unit) is lower, thus an appropriate design-based solution is to aim for relatively high probabilities of detection either on a per survey basis (e.g. by conducting 10-minute rather than 5-mintue surveys) or by increasing the number of repeat surveys. By designing studies to have relatively high overall probabilities of detection, the data will be more robust to natural heterogeneity and less reliance is placed on appropriately accounting for imperfect detection with statistical models when making inferences about occupancy. People who would rather avoid having to worry about accounting for imperfect detection, whether it be for estimating abundance, PAO, or species richness, often point out that any model-based approach assumes no unmodelled heterogeneity, an assumption that is almost surely violated, therefore the methods are unreliable and should not be used. However, if the same type of data are being used with another method that does not account for detection at all (e.g. logistic regression), detection heterogeneity is still going to be problematic; an assumption does not have to be explicitly stated in order to be violated. Hence, while it is acknowledged that unaccounted for detection heterogeneity will lead to biased estimates of PAO, those estimates are likely to be more reliable than estimates that do not account for imperfect detection at all.

Like many statistical methods, it is assumed that observations are independent. An interpretation of independence is that the outcome of one event does not influence the outcome of another event. For example, if the probability of detection in the first survey of a unit is 0.6 and due to disturbance the probability of detection is 0.3 on the second survey, because the probability of detection in the second does not actually depend on the outcome of the first survey (detection/non-detection), the two surveys are independent (i.e. regardless of whether or not the species was detected in the first survey, the disturbance has still occurred, thus reducing detectability). If, however, the probability of detection in the second survey was 0.3 if the species was detected in the first survey, but 0.6 if the species was not detected in the first survey, the surveys are not independent (e.g. detection probability in the second survey is different if the species was detected in the first survey – similar to a trap response in capture-recapture). A lack of independence of this type may create bias in the estimates of PAO, but it is possible to allow the probability of redetection to be different to the probability of first detection to account for it. Another situation where surveys are not independent is surveying two nearby units simultaneously and the same event (e.g. a bird call) is registered as a detection at both locations. Such instances will create a pair of detection histories that are very similar and is a form of overdispersion (i.e. more variation in the data than predicted by the model given the sample size). In this case, the actual estimate of PAO is likely to be unbiased, but the reported standard error is too small and should be inflated. A final situation where people may get unnecessarily concerned about independence is spatial correlation; values at nearby units are more similar than values from units further apart. Spatial correlation is likely to be a reality in many ecological situations; however, it is important to realize that the level of spatial correlation depends on the scale at which the landscape is being assessed (e.g. grid cell size), and it may also be well explained by covariates or predictor variables (e.g. habitat type or elevation). While accounting for spatial correlation may be useful for unit-level predictions about the presence or absence of a species (e.g. for creating distributional maps), not accounting for it will not invalidate estimates of PAO using these methods provided the units are selected for surveying independently (e.g. randomly), even though the underlying value may be correlated. For example, suppose two grid cells are randomly selected. The target species is either present or absent from those grid cells and this has been determined prior to their selection. Therefore, the outcome from surveying one grid cell cannot influence the outcome at the other because it has already been predetermined, hence the independence assumption is actually being met. However, if other sampling strategies are used where landscape units are not selected independently of one another, such as systematic or adaptive sampling, spatial correlation may introduce some bias. Hence random sampling, while often frowned upon by field workers, increases the robustness to estimation procedures. There are some methods currently available that could be used to account for spatial correlation if necessary (e.g. Magoun *et al.*, 2007), and others are currently being developed.

Species richness

Species richness may be defined as the number of species (possibly restricted to specific taxa or other groups of species) present at a location or within a region. Certain values may also be assigned to each species hence the overall value of the community is of prime concern rather than overall richness (Yoccoz *et al.*, 2001). However, not all species within a region are likely to be detected during the surveys, therefore overall richness (or value) must be estimated. Mark-recapture (Nichols and Conroy, 1996; Boulinier *et al.*, 1998; Nichols *et al.*, 1998; Cam *et al.*, 2000) and coverage estimators (Chao and Lee, 1992) have both been used to estimate species richness previously, although the above occupancy models can also be used with certain advantages (MacKenzie *et al.*, 2006). Mark-recapture and coverage estimators work in a similar manner where the frequency of observations of species detected at least once is used to estimate the number of species that may have been in the area but were never detected. There is no specification of which species may have been present, but unseen, just how many. The main difference with using occupancy models is that a list of potential species of interest must be defined prior to analysis, preferably prior to data collection. There are two main advantages to defining a species list: (i) it can address questions about particular species that were on the list but never detected; and (ii) covariate information about all species on the list (e.g. size, coloration, calling behaviour, microhabitat preference) can be included in the analysis regardless of whether they were detected or not. Covariate information could also be included using mark-recapture techniques (i.e. Huggins, 1991), but only for those species that were detected.

If species richness is of interest at just a single location (or small number of locations), the occupancy models described above can be applied without modification, but instead of sampling landscape units we are sampling species, hence a detection history is at the scale of a single species rather than individual units. Repeat surveys are again required to account for imperfect detection and these may be either temporal or spatial (e.g. surveying five consecutive nights or five different plots within a study site).

If operating over a larger area, with data being collected at a greater number of locations (e.g. 20+ locations), another manner in which occupancy models could be applied to estimate species richness is to essentially fit an occupancy model to each of the species of interest. Species richness can then be calculated at both an individual unit level and also for the greater area (Dorazio *et al.*, 2006).

These developments are relatively recent and interested readers are directed to chapter 9 of MacKenzie *et al.* (2006) for further details and ideas.

Conservation management of red slender loris in Sri Lanka

Sri Lankan management of biodiversity has relied on the designation of protected areas and legislation designed to minimize direct exploitation of biological resources; however, as pressure from anthropogenic drivers increases, management intervention and an understanding of biodiversity processes will be essential to achieve security for the remaining biological resources. A conservation initiative for the red slender loris has functioned as a pilot programme for researchers and conservationists in Sri Lanka, with the intention of developing the strategic and technical capacity to design informed management for security of the species. As stated above, their threat status and the growing pressures on remaining forests means that action is required to identify status of the species across its range, and to consider that status against the distinctive forest ecotypes and the management regimes under which they persist. By conducting such an investigation it is intended that the threat classification of the species can be reviewed and refined, management units identified with regard to preferred/optimal habitat, and most importantly, current management practices through forest legal designation can be assessed for their loris conservation efficacy.

Given the practical difficulties in trying to estimate abundance for the red slender loris across any meaningful spatial scale, PAO has been suggested as a method to quantify the current distribution of red slender loris across southwestern Sri Lanka. Recently, a study has been implemented to trial field methodologies and to establish the current distribution.

The intent is to identify forest stands across southwest Sri Lanka based upon available maps and field knowledge. Forest stands are defined as contiguous patches of primary, secondary, or modified forest (i.e. home gardens), confined within a single ecoregion. Hence, large contiguous forest patches that span multiple ecoregions (e.g. due to large within-patch changes in elevation) are regarded as multiple stands. Stands will be randomly selected and a 2-km transect established within the boundaries of the stand. The transect is to be searched with 2-hour morning and evening surveys during a week, with 12 surveys in total. In the absence of prior knowledge of detection probability of individual loris in the various forest systems, search effort was defined as that necessary to detect loris at the lowest frequencies in previous transect search studies (R. Gamage, unpublished data). Surveys are conducted by a pair of observers, each searching one side of the transect line (alternating sides every 30 minutes). Loris are extremely cryptic but easily detected by eyeshine, hence the sub-canopy is searched using headlamps with dimmed and red-filtered LEDs.

This study is still gathering data but it is noteworthy that in the space of 4 months of discussing the conservation and research needs with a team of biologists, conservationists, and field workers, none of whom were familiar with the concepts of occupancy analysis, nor had benefited from advanced statistical training, a dynamic and informed programme had been established that was able to generate results from preliminary analyses after only 10 weeks in the field. Such an achievement would not be easy using classical abundance or density monitoring methods, and neither would the potential results be as directly interpretable for systems-level management decision-making.

References

Anderson, D.R. (2001) The need to get the basics right in wildlife field studies. *Wildlife Society Bulletin*, 29, 1294–1297.

Allendorf, F.W. and Luikart, G.H. (2006) *Conservation and the Genetics of Populations*. Blackwell Publishing, London.

Boulinier, T., Nichols, J.D., Sauer, J.R., Hines, J.E., and Pollock, K.H. (1998) Estimating species richness: the importance of heterogeneity in species detectability. *Ecology*, 79, 1018–1028.

Buckland, S.T., Anderson, D.R., Burnham, K.P., and Laake, J.L. (1993) *Distance Sampling: Estimating Abundance of Biological Populations*. Chapman & Hall, London.

Bunge, J. and Fitzpatrick, M. (1993) Estimating the number of species: A review. *Journal of the American Statistical Association*, 88, 364–373.

Burnham, K.P. and Overton, W.S. (1979) Robust estimation of population size when capture probabilities vary among animals. *Ecology*, 62, 927–936.

Cam, E., Nichols, J.D., Sauer, J.R., Hines, J.E., and Flather, C.H. (2000) Relative species richness and community completedness: avian communities and urbanization in the mid-Atlantic states. *Ecological Applications*, 10, 1196–1210.

Chao, A. and Lee, S.M. (1992) Estimating the number of classes via sample coverage. *Journal of the American Statistical Association*, 87, 210–217.

Cincotta, R.P., Wisnewski, J., and Engleman, R. (2000) Human population in the biodiversity hotspots. *Nature*, 404, 990–992.

Dorazio, R.M., Royle, J.A., Soderstrom, B., and Glimskar, A. (2006) Estimating species richness and accumulation by modeling species occurrence and detectability. *Ecology*, 87, 842–854.

Green, M.J.B., How, R., Padmalal, U.K.G.K., and Dissanayake, S.R.B. (2009) The importance of monitoring biological diversity and its application in Sri Lanka. *Tropical Ecology*, 50, 41–56.

Gunawardene N.R., Dulip Daniels, A.E., Gunatilleke, I.A.U.N., *et al.* (2007) A brief overview of the Western Ghats – Sri Lanka biodiversity hotspot. *Current Science*, 93, 1567–1572.

Huggins, R.M. (1991) Some practical aspects of conditional likelihood approach to capture experiments. *Biometrics*, 47, 725–732.

MacKenzie, D.I. (2005) Was it there? Dealing with imperfect detection for species presence/absence data. *Australian and New Zealand Journal of Statistics*, 47, 65–74.

MacKenzie, D.I. (2009) Getting the biggest bang for our conservation buck. *Trends in Ecology and Evolution*, 42, 175–177.

MacKenzie, D.I. and Kendall, W.L. (2002) How should detection probability be incorporated into estimates of relative abundance. *Ecology*, 83, 2387–2393.

MacKenzie, D.I. and Royle, J.A. (2005) Designing occupancy studies: general advice and allocation of survey effort. *Journal of Applied Ecology*, 42, 1105–1114.

MacKenzie, D.I., Nichols, J.D., Lachman, G.B., Droege, S., Royle, J.A., and Langtimm, C.A. (2002) Estimating site occupancy rates when detection probabilities are less than one. *Ecology*, 83, 2248–2255.

MacKenzie, D.I., Nichols, J.D., Hines, J.E., Knutson, M.G., and Franklin, A.B. (2003) Estimating site occupancy, colonization, and local extinction probabilities when a species is detected imperfectly. *Ecology*, 84, 2200–2207.

MacKenzie, D.I., Nichols, J.D., Royle, J.A., Pollock, K.H., Bailey, L.A., and Hines, J.E. (2006) *Occupancy Modeling and Estimation*. Elsevier, San Diego, CA.

Magoun, A.J., Ray, J.C., Johnson, D.S., Valkenburg, P., Dawson, F.N., and Bowman, J. (2007) Modeling wolverine occurrence using aerial surveys of tracks in snow. *Journal of Wildlife Management*, 71, 2221–2229.

Martin, J., Runge, M.C., Nichols, J.D., Lubow, B.C., and Kendall, W.L. (2009) Structured decision making as a conceptual framework to identify thresholds for conservation and management. *Ecological Applications*, 19, 1079–1090.

Mills, L.S. (2006) *Conservation of Wildlife Populations: Demography, Genetics and Management*. Blackwell Publishing, London.

Mittermeier, R.A., Myers, N., Gil, P.R., and Mittermeier, C.G. (2007) *Hotspots: Earth's Biologically Richest and Most Endangered Terrestrial Ecoregions*. Cemex, Conservation International and Agrupacion Sierra Madre, Monterrey, Mexico.

Moilanen, A. (2002) Implications of empirical data quality for metapopulation model parameter estimation and application. *Oikos*, 96, 516–530.

Myers, N., Mittermeier, R.A., Mittermeier, C.G., da Fonseca, G.A.B., and Kent, J. (2000) Biodiversity hotspots for conservation priorities. *Nature*, 403, 853–858.

Nekaris, K.A.I. (2003) Rediscovery of the slender loris in Horton Plains National Park, Sri Lanka. *Asian Primates*, 8, 1–7.

Nekaris, K.A.I., Blackham, G.V., and Nijman, V. (2008) Conservation implications of low encounter rates of five nocturnal primate species (*Nycticebus* spp.) in Asia. *Biodiversity and Conservation*, 17, 733–747.

Nichols, J.D. and Conroy, M.J. (1996) Estimation of species richness. In: Wilson, D.E., Cole, F.R., Nichols, J.D., Rudan, R., and Foster, M. (eds), *Measuring and Monitoring Biological Diversity: Standard Methods for Mammals*. Smithsonian Institution Press, Washington, DC.

Nichols, J.D. and Williams, B.K. (2006) Monitoring for conservation. *Trends in Ecology and Environment*, 21, 668–673.

Nichols, J.D., Boulinier, T., Hines, J.E., Pollock, K.H., and Sauer, J.R. (1998) Estimating rate of local extinction. colonization and turnover in animal communities. *Ecological Applications*, 8, 1213–1225.

Nichols, J.D., Hines, J.E., Sauer, J.R., Fallon, F.W., Fallon, J.E., and Heglund, P.J. (2000) A double-observer approach for estimating detection probability and abundance from counts. *Auk*, 117, 393–408.

Otis, D.L., Burnham, K.P., White, G.C., and Anderson, D.R. (1978) Statistical inference from capture data on closed animal populations. *Wildlife Monographs*, 62.

Pearce, D.W. and Moran, D. (1995) *The Economic Value of Biodiversity*. IUCN – The World Conservation Union.

Royle, J.A. and Kéry, M. (2007) A Bayesian state-space formulation of dynamic occupancy models. *Ecology*, 88, 1813–1823.

Skalski, J.R., Robson, D.S., and Simmons. M.A. (1983) Comparative census procedures using single mark-recapture methods. *Ecology*, 64, 752–760.

Thompson, W.L., White, G.C., and Gowan, C. (1998) *Monitoring Vertebrate Populations*. Academic Press, San Diego, CA.

Tyre, A.J., Tenhumberg, B., Field, S.A., Niejalke, D., Parris, K., and Possingham, H.P. (2003) Improving precision and reducing bias in biological surveys by estimating false negative error rates in presence–absence data. *Ecological Applications*, 13, 1790–1801.

Werner, W.L. (1984) *Die Höhen- und Nebelwälder auf der Insel Ceylon (Sri Lanka). Tropische und subtropische Pflanzenwelt 46*. Steiner, Wiesbaden.

Williams, B.K., Nichols, J.D., and Conroy, M.J. (2002) *Analysis and Management of Animal Populations*. Academic Press, San Diego, CA.

Wintle, B.A., McCarthy, M.A., Parris, K.M., and Burgman, M.A. (2004) Precision and bias of methods for estimating point survey detection probabilities. *Ecological Applications*, 14, 703–712.

Yoccoz, N.G., Nichols, J.D., and Boulinier, T. (2001) Monitoring of biological diversity in space and time. *Trends in Ecology and Evolution*, 16, 446–453.

12

Monitoring and Evaluating the Socioeconomic Impacts of Conservation Projects on Local Communities

Katherine Homewood

University College London, London, UK

Introduction

Since the late nineteenth century, the response to rapidly increasing biodiversity loss has been the establishment of protected areas (PAs), which now form the core of most national or regional biodiversity conservation strategies, with currently more than 130 000 designated PAs worldwide (Chape *et al.*, 2008; Butchart *et al.*, 2010). There is no doubt that PAs have played a key role in protecting biodiversity across the globe. PAs are, however, associated with major social and economic challenges: depending on their level of protection, PAs by definition have strict rules that exclude human activities (Salafsky and Wollenberg, 2000). Their establishment and management have significant and generally negative effects on local communities, through, for example, economic opportunity costs, population displacement and the associated risk of impoverishment, alongside increased human–wildlife conflicts (West *et al.*, 2006; Adams and Hutton, 2007).

As an alternative to 'fortress conservation' (Brockington, 2002), approaches aspiring to combine socioeconomic development with conservation goals have proliferated (Hughes and Flintan, 2001; Roe *et al.*, 2009). While emanating from a pragmatic desire for conservation to be sustainable (ecologically, economically, and socially), they have been driven by the conviction that biodiversity conservation and poverty alleviation are linked aims (CBD, 2010). The interaction between the two is complex (Fischer and Christopher, 2007; Wittemeyer *et al.*, 2008; Igoe *et al.*, 2008; CBD, 2010). Following

---

Biodiversity Monitoring and Conservation: Bridging the Gap between Global Commitment and Local Action, First Edition. Edited by Ben Collen, Nathalie Pettorelli, Jonathan E.M. Baillie and Sarah M. Durant.
© 2013 John Wiley & Sons, Ltd. Published 2013 by John Wiley & Sons, Ltd.

a recent typology (Adams *et al.*, 2004), some initiatives treat conservation and development as separate policy realms that should not overlap (Gartlan, 1997; Oates, 1999); others aim to enhance socioeconomic conditions of communities surrounding areas of conservation interest, as key to achieving effective biodiversity conservation (Rands *et al.*, 2010); a third group pursue environmental conservation as being in itself key to improving socioeconomic conditions (Millennium Ecosystem Assessment, 2005), and a final group of initiatives see both goals as equal moral imperatives that are inextricably intertwined and to be pursued hand in hand (Brechin *et al.*, 2002). Apart from the first group, these approaches generally aim for a mix of conservation and development objectives and seek to provide appropriate development opportunities, emphasizing local community involvement, adopting shared management, ensuring local autonomy, guaranteeing rights to harvest, promoting knowledge, awarding cash compensation, and encouraging tourism (Brooks *et al.*, 2006). In evaluating social and economic impacts, a first framework of reference must be what the project sets out to do initially, and whether the initiative can legitimately claim to be working towards community development as well as conservation, and with what relative emphasis.

Community-based conservation (CBC) emerged during the 1970s in individual sites such as Amboseli in Kenya (Western, 1994) and was first widely advocated at the 1982 World National Parks Congress with the promotion of the biosphere reserves model. It seeks to protect large areas of land by encouraging local stewardship and integrating social and environmental priorities. This people-oriented approach has acquired the slightly misleading[1] name of 'community-based' approaches, and is applied in a wide range of natural resource management contexts including community-based wildlife management (CWM); community-based fisheries management (CFM); community-based forest management (CBFM); community-based natural resource management (CBNRM); and community-based ecotourism. All aspire or at least claim to be participatory, community-led, collaborative, joint and/or popular forms of natural resource management, and all at some level aim for triple objectives of poverty reduction, natural resource conservation, and good governance, through collective management of ecosystems, empowering communities to manage their own resources sustainably (e.g. Rotha, 2005; Fabricius and Collins, 2007; Roe *et al.*, 2009). Integrated conservation and development projects (ICDPs) emerged as an early and particular type of initiatives promoting the flow of conservation-derived benefits to PA-adjacent dwellers in order to complement strategies for strengthening law enforcement inside PAs (Gubbi *et al.*, 2009). However, the meaning of the term has shifted through time and ICDPs are perhaps simply best seen as being on the spectrum of CBC initiatives,

---

[1] Misleading in that these initiatives are almost invariably conceived and driven at least initially by agencies and interests external to local communities.

with all the site- and case-specific flexibility of meanings that this entails (Hughes and Flintan, 2001).

The social and economic impacts of conservation projects on local communities in the developing world have been the subject of comprehensive reviews (West *et al.*, 2006; Adams and Hutton, 2007), form a central dimension within current systematic reviews (Bowler *et al.*, 2010; Waylen *et al.*, 2010) and have triggered ongoing polarized debate both around PA/fortress conservation (Brockington, 2003; Schmidt-Soltau and Brockington, 2004; Curran *et al.*, 2009) and community-based initiatives (Blaikie, 2006; Brockington, 2007; Roe *et al.*, 2009). Rather than revisiting the qualitative detail of such impacts, I focus on three specific questions central to socioeconomic monitoring for conservation initiatives. These are:

- Why carry out such monitoring?
- What are the main dimensions to be monitored?
- What methods are available, and what are the main pitfalls in monitoring social and economic impacts?

In addressing these questions, I draw not only on the experience of past conservation initiatives, but on the now considerable experience of social and economic monitoring associated with much wider development and resource use contexts (see, e.g., Cernea and McDowell, 2000; Goldman, 2000). The conservation literature suggests much creative effort goes into inventing new instruments and indicators (Schreckenberg *et al.*, 2010; Malleret-King, 2000; Wilder and Walpole, 2008; BirdLife International, 2011). Most are variants of the tried and tested basic set: key informant (KI), semi-structured interview, focus group, and household survey (Russell Bernard, 2006). I exclude from this chapter methods focusing on evaluating the *conservation* outcomes of interventions addressing livelihoods and institutions, though these are undergoing a parallel evolution (Kapos *et al.*, 2008). The current explosion of individual approaches to evaluating the social and economic impacts of conservation carries the risk not only of rediscovering the wheel but, more seriously, of falling prey to known pitfalls. Conservation workers are keen to adopt new tools as the most up-to-date methods for assessing impacts, but are rarely sufficiently aware of the central issue of data validity and how easily this may be compromised in social science research. Briefly, in highly political situations involving significant shifts in who wins or loses access to and control of resources, it may be easy to record what people say, but extremely difficult to get at what people really think and do (and why). It is therefore not so much the tools themselves, but rather the qualitative subtleties of the ways in which those tools are applied, which determine whether the data that result are both meaningful and useful in understanding impact. Those seeking to monitor the social and economic impacts of conservation interventions do well to learn from and build upon established social impact assessment experience.

Why monitor the socioeconomic impacts of conservation projects?

There are several reasons to monitor the social and economic impacts of conservation initiatives, as considered below.

Pragmatic

It is self-evident that conservation is likely to work better where it has successfully enlisted the support of local people and other stakeholders. Yet we are all familiar with conservation initiatives that look good on paper and work significantly less well in practice, in ways that suggest local social, political, and economic realities have got in the way of the best-laid plans. Conservation workers may have genuinely tried to ensure that benefits flow to the community and that those benefits are somehow commensurate with the costs of lost access to and use of the resources being conserved. The conservation initiative may well have tried to understand local people's priorities and problems, to engage their participation, to address and resolve tensions and conflicts, but something has not worked: ultimately ecological as well as economic outcomes are disappointing, and tensions and conflicts abound. Properly conceived and implemented, social and economic monitoring can both engage local people and also inform adaptive management in ways that will address or circumvent many of these problems. Where it is not only well designed but also actively mainstreamed into the ongoing, longer-term structure and delivery of the intervention, socioeconomic monitoring can become the basis for robust, participatory adaptive management institutions able to identify and deal with problems as they arise, and before serious damage is done.

Legal

In Western and industrialized economies, legislation imposes not only the requirement to carry out social impact assessment for any major intervention (conservation or other), but also the need for that impact assessment to conform to certain standards. Even primarily profit-making financial institutions such as the World Bank and the Organization for Economic Co-operation and Development (OECD) have increasingly rigorous guidelines on social impact assessment: some are widely cited in the conservation literature (see, e.g., Cernea and McDowell, 2000). Increasingly, developing countries have comparable legislation, and though its implementation may be poorly resourced and even more poorly regulated, even minimal standards of social responsibility on the part of conservation agencies should ensure that projects and workers observe these requirements, hopefully to a standard that they would expect in their often Western base countries.

Increasingly, reporting on the socioeconomic impacts of conservation projects has also become a legal financial accounting prerequisite for the funding scheme supporting such projects. Reducing Emissions from Deforestation and Degradation plus Poverty alleviation (REDD+) is rapidly becoming a salient case in point. It is set out in more detail here as an example of the growing legal and financial importance of social and economic monitoring, associated with a global environmental conservation policy translating into numerous local projects. Deforestation and forest degradation lead to biodiversity reduction, loss of ecosystem services, and increased emission of greenhouse gases (GHG; IPCC, 2007). Reducing Emissions from Deforestation and forest Degradation (REDD) can impact biodiversity loss through (i) reducing habitat loss, fragmentation, and degradation; and (ii) limiting the negative impacts of climate change on biodiversity (Campbell *et al.*, 2009). A range of mechanisms, funds, and initiatives have been developed to promote reductions in GHG emissions at various scales (i.e. international, country, and project levels) through REDD-relevant activities (Forest Investment Review, 2009). Such mechanisms, funds, and initiatives are linked to associated criteria and/or recommendations that need to be met by projects for them to continue being funded (or for them to be able to sell their carbon offsets). REDD+ addresses social as well as environmental sustainability, particularly poverty alleviation, and thus several of these criteria/recommendations are directly linked to the socioeconomic impacts of the project, promoting sustainable development and rural livelihood improvement (see, e.g., Plan Vivo, 2008; CCB, 2011). REDD/REDD+ is widely hailed by the climate change community as one of the few – if not the only – emerging mechanism that could genuinely address mitigation of climate change processes. Yet because it involves negotiations between international- and national-level agencies, and because there is often doubt about levels of transparency and accountability between national and local levels over payments for such services (Burnham, 2011), it will be doubly important not only to monitor social and economic impacts on local communities but also to ensure meaningful data are collected in ways that truly represent experienced change.

Ethical accountability

Most conservation workers would subscribe to at least a minimum ethic of 'do no harm' as far as their environmental practices are concerned. Increasingly, conservation agencies and projects aspire to work with communities for socially as well as environmentally sustainable conservation, and present mission statements that enshrine concern for community welfare and development. Even if community welfare is not the primary consideration of a conservation initiative, this awareness and its proclamation in the conservation literature (not least its widespread use on conservation websites as part of their fundraising and public profile) entails an inescapable ethical responsibility towards the communities involved. At minimum, that responsibility is to do no harm, and requires that outcomes be monitored. Accountability requires that any claims to achieving positive outcomes for communities – again increasingly common in the conservation literature and on websites – be backed up by valid data. Any intervention will have winners and losers: only monitoring can make it possible to improve systems so as to move towards achieving better results for any who have been initially disadvantaged by the conservation process.

What to monitor?

This section looks first at the broad framing of monitoring studies. It starts with the often overlooked issue of study context, and of potentially confounding external factors driving change. It goes on to outline the broad set of dimensions that need to be considered. Past monitoring attempts have tended to focus on economic outcomes, but these represent only one of several crucial dimensions in social and economic monitoring. While there are many ways to categorize impacts (see, e.g., Goldman, 2000) and detail is case-specific, the sustainable livelihoods framework (Carney, 1998, 1999; Ashley and Carney, 1999) provides one possible approach to keeping the broad range of significant dimensions in view. This section goes on to look at the basic methods available to collect such data, and the approaches needed to ensure they are both representative and of reliable quality.

Background to monitoring: context and confounding factors, scale and balance

Experience shows that in designing social impact assessment it is necessary to consider from the very start (i) the wider context and potential confounding factors driving observed change in the population affected by the conservation initiative; (ii) the scale (individual, household, community . . .) at which impacts may be felt; and (iii) robust ways to ensure 'expert' scientific analysis is balanced with local knowledge in the formulation of monitoring (and in ongoing management).

External factors and the wider context

The conservation project will not be the only driver of change, and its effects must be distinguished from those of other factors. For example, in a recent study of bushmeat hunting in the Democratic Republic of Congo, logging initiatives had vastly greater economic impacts on behaviour and economic welfare than did the long-established ICDP and associated livelihoods projects being evaluated (Riddell, 2010). Such contextual factors may operate at any level from local micro-conditions through to national and international circumstances. For example, a period of global economic recession, political instability affecting tourism, an extreme climatic event, or sudden changes in global market demand for a particular locally sourced non-timber forest product will all affect the economic outcomes being measured as much as will the local start-up or withdrawal of particular non-conservation enterprises. However, they may have no direct relation to the intervention being evaluated and it is essential to tease apart relevant change from that caused by unrelated contextual factors.

It is also important to consider external factors and context when making comparisons between areas with and those without a particular conservation intervention. Such comparative studies are often taken as providing the next best thing to a controlled experiment comparing situations with/without intervention (Chatre and Agrawal, 2009). But without detailed baseline data on context, studies may be comparing, in terms of outcomes with/without intervention, two categories of areas that were different from the outset. Any differences between the two are then potentially interpreted as products of presence/absence of the conservation intervention, when in reality they may be products of pre-existing differences.

It is rarely possible in the context of social and economic monitoring of conservation initiatives to set up a truly balanced before/after, control/intervention comparison (Bowler *et al.*, 2010). Nonetheless any assessment of social and economic impact must establish a study design that gives due weight to context and that controls for/distinguishes the effects of external factors. This can only be done through in-depth familiarity with the area – gained through, for example, ethnographies, grey literature from development and other reports, media commentary, discussion with informed individuals from as wide a range of perspectives as possible – local, national, and international, before designing any monitoring.

Scale at which impacts may be felt

Individual, household, and community levels may be impacted differently. For example, in Tanzanian Wildlife Management Areas, some revenues were paid to village governments and at least potentially supported community-level enterprises such as schools or transport infrastructure (Nelson, 2007; Nelson *et al.*, 2007, 2009). However, no revenues were paid to or detectable at individual or household levels, though interviewees made it clear that they valued benefits at these levels more highly than they valued those accruing to the wider community (Nelson *et al.*, 2009; Homewood *et al.*, 2009). Documenting the impacts on different levels is likely to form an important dimension in understanding and ultimately revealing avenues for moderating differential impacts and the tensions and conflicts these may drive. It is important to collect these data in as disaggregated a form as possible to make such differentiated analyses possible.

Balancing 'expert' analysis with local knowledge

Conservation interventions – even 'community-based' ones – have tended to be conceived in a Western and/or international forum before being transplanted into less developed country (LDC) contexts, where they may be alien to local concerns and priorities. Their success may be correspondingly limited. This has often happened despite the now standard 'participatory' consultation exercises, which are intended to give the opportunity for local knowledge to be integrated into the intervention, and for local people to engage with decision-making, but which may often fail to achieve the desired result (see below).

In the section that follows, local environmental knowledge and other forms of local knowledge (political, social) are seen as making up a comprehensive knowledge system or world view that is commonly dominated and to some extent displaced by the knowledge systems of powerful outsiders – with Western scientific knowledge being only one dimension of the incoming, overruling world view. Local knowledge cannot be neatly partitioned into 'useful' objective environmental insights and 'dispensable' subjective knowledge superseded by superior Western understanding. 'Participatory' consultations are commonly conceived and implemented by people situated firmly within the incoming, overriding knowledge system, with its own subjectivities interwoven with objectively demonstrable known truths, and this may make it difficult for them to connect with local environmental knowledge or indeed engage effectively with local political understanding and support.

Historically there has been a widespread assumption that scientific expertise trumps local political and ecological knowledge. But few would question that conservation interventions in industrialized nations demand early and in-depth engagement and consultation with an informed public, and that this local knowledge can influence scientific understanding of the workings of the system, and help develop more appropriate management applications. Increasingly it is recognized that conservation interventions in LDC contexts are no different. Without succumbing to the myth of the 'noble savage' living in harmony with the environment (Alvard, 1993; Smith and Wishnie, 2000; Hames, 2007), examples have proliferated to show that western scientific models have often been less than robust, and less effective, where they have ignored local knowledge as a valid basis for understanding and managing LDC ecosystem dynamics (Laris and Klepeis, 2006); examples from the literature include: pastoralist use of arid and semi-arid rangelands (Homewood, 2008, Roba and Oba, 2009); fire ecology in Australia (Bradstock *et al.*, 2002, Dennis, 2003, Bird *et al.*, 2008) and in West Africa (Wardell *et al.*, 2004); forest dynamics in Guinean savanna (Fairhead and Leach, 1996); and soil fertility management in West Africa.

Not only environmental understanding, but also other aspects of local knowledge are key to managing local systems, both in terms of establishing effective communication between the different stakeholders, and in terms of establishing sustainable decision-making and management structures. From the most industrialized Western context to the most rural and marginalized developing country populations, long experience with social impact analysis shows that interventions overly dominated by either scientific and technical 'rationality' or indeed by political ideology, are less likely to work in the long run than those developed through balanced interaction between political and scientific experts challenged and moderated by ongoing discussion with an informed public able to express local knowledge on different dimensions of the project and apply it in evaluating likely outcomes (Goldman, 2000). This applies to the design of monitoring plans as much as to the intervention itself. Local knowledge, perceptions, priorities, and representatives all need to be central to the design and implementation of monitoring, fostering meaningful engagement and participation, giving legitimacy to the results, and ensuring buy-in to future developments (Brechin et al., 2002).

This understanding has begun to be expressed in an increasingly sophisticated grasp of (i) what is meaningful as opposed to widely observed token participation (Guijt, 1999; Paudel, 2005; Lund et al., 2009); (ii) of the institutions necessary to establish meaningful participation and the markers that these are working successfully (Lund et al., 2009); and (iii) of the political and sociocultural implications of conservation agencies' and workers' choices as to which local groups to engage as participating partners (Ribot, 2009), building or conversely undermining local grassroots democracy in ways that are critical to fostering or corroding longer-term success.

Early and in-depth local engagement defining local priorities for social and economic monitoring, *prior to* the design of any monitoring data collection, is crucially important (Goldman, 2000; Browne-Nunez and Jonker, 2008; Drury et al., 2011). In remote rural LDC areas with little effective official administration, social impacts monitoring bodies established with strong local representation through such engagement, as well as government alongside conservation agency representatives, can become effective management institutions, well placed to identify and respond in a timely way to potentially problematic project impacts, and to develop solutions with wider legitimacy and buy-in. Where meaningful participation is not built into planning monitoring and project management in ways locally accepted as legitimate, conservation workers will find themselves caught up with *post facto* problem-solving, dealing with protest and advocacy, conflict and division, as opposed to equitable and sustainable development outcomes (Goldman, 2000; Sullivan, 2003).

Dimensions to be considered in social and economic impact monitoring

Conservation initiatives increasingly document economic dimensions of project impacts. Potential economic benefits include the extent to which the affected population has received opportunities for employment, alternative livelihoods (substituting for activities discouraged by the intervention), and new income streams. Less commonly, but increasingly, projects look at the flipside to benefits: the costs of conservation restrictions – lost production through restrictions on grazing, gathering, or hunting. Relatively rarely have projects considered the implications of displacing residence and loss of mobility or through-passage.

In the wider context of development-induced displacement, Cernea and McDowell (2000), working for the World Bank, identified as central costs: landlessness; homelessness, joblessness, risk of marginalization, risk of food insecurity, risk of increased morbidity and mortality, risk of exclusion from common property resources, and risk of social disarticulation (Cernea, 2000). These are reiterated elsewhere (Adams and Hutton, 2007). World Bank guidelines now make clear that restrictions on access to resources should be considered a form of displacement. More recently Cernea discussed the application of these criteria in the case of conservation-induced displacement (Cernea and Schmidt-Soltau, 2003), triggering a strongly polarized debate over the scale and importance of conservation-induced displacement (Brockington and Igoe, 2006; Curran *et al.*, 2009).

Comparable dimensions emerge across much of the social impact analysis literature. Goldman (2000), reviewing social impact analyses of development interventions in a variety of first-world as well as LDC settings, cites changes in population, employment, access, land use, displacement/relocation, community organization and amenities, health and safety, leisure and recreation, and multiple dimensions of cultural and psychological as well as social and economic disruption (loss of identity, valued spiritual and aesthetic aspects) as all having emerged in individual case studies as being impacts of central importance.

Other approaches such as the Sustainable Livelihoods Assets framework (SLA; Carney, 1998; Scoones, 1998) allow comparable mapping of changes in broad context and along essential dimensions (Figure 12.1). The five categories of assets can be summarized as:

- *Natural* – access to natural resources as the basis of production and consumption.
- *Physical* – changes in infrastructure, transport, communications, etc.
- *Financial* – financial and in-kind income and assets also known as stocks and flows.
- *Social* – change in social networks, social cohesion or conversely tension/conflict.
- *Human* – population composition, educational attainment, health, food security, living standards.

Alongside these assets the framework flags up contexts, trends, transforming processes, institutional processes, and organizations, making the livelihoods approach a potentially comprehensive framework guiding consideration of what to monitor, and how to interpret the findings.

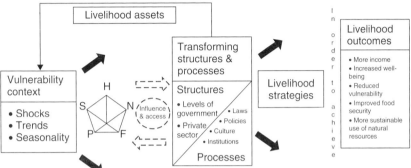

Figure 12.1 Sustainable livelihoods framework. The sustainable livelihoods assets framework (SLF) was developed as a way of making clear the complexity and multidimensional nature of poverty, going well beyond the limitations of the datum line dollar-per-person-per-day approach. SLF has become a central, if complex, tool in understanding the livelihoods of poor people and the ways in which different factors impact on those livelihoods. The five categories of assets can be summarized as: (i) *natural* (access to natural resources as the basis of production and consumption); (ii) *physical* (infrastructure, transport, communications, etc.); (iii) *financial* (financial and in-kind income and assets, also known as stocks and flows); (iv) *social* (social networks, patterns of social cohesion or conversely tension/conflict); and (v) *human* (population composition, educational attainment, health, food security, living standards). The dimensions of poverty – and therefore the dimensions on which interventions may impact – are comprehensively captured in these 'assets'. Alongside assets the SLF flags up context and trends (e.g. climate change driving increased frequency of extreme events); and transforming structures and processes (such as enabling or restrictive policies and market conditions, institutional processes fostering or constraining participation). Context, and transforming structures and processes effect changes in those assets and in the livelihoods, activities, and strategies of people characterized by those assets. The SLF thus represents a potentially comprehensive framework guiding consideration of what to monitor, and how to interpret the findings. Adapted from Ashley and Carney (1999).

The SLA framework has informed a wide range of rapid participatory methodologies (see Schreckenberg *et al.*, 2010, for a comprehensive catalogue). However, there are problematic trade-offs between the comprehensive nature of the framework and the quick and dirty nature of Participatory Rural Appraisal (PRA). Many of these rapid methodologies inevitably suffer from significant data quality issues (see next section) as well as often being framed in ways that limit their usefulness (Schreckenberg *et al.*, 2010). Overall, they tend to focus on economic at the expense of political dimensions (Ribot, 2009) including the long-term, ongoing processes of institutional evolution (Brechin *et al.*, 2002), and play into the hands of local and national elites (Kaswamila and Songorwa, 2009). They also tend to neglect social and cultural aspects.

Given the range of different dimensions set out above that need to be considered, and the pitfalls of trying to do too much with limited resources and sacrificing data quality and validity as a result, it may help to conclude this section with two guiding principles. First, local people's priorities, and local interest in contributing to and making use of the results, should guide, define, and operationalize the dimensions on which monitoring and evaluation should focus. Second, it is better to get a good qualitative understanding of impacts, than to get a faulty quantitative description. These principles are further developed in subsequent sections.

How to monitor? What are the pitfalls?

This section sets out some principles of methods in monitoring. It does not attempt to present cookbook recipes for individual methods. This is because these are extremely well covered in applied textbooks (Goldman, 2000; Russell Bernard, 2006). It does, however, seek to make clear an important warning. Conservation workers will be naturally not just aware of but drawn to reports of socioeconomic impact monitoring methods devised and used by other conservation scientists and published in the conservation literature (Schreckenberg *et al.*, 2010). However, as is the case with surveys of attitudes and perceptions (see below), many attempts by conservation workers to monitor social and economic impacts have been designed and implemented with limited grasp of fundamental social science principles essential to ensuring data validity.

Most of the potential problems hinge on the fact that collecting social data usually involves talking to people and relying on (i) them understanding the question in the same light as the person asking it, and (ii) the truth of their answers. But local populations may have a very different attitude to conservation than do people from the industrial West, or even in-country nationals whose education and professional training make conservation their livelihood. If the questions making up the interview or questionnaire have been formulated by an outsider, from that outsider's perception of the local situation, which assumes that conservation is of central and paramount importance, and embodies a Western approach to the dimensions under investigation, it is quite possible that the interviewee hears something that is at best quite different, and at worst, meaningless. This problem is made far more acute where the person formulating the question is not fluent in the local language, and is effectively formulating questions not just from an alien world view but also in an alien language, which does not map onto local means of expression.

This problem extends to cover, for example, questions asking people to perform alien types of computation. These could be questions as to how much was earned by an individual in the previous year, where cash plays a limited part in the economy, and where there is no regular employment, but rather a seasonal cycle in activity patterns with very irregular incomes. Similarly questions about household income may be impossible to answer in situations where individuals do not share information about their respective earnings. A different kind of potentially unrealistic computation is demanded of interviewees when rapid appraisal exercises ask people to rate relative changes in well-being (Schreckenberg et al., 2010), in context where people's lives have been impacted by multiple external local and personal factors among which the conservation initiative may or may not be salient.

Even if the meaning of the question is crystal clear to the interviewee, does the interviewer understand the answer? People are political animals, and conservation initiatives affecting access to resources are highly political interventions. People's answers are extremely sensitive to the context of the interview, to the power relations between interviewer and interviewee, to the perceived allegiances of the interviewer, and to the perceived possibilities for influencing material and political outcomes by using strategic answers. Many rural populations in LDCs are extremely hierarchical, with large inequalities between the economic and political elite and the most marginalized, who may be women and young people, or a marginalized ethnic or migrant group, or simply the economically disadvantaged. The same inequalities apply between local people and relatively elite district, regional, or national government officials, university students or other researchers, including project employees. Under those circumstances, it is extremely unlikely that marginalized people will be able to articulate their priorities and concerns freely. The researcher is only there for a brief period: the marginalized person has to be politic so as not to bring down on their own heads still worse problems through displeasing the elite. It is naive to assume that by holding participatory rapid appraisal exercises in groups of rich, or poor, or women, or young people, such issues are necessarily resolved. The fact is that there are many issues around resource use and around conservation impacts that cannot be raised in public, or indeed at all, even with a trusted, known individual. Similarly, individuals engaged in corrupt or illegal practices are not going to be forthcoming about their activities and may be in a position to make life difficult for anyone else who is.

Another aspect of the problem involves interviewees trying to display courtesy to the interviewer by second guessing and saying what they want to hear, or through a 'prestige effect' giving answers that do not accurately reflect day-to-day conditions but rather emphasize conspicuous consumption. The problem with many social and economic impact assessment data collection exercises devised and implemented by conservation scientists is that they assume that information can be elicited in a straightforward way by direct questioning and that answers can be taken at face value.

Time, money, resources

Having taken on board the need to keep the broad framework in view, and the multidimensional nature of monitoring, as well as the perennial problem of ensuring the data are valid and meaningful, limitations of time, money, and resources will inevitably make it necessary to focus down to a minimum set of achievable and relevant data collection activities.

First, it is vital to make full use of the *secondary data* available. Researchers planning a monitoring exercise need to read every bit of information they can lay their hands on about the area, its people, and past interventions, even if this information comes in a form that is alien to their usual mode of publication. After reading all the available background material, researchers will be in a position to use those secondary data, alongside discussion with as wide a range as possible of informed people, to establish which different stakeholder groups comprise the affected population (including stakeholders who are not necessarily residents, but are affected), bearing in mind the classic fault lines (age, gender, wealth, ethnicity, migration/residence status, religion/culture, occupation). This is necessary in order to set up a *sampling* schedule that will take account of the full array of those affected stakeholders, in order to establish the dimensions that matter to the various groups. This sample frame may be the basis for either qualitative or quantitative data collection.

At the same time, researchers need to establish what matters to different groups affected by the intervention. While scientists may have strong preconceived ideas as to what they can measure, and income tends to be an obvious variable to pursue, relevance and therefore usefulness of monitoring can only be ensured if priorities as perceived by the affected population are made central to the data collected. The background reading and the discussions will have begun to convey in a preliminary way the priorities for monitoring as held by different stakeholders affected by the intervention. That preliminary scoping needs to be carried forward through qualitative methods in order to establish a meaningful study design that encompasses locally meaningful dimensions of impacts.

Mixed methods: qualitative and/or quantitative data?

As must be clear from the emphasis on context and confounding factors set out above, as well as the breadth of dimensions of likely importance, socioeconomic monitoring must rely on qualitative as much as (often more than) quantitative data. Conservation scientists are commonly trained as natural scientists, with a strong desire to quantify, and a certain suspicion of or even contempt for qualitative data. Yet that mistrust of the value of qualitative data is misplaced: such data should be seen both as a necessary prerequisite to good quantitative data collection, and also as a type of data valid in its own right. Such qualitative work is likely to rely on key informants, individuals with particular knowledge and/or representing particular perspectives (official and higher-level as well as local and grassroots), and on low-key discussions of the important dimensions of monitoring, allowing the KI to lead the discussion and open unforeseen avenues, rather than seeking a statistically representative sample. Qualitative work may also use focus groups and semi-structured interviews, and participant observation, with low-key discussions as well as passively witnessed exchanges potentially yielding important insights. Media reports and local or district archives may also be important independent sources of information.

It is quite often the case that the drive to quantify may be counterproductive. For example, there are increasing numbers of surveys quantifying changes in attitudes and perceptions as an outcome of conservation interventions. The major methodological problems commonly biasing many if not most of these studies have been dealt with by other commentators (Browne-Nunez and Jonker, 2008). Without going into detail here, it is extremely unlikely that any such survey could collect valid or representative data without considerable prior qualitative work, and without the person collecting the data possessing strong local language skills, as well as enjoying complete acceptance by the affected population as being personally well trusted (through long acquaintance) and having independence from the conservation initiative that is beyond question (Drury *et al.*, 2011). These conditions are unlikely to be met. If they cannot be satisfied it is best not to attempt such a quantitative survey of attitudes and perceptions. Careful qualitative work meeting some of these conditions could, however, give important insights into the reactions of local people and the way these relate to the different circumstances of different stakeholders (Drury *et al.*, 2011).

Box 12.1 **Quick-and-dirty Participatory Rural Appraisal (PRA) quantification, or qualitative Key Informant (KI) data?**

It is not uncommon for conservation organizations seeking to evaluate a community-based conservation (CBC) project to schedule a set of participatory rural appraisal exercises with groups of people possibly chosen to represent old/young, male/female, and other categorizations. They will be asked to rank sources of income, including those from tourism, or levels of use of a resource, and to estimate change before/after the conservation intervention.

But such public groups make a poor forum in which to articulate sensitive issues around displacement impacts, illegal activities, or elite capture. It is also unlikely that they will elicit comments as to perhaps ill-informed or inappropriate aspects of the project, as these comments would constitute public discourtesy towards an important outsider with power to impact on local affairs. Yet poor qualitative validity of the data may not be immediately apparent to conservation workers.

Also, the PRA groups do not constitute statistically representative samples, but it is hard for conservation workers to resist some simple quantification, irrespective of its acknowledged lack of statistical validity – and the results may ultimately feed into sophisticated statistical analyses, themselves methodologically rigorous, but based on dubious data.

Finally, PRA exercises can morph from being a potentially enjoyable group social occasion with appropriate hospitality laid on, to an outright imposition on hard-pressed people's time, with sometimes uncomfortable overtones through problematic exercise of local government power and local elite control.

In such situations, it is better to use a low-key approach, respecting the age and experience of elders, or the innovative/entrepreneurial approach of some individuals, and seeking quiet, extended conversations with such individuals identified through earlier enquiries as knowledgeable people, at their convenience, in a place chosen by them, with appropriate privacy, and with appropriate tokens of respect and/or hospitality offered by the researcher. Such conversations can be used to elicit individual resource use histories, or histories of changing structures that have governed access to resources across the community and the implications for the individual, which will in the long run convey much more valid if qualitative information than the supposedly participatory, cookery book PRA exercise.

In the conservation literature, the term 'key informant' (KI) is commonly applied to prominent individuals – such as government officials and local leaders. In the social science literature, KI means a knowledgeable, reliable, and disinterested individual unlikely to have significant gatekeeping powers. Such KIs are identified only after a long period of the researcher's settling into the background, becoming accepted as being of neutral status, building trust, developing insight into local institutions and hierarchies, and coming to an understanding of who is a knowledgeable and reliable witness. Conservation workers cannot establish neutral status – their affiliation to and involvement in the conservation intervention is obvious to all, and local individuals with vested interests will put themselves forward as KIs. Conservation researchers have to use their best judgment to ensure they capture contrasting (and dissenting)

> viewpoints by (i) satisfying local protocol by interviewing prominent people, (ii) interviewing people who are not gatekeepers, (iii) putting as much weight on informally observed opportunistic interactions as on information from formal PRA/interview data collection exercises, and (iv) interpreting/weighting the results according to the known context (kin relations, political affiliations, economic interests, etc.) and position of those consulted.

Quantitative survey

Once such qualitative information is well established, cross-checked, tested, and validated through multiple sources and also through independent lines of evidence (e.g. administrative archives, project records, previous research or development work), it may be appropriate to carry out quantitative work. This would commonly take the form of exercises such as household survey (Russell Bernard, 2006; Schreckenberg *et al.*, 2010), using questionnaires documenting, for example, demographic and socioeconomic variables such as household size, composition and inter-relations, household wealth and assets, people's residence status, their primary, secondary, and tertiary economic activities, and their educational attainment.

Depending on the particular impacts in question, such a survey might then gather detail on particular activities that may relate to project initiatives. It may be relevant to enquire about investment of labour, cash, and other resources in different activities – farming, herding, fishing, hunting, trading, wage labour, remittances – and into returns from these activities. It may be relevant to look at people's food systems, with data on food produced on farm or harvested wild, what is consumed locally, and what is sold on or given or exchanged. These types of data often require recall, and it is important to consider whether the categorizations, and the period of time over which recall is being asked, will make good sense to the interviewee (Russell Bernard, 2006). Some surveys attempt to quantify changes in morbidity, mortality, and migration as correlates of conservation interventions. Each of these is technically extremely difficult to manage, and would require consideration of the specialist literature (Igoe *et al.*, 2008). More generally, careful thought needs to go into not only formulating survey questions (based on in-depth discussion with local people as to priority issues, with local language speakers as to appropriate wording, and using translation and back translation to ensure meaning is clear and consistent) but also their order (most neutral, least sensitive first) and whether some are best asked differently – or not at all (Browne-Nunez and Jonker, 2008).

Given the time and expense of establishing an appropriate sample frame where no electoral roll or similar comprehensive listing exists, and of administering such surveys (expensive in terms of interviewers, in terms of logistics, and in terms of local people's time, as well as potentially hard to schedule with respect to interviewees' other commitments); given also the ethical obligation not to burden poor people unnecessarily (Hirotsuga, 2007; Alcser et al., 2010) or to gather exhaustive data that then cannot be used, such surveys should not be undertaken casually. Designing monitoring and evaluation will be best achieved through discussion with local people as to the best way to structure and implement any survey. It is also important to consider how to temper the agreed process: local hierarchies may give little weight to the time or rights of particular groups – women, youth, marginalized ethnic groups, poor people. The scale of transaction costs can best be anticipated and minimized by thinking through what an interviewee of a particular age/sex class and from a particular economic occupation will forgo to complete the interview, and ensuring compensation that is locally relevant while making fair return. In some rural LDC cases, small household gifts of food or soap or tea and sugar may be appropriate; in others, employment or cash (e.g. some proportion of a day's wage) may be expected.

As with any social science research, there is a need for continual reflexive awareness and evaluation of issues and conflicting demands. Professional ethical guidelines (American Anthropological Association, 1998; Association of Social Anthropologists, 2011) invoke guiding principles of respect for research subjects, with the fundamental imperative being to do no harm. Researchers are bound to gain free, full, and informed consent prior to beginning the process, not simply blanket access granted by some national or local gatekeeper. No harm should ever come to people through the monitoring and evaluation process, whether in terms of personal or professional standing, or through waste of time or resources. The researcher should be focused, and bound by the professional duty to make sure that any information sought is necessary and relevant. Researchers need to ask themselves who are they working for – balancing their own career interests, the progress of the conservation intervention, and the interests of local people; the latter are, ethically speaking, a researcher's first responsibility in any social science research.

If, after all these considerations, quantitative surveys are undertaken and maintained over the long term, periodic repeat round surveys can be extremely useful in demonstrating the scale, importance, and trends of, for example, economic impacts. However, their usefulness is determined by their qualitative validity, and the validity of the exercise is at once the hardest dimension to ensure, and the least well considered. Subject fatigue easily sets in with repeat round surveys, especially if they are felt to be extracting information for no return. In addition to the caveats over data quality set out earlier, there is an overriding caveat on the issue of independence, dealt with below.

Independence of social and economic monitoring

In many developing country situations there are few or no local or national government resources for social and economic monitoring. It is the rule, not the exception, for conservation projects to fund or even undertake such monitoring themselves as part of their own portfolio. This lack of independence is truly problematic. Most conservation projects are limited in terms of time, money, and personnel. The primary concern for most is conservation not development, however much they may subscribe to the two goals being in some way linked. There is rarely either the absolute requirement, or the opportunity, to carry out an in-depth social and economic assessment. Sometimes there is limited motivation to uncover problems; there is rather an expectation that social impact assessment will give the required clean bill of health. This expectation can translate into considerable pressure, conscious or unconscious, for the results to be favourable, rather than for the monitoring to be an opportunity to identify and address potential problems so as to help make things work better in the longer term. Recent work by Sachedina (2008) distinguished a conservation NGO's downward and upward accountability. Upward accountability to the donors was taken extremely seriously: the effort and strategic thought put into reporting upwards to donors was so successful that the organization's fundraising turnover and size grew by orders of magnitude over a short period of time. However, downward accountability to the local people on the ground was neglected, glossing over legitimate problems to portray the initiatives as unqualified successes and persistently ignoring the growing tension and conflicts that resulted. In another example, Sullivan (2003) documented the outright intimidation of dissenters to a community-based conservation project in Namibia.

Only independently designed, implemented, and reported monitoring can circumvent such pressures; only such independent monitoring can give a robust basis for identifying and addressing problems. Though there will always be the temptation to carry out the monitoring in-house, and there will often be associated pressures to focus on favourable outcomes where less favourable ones may in reality exist, repeated failures in conservation projects show that misrepresenting reality for purposes of political expediency only undermines sustainability in the long term.

Social and economic monitoring is not a subsidiary exercise nor a simple box to be ticked. Whatever the pressures on time and resources, however pressing the conservation needs, the fact is that if unbiased, inclusive channels of representation are not established, to report accurately and independently on socioeconomic problems and perceptions associated with the intervention, the long-term outcome is likely to be tension, conflict and failure. Far from it being problematic to return monitoring reports that articulate problems and dissent, such a report is the first step towards getting things right. Conservation organizations need to put resources into independent evaluation, creating a safe and inclusive forum for people to flag up issues, and inviting independent assessment. Inviting social scientists to work alongside local people in establishing participatory monitoring could be an affordable way to do this, as set out below.

Participatory monitoring

Participatory monitoring is increasingly advocated and used as an opportunity for local people to engage in the overall project, and because it can moderate perspectives driven primarily by the interests or biases of internal project staff. However, as with all processes labelled 'participatory', a wide range of approaches are subsumed under the term. At one end of the scale are monitoring processes that are participatory only in the sense that local people identified and selected by the conservation agency are used as sources of information on social and economic variables chosen by the conservation agency as being of importance (e.g. Jones *et al.*, 2008). 'In some cases, no beneficiaries have been involved but fieldworkers have helped senior management to design the MandE [monitoring and evaluation], thus making it "participatory"' (Guijt, 1999). At the other end of the scale are monitoring systems that genuinely establish channels of representation for a wide cross-section of local people, who choose to do so to record and report on issues selected by them as being of importance (e.g. Lewis, 2007). Understanding the difference between token and meaningful participation, and ensuring the latter, through establishing the institutions and technologies making it possible, is key to making the process effective (see, e.g., Guijt, 1999; Lund *et al.*, 2009).

Increasingly, technology is expanding the possibilities for citizen science, with local people reporting directly to the highest levels and the widest audiences in ways that forestall their being penalized for making the information clear to outsiders. Examples include the use of YouTube to alert a wider public to violation of human rights and social justice associated with wildlife hunting leases in northern Tanzania,[2] through to handheld GPS with icon displays for pre-literate forest peoples to flag sites of special significance, warn off loggers and/or to report infringements of the rules of access and use (Lewis, 2007). The explosive rise of social media and their potential for conveying otherwise suppressed events, opinions, wants, and needs, is forging an increasingly significant channel for the voices of hitherto poorly represented groups.

The role of the conservation organization then becomes to facilitate and follow such media, to take note of the issues raised, to find ways to address concerns that are raised, and avoid the temptation to seek counterarguments to justify or dismiss them.

---

[2] http://www.youtube.com/watch?v=i-FP2gRvziw; accessed 14 February 2011.

Conclusions

There is no silver bullet for social and economic monitoring; no single tool, no cookery book solution. Above all, social and economic monitoring requires what Bill Adams has called 'thinking like a human' (Adams, 2007). We need to be clear that this does not immediately come naturally to the natural sciences-trained conservation worker, whatever their overwhelming strengths in field biology, ecological theory, and quantitative and statistical data management. Conservation workers are not the best people to design or carry out monitoring of social and economic surveys, and should refrain outright from doing so on their own conservation initiatives, by the requirements of independence. Even so, there are pressing reasons – pragmatic, legal, ethical – for conservation workers to have a clearer understanding of the basic principles of social and economic monitoring of the impacts of conservation initiatives on local communities, of the importance of meaningful rather than token participation and representation in monitoring and evaluation (MandE), and of the need for independence of feedback. If conservation workers do not understand the importance not only of monitoring social and economic impacts, but also of doing it well, they cannot expect to see long-term conservation success.

References

Adams, W.M. (2007) Thinking like a human: social science and the two cultures problem. *Oryx*, 41, 275–276.
Adams, W. and Hutton, J. (2007) People, parks and poverty: political ecology and biodiversity conservation. *Conservation and Society*, 5, 147–183.
Adams, W., Aveling, R., Brockington, D., *et al.* (2004) Biodiversity conservation and the eradication of poverty. *Science*, 306, 1146–1149.
Agrawal, A. and Redford, K. (2006) Poverty, development and biodiversity conservation: shooting in the dark? *WCS Working Paper 26*. Wildlife Conservation Society, New York.
Alcser, K., Antoun, C., Bowers, A., Clemens J., and Lien, C. (2010) Ethical considerations in surveys. In: Cross-Cultural Survey Guidelines. Available at: http://ccsg.isr.umich.edu/pdf/03EthicalConsiderationsInSurveysNov2010.pdf. Accessed 10 February 2011.
Alvard, M. (1993) Testing the "ecologically noble savage" hypothesis: interspecific prey choice by Piro hunters of Amazonian Peru. *Human Ecology*, 21, 355–387.
American Anthropological Association (1998) Code of ethics. Available at: http://www.aaanet.org/committees/ethics/ethcode.htm.
Ashley, C. and Carney, D. (1999) *Sustainable Livelihoods: Lessons from Early Experience*. DFID, London.
Association of Social Anthropologists (2011) Ethical guidelines for good research practice. Available at: http://www.theasa.org/downloads/ASA%20ethics%20guidelines%202011.
Benjaminsen, T. and Lund, C. (2002) Formalisation and informalisation of land and water rights in Africa: an introduction. *European Journal of Development Research* 14, 1–10.

Bird, R.B., Bird, D.W., Codding, B.F., Parker, C.H., and Jones, P.G. (2008) The "fire stick farming" hypothesis: Australian Aboriginal foraging strategies, biodiversity, and anthropogenic fire mosaics. *Proceedings of the National Academy of Sciences of the USA*, 105, 14796–14801.

BirdLife International (2011) Ecosystem services demonstrate the socio-economic value of Important Bird Areas. Presented as part of the BirdLife State of the world's birds website. Available from: http://www.birdlife.org/datazone/sowb/casestudy/238. Accessed 29 March 2011.

Blaikie, P. (2006) Is small really beautiful? Community-based natural resource management in Malawi and Botswana. *World Development*, 34, 1942–1957.

Bowler, D., Buyung-Ali, L., Healey, J.R., Jones, J.P.G., Knight, T., and Pullin, A.S. (2010) *The Evidence Base for Community Based Forest Management as a Mechanism for Supplying Global Environmental Benefits and Improving Local Welfare*. Centre for Evidence-Based Conservation, Bangor, UK. Available at: http://www.unep.org/stap/Portals/61/pubs/STAP%20CFM%20document%202010.pdf. Accessed 9 February 2011.

Bradstock, R.A, Williams, J.E., and Gill, A.M. (2002) *Flammable Australia: the Fire Regimes and Biodiversity of a Continent*. Cambridge University Press.

Brechin, S.R., Wilshusen, P.R., Fortwangler, C., and West, P. (2002) Beyond the square wheel: toward a more comprehensive understanding of biodiversity conservation as social and political process. *Society and Natural Resources*, 15, 41–64.

Brockington, D. (2002) *Fortress Conservation. The Preservation of the Mkomazi Game Reserve, Tanzania*. James Currey, Oxford, UK.

Brockington, D. (2003) Injustice and conservation – is 'local support' necessary for sustainable protected areas?' *Policy Matters*, 12, 22–30.

Brockington, D. (2007) Devolution, community conservation and forests. On local government performance and Village Forest Reserves in Tanzania. *Society and Natural Resources*, 20, 835–848.

Brockington, D. and Igoe, J. (2006) Eviction for conservation: a global overview. *Conservation and Society*, 4, 424–470.

Brooks, J.S., Franzen, M.A., Holmes, C.M., Grote, M.N., and Borgerhoff Mulder, M. (2006) Development as a conservation tool: Evaluating ecological, economic, attitudinal, and behavioural outcomes. Collaboration for Environmental Evidence. Available at: http://environmentalevidence.org/SR20.htm. Accessed 9 February 2011.

Browne-Nunez, C. and Jonker, S.A. (2008) Attitudes toward wildlife and conservation across Africa: a review of survey research. *Human Dimensions of Wildlife*, 13, 47–70.

Burnham, P. (in press) Climate change and forest conservation: a REDD flag for Central African forest people? In: Brokensha, D., Castro, P., and Taylor, D. (eds), *Climate Change and Indigenous Knowledge: Perceptions and Adaptations of Threatened Communities*. Practical Action Publications.

Butchart, S.H.M., Walpole, M., Collen, B., *et al.* (2010) Global biodiversity: indicators of recent declines. *Science*, 328, 1164–1168.

Campbell, A., Miles, L., Dunning, E., Rashid, M., Teobaldelli, M., and Dickson, B. (2009) *Issues and Potential Outcomes from UNFCCC's COP 15 Meeting. Stage 1 report*. UNEP World Conservation Monitoring Centre, 50 pp.

Carney, D. (1998) Implementing the sustainable rural livelihoods approach. In: Carney, D. (ed.), *Sustainable Rural Livelihoods: What Contributions Can We Make?* DFID, London.

Carney, D. (1999) Approaches to sustainable livelihoods for the rural poor. *ODI Poverty Briefing No.2.* ODI, London.

CBD (2010) Linking biodiversity conservation and poverty alleviation: a state of knowledge review. *CBD Technical Series No. 55.* Convention on Biological Diversity. Available at: http://www.cbd.int/doc/publications/cbd-ts-55-en.pdf. Accessed 9 February 2011.

CCB (2011) Climate, Community and Biodiversity Alliance home page. Available at: http://www.climate-standards.org. Accessed 25 January 2011.

Cernea, M.M. (2000) Risks, safeguards and reconstruction: a model for population displacement and resettlement. In: Cernea, M.M. and McDowell, C. (eds), *Risk and Reconstruction: Experiences of Resettlers and Refugees.* World Bank, Washington, DC.

Cernea, M.M. (2006) Re-examining "displacement": a redefinition of concepts in development and conservation policies. *Social Change,* 36, 8–35.

Cernea, M.M. and McDowell, C. (2000) *Risk and Reconstruction: Experiences of Resettlers and Refugees.* World Bank, Washington, DC.

Cernea, M.M. and Schmidt-Soltau, K. (2003) The end of forcible displacements? Conservation must not impoverish people. *Policy Matters,* 12, 42–51.

Chape, S., Spalding, M., and Jenkins, M. (2008) *The World's Protected Areas: Status, Values and Prospects in the Twenty-first Century.* University of California Press, Berkeley.

Chatre, A. and Agrawal, A. (2009) Tradeoffs and synergies between carbon storage and livelihood benefits from forest commons. *Proceedings of the National Academy of Sciences of the USA,* 106, 17667–17670.

Curran, B., Sunderland, T., Maisels, F., *et al.* (2009) Are Central Africa's protected areas displacing hundreds of thousands of rural poor? *Conservation and Society,* 7, 30–45.

Dennis, C. (2003) Burning issues. *Nature,* 421, 204–206.

Dietz, T., Ostrom, E., and Stern, P. (2003) The struggle to govern the commons. *Science,* 302, 1907–1912.

Drury, R., Homewood, K., and Randall, S. (2011) Less is more: the potential of qualitative approaches in conservation research. *Animal Conservation,* 14, 18–24.

Emerton, L. (2001) The nature of benefits and the benefits of nature. In: Hulme, D. and Murphree, M. (eds), *African Wildlife and Livelihoods: The Promise and Performance of Community Conservation.* James Currey, Oxford.

Fabricius, C. and Collins, S. (2007) Community-based natural resource management: governing the commons. *Water Policy,* 9, 83–97.

Fairhead, J. and Leach, M. (1996) Rethinking the forest-savanna mosaic: colonial science and its relics in West Africa. In: Leach, M. and Mearns, R. (eds), *The Lie of the Land: Challenging Received Wisdom in African Environmental Change and Policy.* James Currey, London, pp. 63–80.

Fischer, B. and Christopher, T. (2007) Poverty and biodiversity: measuring the overlap of human poverty and the biodiversity hotspots. *Ecological Economics,* 62, 93–101.

Forest Investment Review (2009) Available at: http://www.forumforthefuture.org/projects/forest-investment-review. Accessed 25 January 2011.

Gartlan, S. (1997) Every man for himself and God against all: history, social science and the conservation of nature. In: Worldwide Fund for Nature Annual Conference on People and Conservation.

Goldman, L. (2000) *Social Impact Analysis: An Applied Anthropology Manual*. Berg Press, Oxford. See Introduction (Chapter 1) and Chapters 2, 9, and 10.

Gubbi, S., Linkie, M., and Leader-Williams, N. (2009) Evaluating the legacy of an integrated conservation and development project around a tiger reserve in India. *Environmental Conservation*, 35, 331–339.

Guijt, I. (1999) *Participatory Monitoring and Evaluation for Natural Resource Management and Research*. Socio-economic Methodologies for Natural Resources Research. Natural Resources Institute, Chatham, UK. Available at: http://www.nri.org/publications/bpg/bpg04.pdf.

Halme, P., Toivanen, T., Honkanen, M., *et al.* (2010) Flawed meta-analysis of biodiversity effects of forest management. *Conservation Biology*, 24, 1154–1156.

Hames, R. (2007) The ecologically noble savage debate. *Annual Review of Anthropology*, 36, 177–190.

Hirotsuga, A. (2007) Bombarding people with questions: a reconsideration of survey ethics. *Bulletin of WHO*, November, 823–824.

Homewood, K. (2008) *Ecology of African Pastoralist Societies*. James Currey.

Homewood, K., Kristjanson, P., and Trench, P.C. (2009) *Staying Maasai? Livelihoods, Conservation and Development in East African Rangelands*. Springer Verlag, New York.

Hughes, R. and Flintan, F. (2001) *Integrating Conservation and Development Experience: A Review and Bibliography of the ICDP Literature*. International Institute for Environment and Development, London. Available at: http://www.oceandocs.org/bitstream/1834/805/1/HUGHES,%20R1-24.pdf. Accessed 9 February 2011.

Igoe, J., Brockington, D., Randall, S., and Schofield K. (2008) Lessons to be learned about migration around protected areas. *Science*, E-Letter, 11 December.

IPCC: Solomon, S., Qin, D., Manning, M., *et al.* (eds) (2007) *Climate Change 2007: The Physical Science Basis*. Contribution of Working Group I to the Fourth Assessment Report of the Intergovernmental Panel on Climate Change. Cambridge University Press.

Johnson, C. (2004) Uncommon ground: the 'poverty of history' in common property discourse. *Development and Change*, 35, 407–433.

Jones, J.P.G., Andriamarovololona, M.M., Hockley, N., Gibbons, J.M., and Milner-Gulland, E.J. (2008) Testing the use of interviews as a tool for monitoring trends in the harvesting of wild species. *Journal of Applied Ecology*, 45, 1205–1212.

Kapos, V., Balmford, A., Aveling, R., *et al.* (2008) Calibrating conservation: new tools for measuring success. *Conservation Letters*, 1, 155–164.

Kaswamila, A. and Songorwa, A. (2009) Participatory land-use planning and conservation in northern Tanzania rangelands. *African Journal of Ecology*, 47, 128–134.

Laris, P. and Klepeis, P. (2006) Degradation narratives. In: Geist, H. (ed.), *The Earth's Changing Land: An Encyclopedia of Land-Use and Land-Cover Change*. Greenwood Publishing Group, pp. 161–164.

Lewis, J. (2007) Enabling forest people to map their resource and monitor illegal logging in Cameroon. *Before Farming*, 2007/2, article 3. Available at: http://www.waspress.co.uk/journals/beforefarming/journal_20072/news/2007_2_03.pdf. Accessed 10 February 2011.

Lund, J.F. and Treue, T. (2008) Are we getting there? Evidence of decentralized forest management from the Tanzanian miombo woodlands. *World Development*, 36, 2780–2800.

Lund, J.F., Balooni, K., and Casse, T. (2009) Change we can believe in? Reviewing studies on the conservation impact of popular participation in forest management. *Conservation and Society*, 7, 71–82.

Malleret-King, D. (2000) *A Food Security Approach to Marine Protected Area Impacts on Surrounding Fishing Communities: the Case of Kisite Marine National Park in Kenya*. Cuvillier-Verlag, Göttingen.

Mortimore, M. (1998) *Roots in the African Dust: Sustaining the Sub-Saharan Drylands*. Cambridge University Press, Cambridge.

Millennium Ecosystem Assessment (2005) *Ecosystems and Human Well Being: Synthesis*. World Resources Institute, Washington, DC.

Nelson, F. (2007) *Emergent or Illusory? Community Wildlife Management in Tanzania*. International Institute for Environment and Development, London.

Nelson, F., Nshala, R., and Rodgers, W.A. (2007) The evolution and reform of Tanzanian wildlife management. *Conservation and Society*, 5, 232–261.

Nelson, F., Gardner, B., Igoe, J., and Williams, A. (2009) Community-based conservation and Maasai livelihoods in Tanzania. In: Homewood, K., Kristjanson, P., and Trench, P. (eds), *Staying Maasai? Livelihoods, Conservation and Development in East African Rangelands*. Springer, New York.

Oates, J.F. (1999) *Myth and Reality in the Rain Forest: How Conservation Strategies are Failing in West Africa*. University of California Press, Berkeley, CA.

Ostrom, E. (2009) A general framework for analyzing sustainability of social-ecological systems. *Science*, 325, 419–422.

Ostrom, E., Burger, J., Field, C., Norgaard, R., and Policansky, D. (1999) Revisiting the commons: local lessons, global challenges. *Science*, 284, 278–282.

Paudel, N. (2005) Conservation and livelihoods: exploration of local responses to conservation intervention in Royal Chitwan National Park, Nepal. PhD Thesis, University of Reading.

Peet, R., Robbins, P., and Watts, M. (2010) *Global Political Ecology*. Taylor & Francis Ltd.

Plan Vivo (2008) The Plan Vivo Standards. Plan Vivo Foundation. Available at: http://www.planvivo.org/documents/standards.pdf. Accessed 25 January 2011.

Ramisch, J. (1999) *In the Balance? Evaluating Soil Nutrient Budgets for an Agro-Pastoralist Village of Southern Mali*. Managing Africa's Soils 9. Drylands Programme, International Institute for Environment and Development, London.

Rands, M.R.W., Adams, W.M., Bennum, L., *et al.* (2010) Biodiversity conservation: challenges beyond 2010. *Science*, 329, 1298–1303.

Ribot, J. (2009) Authority over forests: empowerment and subordination in Senegal's Democratic decentralization. *Development and Change*, 40, 105–129.

Riddell, M. (2010) Hunting, livelihoods and conservation in northern Republic of Congo. PhD thesis, University of Oxford.

Roba, H. and Oba, G. (2009) Efficacy of integrating herder knowledge and ecological methods for monitoring rangeland degradation in Northern Kenya. *Human Ecology*, 37, 589–612.

Roe, D., Nelson, F., and Sandbrook, C. (2009) Community management of natural resources in Africa: impacts, experiences and future directions. *Natural Resource Issues No. 18*. International Institute for Environment and Development, London.

Rotha, K.S. (2005) Understanding key CBNRM concepts. The development of Community Based Natural Resource Management (CBNRM) in Cambodia. Selected Papers on Concepts and Experiences. CBNRM Learning Initiative.

Russell Bernard, H. (2006) *Research Methods in Anthropology: Qualitative and Quantitative Approaches*. Altamira Press.

Sachedina, H.T. (2008) Wildlife is our oil: conservation, livelihood and NGOs in the Tarangire ecosystem, Tanzania. DPhil thesis, University of Oxford.

Salafsky, N. and Wollenberg, E. (2000) Linking livelihoods and conservation: A conceptual framework and scale for assessing the integration of human needs and biodiversity. *World Development*, 28, 1421–1438.

Schmidt-Soltau, K. and Brockington, D. (2004) The social impacts of protected areas. Available at: http://www.social-impact-of-conservation.net.

Schreckenberg, K., Camargo, I. Withnall, K., *et al.* (2010) Social assessment of conservation initiatives: a review of rapid methodologies. *Natural Resource Issues no. 22*. IIED, London.

Scoones, I. (1998) Sustainable rural livelihoods: a framework for analysis. *IDS Working Paper 72*. Institute for Development Studies, Brighton, UK.

Smith, E. and Wishnie, M. (2000) Conservation and subsistence in small-scale societies. *Annual Review of Anthropology*, 29, 493–524.

Sullivan, S. (2003) Protest, conflict and litigation: dissent or libel in resistance to a conservancy in north-west Namibia. In: Berglund, E. and Anderson, D. (eds), *Ethnographies of Conservation: Environmentalism and the Distribution of Privilege*. Berghahn Press, Oxford, pp. 69–86.

Sullivan, S. (2008/2009) An ecosystem at your service. *The Land*, Winter, 21–23. Available at: http://siansullivan.files.wordpress.com/2010/02/sullivan-ecosystems21-23.pdf. Accessed 9 February 2011.

Sullivan, S. (2009) Green capitalism, and the cultural poverty of constructing nature as service-provider. *Radical Anthropology*, 3, 18–27.

Wardell, D.A., Nielsen, T.T., Rasmussen, K., and Mbow, C. (2004) Fire history, fire regimes and fire management in West Africa: an overview. In: Goldammer, J.G. and de Ronde, C. (eds), *Wildland Fire Management Handbook for Sub-Saharan Africa*. Global Fire Monitoring Center, pp. 350–381.

Waylen, K.A., Fischer, A., McGowan, P.J.K., Thirgood, S.J., and Milner-Gulland, E.J. (2010) Effect of local cultural context on the success of community-based conservation interventions. *Conservation Biology*, 24, 1119–1129.

West, P., Igoe J., and Brockington, D. (2006) Parks and peoples: the social impact of protected areas. *Annual Review of Anthropology*, 35, 251–277.

Western, D. (1994) Ecosystem conservation and rural development: the case of Amboseli. In: Western, D. and Wright, R.M. (eds), *Natural Connections: Perspectives in Community-Based Conservation*. Island Press, Washington, DC, pp. 15–52.

Wilder, L. and Walpole, M. (2008) Measuring social impacts in conservation: experience of using the Most Significant Change method. *Oryx*, 42, 529–538.

Wittemeyer, G., Elsen, P., Bean, WT., Coleman A., Burton, O., and Brashares J.S. (2008) Accelerated human population growth at protected area edges. *Science*, 321, 123–126.

Wunder, S. (2005) Payments for environmental services: some nuts and bolts. *CIFOR Occasional Paper No. 42*. Center for International Forestry Research, Bogor Barat, Indonesia. Available at: http://www.cifor.cgiar.org/publications/pdf_files/OccPapers/OP-42.pdf. Accessed 9 February 2011.

13

Science to Policy Linkages for the Post-2010 Biodiversity Targets

Georgina M. Mace[1], Charles Perrings[2], Philippe Le Prestre[3], Wolfgang Cramer[4], Sandra Díaz[5], Anne Larigauderie[6], Robert J. Scholes[7], and Harold A. Mooney[8]

[1]Imperial College London, Centre for Population Biology, Ascot, UK
[2]ecoSERVICES Group, School of Life Sciences, Arizona State University, Tempe, AZ, USA
[3]Institut Hydro-Québec en environnement, développement et société, Université Laval,Québec, Canada
[4]Mediterranean Institute of Biodiversity and Ecology (IMBE), Aix-en-Provence, France
[5]Instituto Multidisciplinario de Biología Vegetal, Universidad Nacional de Córdoba, Cordoba, Argentina
[6]DIVERSITAS, Muséum National d'Histoire Naturelle, Paris, France
[7]CSIR Natural Resources and Environment, Pretoria, South Africa
[8]Department of Biology, Stanford University, Stanford, California, USA

Introduction

It is increasingly widely appreciated that biodiversity loss and ecosystem degradation jeopardize human well-being, both now and in the future. The Convention on Biological Diversity, established at the Earth Summit in 1992, took this issue firmly into its remit in 2002 when governments committed to work towards an international goal to reduce the rate of biodiversity loss by 2010 (Balmford *et al.*, 2005b). The '2010 target' was agreed at the sixth session of the Conference of the Parties to the Convention on Biological Diversity (CBD) and then adopted by the broader international community at the World Summit on Sustainable Development in 2002 in the Johannesburg Plan of implementation. In 2008 the influence of the 2010 target was further enhanced when a target for 'reducing biodiversity loss' was added as a sub-target to the United Nations Millennium Development Goal 7 (MDG7b), which aims to 'ensure environmental sustainability'. The 2010 target was reflected in commitments at national and regional levels too, most notably in Brazil and Europe. It thus became an important political commitment for improved biodiversity conservation and management among almost all countries.

In October 2010, the tenth Conference of the Parties (COP10) adopted a revised strategic plan that included a new formulation of its mission, as well as 20 sub-targets. Whereas in 2002, Parties had resolved 'to achieve by 2010 a significant reduction of the current rate of biodiversity loss at the global, regional and national level as a contribution to poverty alleviation and to the benefit of all life on Earth', they have now pledged to 'take effective and urgent action *to halt* the loss of biodiversity to ensure that by 2020 ecosystems are resilient and continue to provide essential services, thereby securing the planet's variety of life, and contributing to human well-being, and poverty eradication'. This new mission focuses not on results to be achieved, but on actions to be undertaken in the context of a more ambitious objective ('to halt'). The 20 individual targets associated with the five strategic goals identified, however, are themselves largely results-oriented (Table 13.1).

In this context, it is useful to return briefly to the implementation of the 2010 target in order to reflect on the challenges that lie ahead of the new ones. In order to implement such a broad and ambitious target, the CBD identified seven focal areas of particular interest to the Parties, for each of which goals, sub-targets, and indicators were agreed (see http://www.cbd.int/2010-target/focal.shtml). Datasets, measures, and indicators were developed through a collaborative partnership globally (the 2010 Biodiversity Indicators Partnership – 2010 BIP; www.twentyten.net) and within Europe (SEBI2010 – Streamlining European 2010 Biodiversity Indicators; http://ec.europa.eu/environment/nature/knowledge/eu2010_indicators/index_en.htm).

During the period from 2002 onwards, activities to publicize and implement the 2010 target in government and non-governmental organizations proceeded in parallel with growing interest from the scientific community. While the scientific focus was largely on the structure and design of the indicator set (Balmford *et al.*, 2005a; Dobson, 2005), implementation was directed at using the 2010 biodiversity target as a focus for campaigns and actions at local, national, and international level. For example, one of the most high-profile activities, Countdown 2010 (http://www.countdown2010.net), includes a network of partners each of which has committed to specific efforts to tackle the causes of biodiversity loss. It has proved to be effective at raising awareness and promoting actions with clear benefits for biodiversity, although it must be recognized that this is a campaign that aims to facilitate and coordinate action among the partners.

Table 13.1 Twenty biodiversity targets for 2020. From http://www.cbd.int/sp/targets/

Strategic goal	#	Target
DRIVERS Strategic goal A. Address the underlying causes of biodiversity loss by mainstreaming biodiversity across government and society	1	By 2020, at the latest, people are aware of the values of biodiversity and the steps they can take to conserve and use it sustainably
	2	By 2020, at the latest, biodiversity values have been integrated into national and local development and poverty reduction strategies and planning processes and are being incorporated into national accounting, as appropriate, and reporting systems
	3	By 2020, at the latest, incentives, including subsidies, harmful to biodiversity are eliminated, phased out or reformed in order to minimize or avoid negative impacts, and positive incentives for the conservation and sustainable use of biodiversity are developed and applied, consistent and in harmony with the Convention and other relevant international obligations, taking into account national socioeconomic conditions
	4	By 2020, at the latest, governments, business, and stakeholders at all levels have taken steps to achieve or have implemented plans for sustainable production and consumption and have kept the impacts of use of natural resources well within safe ecological limits
PRESSURES Strategic goal B. Reduce the direct pressures on biodiversity and promote sustainable use	5	By 2020 the rate of loss of all natural habitats, including forests, is at least halved and where feasible brought close to zero, and degradation and fragmentation is significantly reduced
	6	By 2020 all fish and invertebrate stocks and aquatic plants are managed and harvested sustainably, legally, and by applying ecosystem-based approaches, so that overfishing is avoided, recovery plans and measures are in place for all depleted species, fisheries have no significant adverse impacts on threatened species and vulnerable ecosystems, and the impacts of fisheries on stocks, species, and ecosystems are within safe ecological limits
	7	By 2020 areas under agriculture, aquaculture, and forestry are managed sustainably, ensuring conservation of biodiversity
	8	By 2020, pollution, including from excess nutrients, has been brought to levels that are not detrimental to ecosystem function and biodiversity
	9	By 2020, invasive alien species and pathways are identified and prioritized, priority species are controlled or eradicated, and measures are in place to manage pathways to prevent their introduction and establishment
	10	By 2015, the multiple anthropogenic pressures on coral reefs, and other vulnerable ecosystems impacted by climate change or ocean acidification, are minimized, so as to maintain their integrity and functioning

(*continued overleaf*)

Table 13.1 (*continued*)

Strategic goal	#	Target
STATE Strategic goal C: To improve the status of biodiversity by safeguarding ecosystems, species, and genetic diversity	11	By 2020, at least 17% of terrestrial and inland water, and 10% of coastal and marine areas, especially areas of particular importance for biodiversity and ecosystem services, are conserved through effectively and equitably managed, ecologically representative, and well-connected systems of protected areas and other effective area-based conservation measures, and integrated into the wider landscape and seascapes
	12	By 2020 the extinction of known threatened species has been prevented and their conservation status, particularly of those most in decline, has been improved and sustained
	13	By 2020, the genetic diversity of cultivated plants and farmed and domesticated animals and of wild relatives, including other socioeconomically as well as culturally valuable species, is maintained, and strategies have been developed and implemented for minimizing genetic erosion and safeguarding their genetic diversity
IMPACTS Strategic goal D: Enhance the benefits to all from biodiversity and ecosystem services	14	By 2020, ecosystems that provide essential services, including services related to water, and contribute to health, livelihoods, and well-being, are restored and safeguarded, taking into account the needs of women, indigenous and local communities, and the poor and vulnerable
	15	By 2020, ecosystem resilience and the contribution of biodiversity to carbon stocks has been enhanced, through conservation and restoration, including restoration of at least 15% of degraded ecosystems, thereby contributing to climate change mitigation and adaptation and to combating desertification
	16	By 2015, the Nagoya protocol on Access to Genetic Resources and the Fair and Equitable Sharing of Benefits Arising from their Utilization is in force and operational, consistent with national legislation
RESPONSES Strategic goal E. Enhance implementation through participatory planning, knowledge management, and capacity building	17	By 2015 each Party has developed, adopted as a policy instrument, and has commenced implementing an effective, participatory, and updated national biodiversity strategy and action plan
	18	By 2020, the traditional knowledge, innovations, and practices of indigenous and local communities relevant for the conservation and sustainable use of biodiversity, and their customary use of biological resources, are respected, subject to national legislation and relevant international obligations, and fully integrated and reflected in the implementation of the Convention with the full and effective participation of indigenous and local communities, at all relevant levels

Table 13.1 (*continued*)

Strategic goal	#	Target
	19	By 2020, knowledge, the science base, and technologies relating to biodiversity, its values, functioning, status, and trends, and the consequences of its loss, are improved, widely shared and transferred, and applied
	20	By 2020, at the latest, mobilization of financial resources for effectively implementing the Strategic Plan 2011–2010 from all sources and in accordance with the consolidated and agreed process in the Strategy for Resource Mobilization should increase substantially from the current levels

Meanwhile, the focus of work by the scientific community was to develop robust and measurable biodiversity indicators based on the target, which could then be translated into monitoring and management systems (Green *et al.*, 2005). One view was that too little attention had been paid to the design of the indicator set with the risk that whatever the outcome in 2010, it would be hard to draw clear conclusions about whether or not the target had been met (Mace and Baillie, 2007). The opportunity to put meaningful baselines in place in practice could be missed. A key problem identified with the 2010 indicator set being implemented through the BIP was the weak relevance of many indicators to the overall target, due in part to the rushed process used to develop measures. Many measures were selected mainly because data were available but many of these also lacked baselines and scales, and most were poorly sampled. In addition to highlighting the importance of making the indicators relevant to the target and to the goals of biodiversity management, there were calls to distinguish measures of pressure, state, and response; to design and validate the measures in context; to ensure that indicators communicate effectively to relevant audiences; to decide when composite indicators would be more useful than multiple independent measures; and to maximize the cost-effectiveness of the process (Green *et al.*, 2005; Mace and Baillie, 2007).

Some of these concerns have been met in the new Strategic Plan, which is centred around five main goals that follow the D-P-S-I-R (Drivers-Pressures-State-Impacts-Responses) model, as well as 20 'headline targets' (*20 for 2020*), which combine scientific and political concerns. For example, the level of each target may be political, but the type of target may have clearer scientific roots (see Table 13.1).

Given the short time between 2002 and 2010 and the breadth of the 2010 target, it is perhaps not surprising that many indicators were not developed in time to deliver clear outcomes (Walpole *et al.*, 2009). In fact, as it turned out, and despite perceived weaknesses in the measures, there was a clear consensus, at COP10, that the target had not been met (Butchart *et al.*, 2010). Several of the indicator measures were only developed at the end of the process, while many others are still under construction, so that the reporting in 2010 was necessarily limited in scope and relevance (Mace and Baillie, 2007; Walpole *et al.*, 2009). Undoubtedly this increases the risk of a loss of commitment to the overall process after 2010. At the same time, however, there is the opportunity to build on the initiative and develop robust targets and processes for the post-2010 period, and to learn lessons, especially about where there could be more profitable links between science and policy. Here we discuss how more effective multilateral biodiversity agreements might be built post-2010, including creating targets that are likely to have more impact, and enhancing the links between science and policy. In this regard, the establishment of a new intergovernmental process, the Intergovernmental Panel on Biodiversity and Ecosystem Services (IPBES), which the UN General Assembly formally approved in December 2010, could play a key role in the development, implementation, and measurement of the new post-2010 biodiversity targets (Larigauderie and Mooney, 2010).

Twenty targets for 2020

Understand the role of targets in international governance

The international governance of the biosphere is structured around the three 'Rio Conventions' on biodiversity, climate, and desertification, together with a large number of more specific multilateral environmental agreements. In many cases these establish broad goals only, but in some cases they also include quite specific targets for the environmental processes under consideration. The Long Range Transboundary Air Pollution Convention (LRTAP), for example, has detailed targets for the reduction of emissions of sulphur dioxide and nitrogen oxides under its eight protocols. Analyses of the effectiveness of multilateral environmental agreements have shown the kinds of targets that may be successfully negotiated, the way in which these vary with the environmental problem concerned, and the number of parties to an agreement (Sandler, 2004). The test of the effectiveness of any agreement is its capacity to support an outcome better than would occur in the absence of the agreement. To date, the multilateral environmental agreements that have been most effective in meeting their stated objectives are those that (i) involve a limited number of signatories; (ii) include commitments that have evolved through repeated renegotiation; and (iii) include effective penalties or disincentives to defect from a precisely defined set of objectives (Sandler, 2004). That is, they can only be effective if they are self-enforcing. The targets a particular agreement is able to agree upon depend on the same things that determine its effectiveness and enforceability. Barrett's analysis of the CBD (Barrett, 1994), for example, concluded that given the number of signatories, the agreement could not offer benefits that were significantly different from the non-cooperative outcome. In other words, any targets it was able to agree on could not be significantly different from the outcome in the absence of the agreement.

A later study of the Helsinki Protocol of the LRTAP, an agreement with many fewer signatories than the CBD, found that the initial abatement targets it contained were what the signatories would have expected to do in the absence of any agreement (Murdoch et al., 1997). The initial targets of the Montreal Protocol of the Vienna Convention on the ozone layer were similarly close to the outcome that would have been expected in the absence of agreement. In both cases, however, while the initial targets negotiated in multilateral environmental agreements represented the non-cooperative outcome, these targets have been progressively strengthened through successive renegotiation (Sandler, 2004).

The implications of this for the post-2010 biodiversity target is that the outcome is likely to be sensitive to the process of target setting. A single target negotiated once and not revisited for a decade is likely to be a statement of what countries expect to happen in the absence of collective action. Initial targets may simply be agreed at levels that countries have already attained under current practices or were set to attain anyway, or need to have for ancillary benefits (e.g. aid, political legitimacy). A more effective approach would involve not just a single target, but a series of intermediate objectives and steps that need to be taken to implement each objective, with the target being re-evaluated on the completion of each step. After 20 years, for example, through a continuous political, scientific, and technical dialogue, the Montreal protocol on ozone-depleting substances, which has been ratified by over 190 countries and the European Union, has progressively expanded the range of substances under control, strengthened Parties' commitments (in the context of the principle of common but differentiated responsibilities), and shortened compliance deadlines. There is a robust monitoring programme that measures the atmospheric concentration of ozone-depleting substances, ozone layer thickness, and trends in production, consumption, emissions, and trade. In fact it is believed that with implementation of the Protocol's provisions the ozone layer should return to pre-1980 levels by 2050 to 2075. By contrast, most of the biodiversity protection targets still lack baseline benchmarks and the kind of regular monitoring that would permit real tracking of trends.

Most current environmental targets aim at improving generic capacities (including adoption of plans, creation of policy frameworks, conducting assessments, and setting priorities), or at reducing pressures (lowering emissions, extraction, or conversion). It is rarer to find targets that aim to reduce drivers or to achieve specific states. The new Strategic Plan has made a welcome effort in this regard, even though the drivers and state targets are couched in very broad terms and often lack specific baselines and indicators. Regional air pollution in Europe is the best-developed example of a targeting process that focuses on environmental states (in this case, levels of deposition relative to critical loads). Such targets have the advantage that they include clear links to societal health and well-being and have therefore greater political leverage.

Adjust targets as external conditions change

Most of the comments in this chapter reflect a scientific view of targets. Scientists tend to evaluate targets in relation to their scientific relevance, how measurable they are, and how those measures meet the policy goal. However, governments are often more sensitive to what it means to miss targets; whether they can become political tools in the hands of domestic or foreign critics; or whether donors may use their achievement or non-achievement to reallocate development assistance. Targets may stimulate action, in part by clarifying objectives, rallying organizations around a common goal, empowering weak bureaucracies, and giving civil society a means to judge and publicize government efforts. However, an overemphasis on formal compliance with targets may in fact detract attention from the original objectives, as when states agree to do something they are already set to achieve. Moreover, efforts to present the achievement of the target in the best light possible may lead to information being presented in a manner that weakens the overall credibility of the policy in question. Indeed, meeting or missing a target often says little about the effectiveness of actions designed to further the general goal of policy.

Many of the political disadvantages of targets stem from the fact that they are seldom framed around external conditions, yet changes in external conditions may impede the achievement of targets. For example, unforeseen climate change impacts or the domino effects of a severe financial crisis may nullify policies and programmes that would otherwise be effective. To avoid the danger of missing targets, policy-makers tend to 'shoot low' and to set targets that are achievable in all conditions. Further, regimes tend to be less effective in a context of high uncertainty regarding the state of scientific knowledge, the distribution of costs and benefits, and the nature of the national interests. Thus, learning becomes a crucial pre-condition of effectiveness and needs to be built into the structure of target-setting itself, as it is in the Montreal Protocol. It would be more effective to set targets that recognize the potential conditionality on the state of knowledge, political coalitions, interests markets, or the physical environment, and allow the process to be more adaptive to such changes.

In a way, the targets of the new strategic plan allow for this in that they do not, except in some instances (as in the land area devoted to protected areas), presume how progress will be measured. That is, they set a desirable state of affairs (whose definition can itself evolve) but allow for change in how it is to be achieved, which takes into account both the evolution of knowledge and different national circumstances. What is unclear, however, is how this learning process will be institutionalized.

Organize an iterative process involving decision-makers and the relevant research community

Developing science-based targets for socially relevant biodiversity indicators will require a much stronger link between the science and policy communities. It will also require stronger links between the social sciences (notably economics, sociology, anthropology, and political science) and the natural sciences, and should connect the targets to policy and decision-making. The key topics are likely to lie at the interface of social and environmental concerns, and can benefit from modelling approaches that help to identify alternative kinds of interventions and then identify costs, benefits, and trade-offs (Alcamo *et al.*, 2005; Hezri and Dovers, 2006; Wätzold *et al.*, 2006; Gottschalk *et al.*, 2007; Carpenter *et al.*, 2009).

Compared to other environmental agreements, the science-policy process by which the CBD develops and implements recommendations is relatively unstructured. The CBD has a regular assessment process called Global Biodiversity Outlook, which produces regular scientific syntheses on the state of biodiversity to CBD policy-makers. COP10, for example, concluded that the 2010 target had been missed based on the evidence presented in the Global Biodiversity Outlook 3 (2010). There is, however, no mechanism for the delivery of science advice to be provided through the kind of credible, legitimate, and salient process that has been successful in other areas of environmental assessment, for example the Intergovernmental Panel on Climate Change (IPCC) and the Millennium Ecosystem Assessment (Reid and Mace, 2003, Watson, 2005). This must confound decision-making processes within the CBD since independent scientific advice is not distinguished from recommendations made to align CBD processes with the priorities of individual organizations or parties. Without independent science advice, the Parties also lose out on technical analyses showing the costs and benefits of alternative policies or targets, and therefore underpinning better decisions. As we have seen, this process can also lead to targets or policies that are either trivial or impossible to achieve in practice.

The creation of the IPBES should help develop a more consistent and well-structured link between the international policy community and science than currently exists. Once established, an IPBES will have the capacity both to undertake regular assessments of biosphere change, and to respond to requests for more rapid assessment of emerging issues of potential importance (Larigauderie and Mooney, 2010).

Set goals and targets in context

A key element of a revised set of targets will be to ensure that the global interest in local biodiversity change is properly represented to avoid biodiversity change that threatens human well-being at all levels. This includes securing the current and future supply of ecosystem services as well as meeting broader needs that society has for biodiversity. For society to move towards realizing such a goal, Parties need to adopt a small set of focused, relevant, efficient, and achievable targets. Each of these targets should have scientifically and socially appropriate outcomes and timescales, support biodiversity's role in human well-being, be linked to legislative and regulatory processes, be relevant at global scales but reflect local and national interests, and be open to accurate and efficient reporting.

Although the number of 'headline targets' has remained about the same from the first to the second Strategic Plan, the 20 targets of the post-COP10 are often much more specific, even when they may not meet all the conditions above. Biodiversity in its broadest sense matters to people in different ways. It directly underpins certain ecosystem functions and services, it contributes to aesthetic and cultural values, and it is a part of the sustainable life support system upon which all life ultimately depends. Urgent versus important priorities under each of these headings will not be the same. Choosing among them will benefit from a clear articulation about why certain choices have been made. Here we define three different categories of targets. These are not mutually exclusive, but can be used to classify targets according to their primary motivation (Perrings *et al.*, 2010b). This classification should help decision-makers to clarify priorities among competing agendas, as well as to focus the science-based management strategies appropriately (Larigauderie *et al.*, 2010; Mace *et al.*, 2010).

- *Red targets* – addressing biodiversity change that is directly harmful to people. Red targets are designed to avoid or avert urgent and unacceptable changes in biodiversity that will be damaging to people in the near term. They largely map onto the biosecurity agenda.
- *Green targets* – conserving biodiversity components valued by society for non-utilitarian purposes. Green targets will focus on long-term priorities for the conservation of biodiversity often focusing on species and habitats. They largely map onto the conservation agenda.
- *Blue targets* – understanding and governing the system. In the long term, sustainable management of the biosphere depends on knowledge of the underpinning processes and an effective system to manage it. Blue targets focus on steps in progress towards this end and map on to the long-term sustainability agenda.

Embed the targets within a continuing process

We recommend that parties to the CBD develop, based on the 20 targets for 2020 agreed upon by CBD COP10 in Nagoya, a small set of focused, relevant, efficient, and achievable targets each of which has socially relevant outcomes and timescales, supports biodiversity's role in human well-being, is linked to legislative and regulatory processes, and is open to accurate and efficient reporting (Mace *et al.*, 2010). In contrast to the first Strategic Plan, which had a fixed 8-year period for all targets and sub-targets, a continuing process has many advantages. As any one specific target is met or if it becomes irrelevant it can be replaced or reformulated (Figure 13.1). Thus, three targets (10, 16, and 17) in the revised strategic plan have a shorter timescale (2015). Targets that necessarily involve longer timescales, because the processes involved

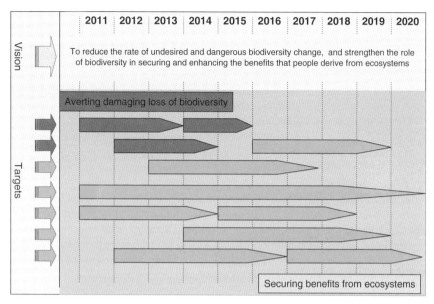

Figure 13.1 A schematic representation for a post-2010 biodiversity target. The long-term vision outlines the ultimate goals. The vision is realized through setting specific short-term targets around averting dangerous or deleterious biodiversity loss and/or securing future benefits from ecosystems. Three different kinds of target are identified (see Mace *et al.*, 2010). Dark-grey targets (arrows 1 and 2 from top) address short-term threats to human well-being from biodiversity change. Mid-grey targets (arrows 3–5 from top) address the conditions needed to ensure long-term sustainability of ecosystem service supplies as well as the governance of biodiversity and for 'access and benefit sharing'. Light-grey targets (arrows 6 and 7 from top) address global conservation goals such as the establishment of protected areas and the prevention of species extinctions. Over the course of time it might be hoped that targets would increasingly focus on proactive measures rather than the reactive ones associated with avoiding dangerous change. Note this image is for illustrative purposes and is not intended to reflect the ideal number of targets nor their timespans.

have longer-term dynamics, can be accommodated, and newly emerging issues and problems can be incorporated without delay.

Improve coordination in the provision and use of biodiversity data

Many biodiversity data still stem from national government sources or from national or international non-governmental organization (NGO) databases. In both cases they have often been collected for other purposes – mostly related to species conservation – leading to spatial, temporal, or taxonomic gaps in global coverage and rendering the information base less reliable for global indicators and less relevant for global assessment, monitoring, and decision-making than if data sampling were more directed or coordinated (Dobson, 2005; Lawler et al., 2006; Pereira and Cooper, 2006). As a result there were important gaps in the data and measures being used for 2010, including key ecosystem processes, functions, and services and the biological taxa (often microorganisms) that underpin them; key drivers such as land-use change, disease, and climate change; or key benefits, such as human health, nutrition, and welfare (Balmford et al., 2005a, 2005b; Mace and Baillie, 2007). The data are also widely dispersed, held in a wide variety of formats, and under diverse ownership with variable accessibility.

Instead of creating new datasets and generating new data-gathering activities, the processes behind emerging specific targets should interact with ongoing international efforts to devise a global biodiversity observation network and associated system (Pereira and Cooper, 2006; Scholes et al., 2008). The Global Earth Observation System of Systems (GEOSS; www.earthobservations.org) is being developed under the aegis of the Group on Earth Observations (GEO) to address just such problems of international data sharing. It has recently established a Biodiversity Observation Network (GEO BON), a new global partnership to collect, manage, analyse, and report on data relating to the status of the world's biodiversity in a more coordinated and strategic way. Rather than creating new databases, the rationale for GEO BON is that it should add value to existing efforts by coordinating and linking the broad range of data-gathering and management that already exists. The links to GEO also provide new opportunities for integrating remote sensing observations with ground-based studies and surveys, and using emerging computational, imaging, and visualization tools to link diverse datasets in a way that is informative for users. The fact that GEO BON is in its early phase of development makes it timely for a discussion of how it could contribute to a new set of targets (www.earthobservations.org/geobon) (Walther et al., 2007; Scholes et al., 2008). The CBD has formally recognized GEO BON as a strategic partner for the implementation of its new strategic plan 2011–2020, and at COP10, tasked GEO BON with evaluating observation capabilities relevant to the 20 targets of the strategic plan.

Exploit the science base

The publication of the Millennium Ecosystem Assessment (2005) and the 2010 target both stimulated a great deal of new work on biodiversity and ecosystem services, building upon existing ecological studies undertaken in model and experimental systems (Carpenter et al., 2006, 2009; Díaz et al., 2006). This had the potential to support enhanced biodiversity assessment, but currently much policy development still lags far behind scientific discovery, and without the 'policy-pull' as well as the 'science-push' this is unlikely to change. As the IPCC has shown, effective interaction with policy can stimulate necessary and useful new science. We give some examples here of potentially useful scientific developments, but note that this is by no means a complete review.

1. *Biodiversity indices* developed in recent years could provide efficient ways for summarizing biodiversity condition and trends. While further work is needed to validate and fine-tune such measures and to assess their utility and efficiency, the Biodiversity Intactness Index (BII) (Scholes and Biggs, 2005; Rouget *et al.*, 2006; Nielsen *et al.*, 2007; Faith *et al.*, 2008) and the Human Appropriation of Net Primary Production (HANPP) (Imhoff *et al.*, 2004; Haberl *et al.*, 2007) are two examples of promising new approaches to the task of summarizing complex information in a way that can contribute to relevant indices (Dobson, 2005).
2. *Biodiversity processes*. Recent work has also demonstrated the important distinction between biological diversity (variability), which is important for the resilience and adaptability of ecological systems, compared to the role of composition (the presence of certain types of species) in many provisioning and some cultural and regulating services (Diaz *et al.*, 2007). Additionally, the emerging evidence is that it is trait variability and composition (called 'functional diversity') that influences both the responses and functions of species within ecosystems, and new compilations of trait values across species and ecosystems offer the potential to assess ecosystem effects that will result from ecosystem transformation or species loss (Diaz *et al.*, 2006, 2007). A communal worldwide repository of plant trait information, called TRY (www.try-db.org), has been established and already contains information on more than 60 000 species and populations. Potentially these emerging datasets could soon point to places or ecological systems where ongoing transformations or loss of biodiversity could be particularly risky or deleterious for specific ecosystem services. For example, we may be able to identify community changes that reduce decomposition rates, which limit productivity or reduce the flood regulatory role of an ecosystem. Quantifying and spatially mapping important ecological and evolutionary processes that drive the distribution and abundance of biodiversity can now explicitly be incorporated into prioritizations (see Klein *et al.*, 2009). Data and methods are now becoming available that attempt to quantify some of these processes at appropriate spatial scales (Pressey *et al.*, 2007; Mackey *et al.*, 2008), but incorporating all known major ecological and evolutionary processes in a target-driven process has not yet been done.
3. *Baselines and thresholds*. Unlike the 2010 target, which had no baselines or reference points, and where declines are treated the same regardless of the starting state of the system and the rate of change, scientific work could identify specific levels or rates of change that should not be exceeded. Several of the 2020 targets are

based around avoiding especially risky levels of exploitation (Milner-Gulland and Akcakaya, 2001); critical population sizes and structures that disproportionately increase extinction risk; and habitat areas or spatial configurations that lead to loss of key functions or local community compositions (Laurance and Williamson, 2001; Scheffer et al., 2001; Laurance et al., 2006; Chapin et al., 2008). True thresholds such as these are far more defensible than arbitrary levels, or targets set at levels that are thought to be achievable or the likely outcome of current policies.

4. *Indicators of state changes.* Another emerging set of studies is investigating state changes and ecosystem shifts (Pace et al., 1999; Scheffer and Carpenter, 2003; Chapin et al., 2010). Such shifts are increasingly documented and are characterized by occurring suddenly and unpredictably in systems under change that may also be exposed to some external shock or pressure. The shifts often take ecosystems to new states and are hard to reverse, even when the shocks are removed and environmental conditions are returned to their original states (Scheffer et al., 2001). New work focused on features of systems that are approaching state changes offers some hope that signals may be identified that might allow interventions to reverse the trend, at least for certain kinds of management (Biggs et al., 2009). More generally, enhanced understanding of the biophysical processes involved will always contribute to more accurate predictions about how ecosystem processes and functions will respond to various kinds of environmental change – a key role for basic science.

5. *Enhanced efforts to understand and monitor new and interacting processes.* Climate change impacts on biodiversity loss, invasive species, and the emergence of novel diseases are processes that are escalating in impact as a result of interactions with a set of anthropogenic processes, of which the most important may be the growth of global trade, transport, and travel (Daszak et al., 2000; McGeoch et al., 2006; Mooney et al., 2009). In some cases, and for some pathogen types, biodiversity loss may lead to the dominance of highly competent reservoir hosts of some human pathogens, and thus a greater disease risk in people (Perrings et al., 2010a, 2010b; Thomas and Ohlemüller, 2010). Thus, conservation of biodiversity may buffer human populations from infectious disease risk. This has been shown using long-term field studies for Lyme disease, a tick-borne pathogen of humans with rodent and other reservoir hosts, and may occur for a number of other vector-borne diseases. In other cases, the link with biodiversity is more complicated. Emergent zoonoses depend on increasingly close contact between species – and especially between wildlife, domesticated livestock, and people (Perrings and Mooney, 2010).

6. *Integrated technologies for biodiversity assessment.* Emerging techniques and tools, for example in molecular genetics, biodiversity models, and remote sensing, may provide new and efficient means for biodiversity monitoring. There are relevant metrics to be developed based on genetic assessments of whole communities, genetic or phylogenetic measures of spatial and temporal turnover, and remote-sensed observations of biodiversity that contribute to global models (Ferrier et al.,

2000; Jennings *et al.*, 2008; Yahara *et al.*, 2010) or to assessments of ecosystem change (DeFries *et al.*, 2005).
7. *Optimization of monitoring.* New research is identifying methods to optimize the allocation of effort in monitoring or measurement of biodiversity in context. While rarely considered, there is much potential for more explicit consideration of why global biodiversity indicators are needed, what characteristics such indicators should have, and how available measures perform with respect to those characteristics compared to new measures that could be developed from scratch (Jones *et al.*, 2011; see also Chapter 15). The optimization techniques can help decision-makers determine when they have enough information to act (Gerber *et al.*, 2005). Such approaches could demonstrably improve the cost-effectiveness of indicator development (Margules and Pressey, 2000; Wilson *et al.*, 2009).
8. *Strengthening of institutions for biodiversity management and the behaviour of agents that drive biodiversity change.* Increasing attention is being paid to the impact of interactions among various levels of governance, and the performance of various multi-scalar governance models, the role of various agents beyond the state (industry, non-profit organizations, private networks), the drivers of norm dissemination and the interactions among competing norms of justice, and the relationship between perceptions, legitimacy, and behaviour (Waddel and Khagram, 2005; Lebel *et al.*, 2006). Governance targets are likely to be amongst the most important of the blue targets discussed above, and their dominance in the emerging 2020 targets is one illustration of this trend (Perrings *et al.*, 2010b).

Conclusions

The development of an efficient set of biodiversity targets and indicators for the post-2010 period is a precondition for the coordination of efforts to address the most serious environmental problem the world faces. Without agreement on global objectives embodied in feasible targets, and supported by adequate measures, national action on biosphere change will continue to be wholly driven by domestic concerns, and will continue to neglect the consequences of national actions for the global public good. We have recommended, here and elsewhere, three sets of targets: 'red' targets to address changes in biodiversity that have imminent and grave consequences for human well-being (the biosecurity agenda); 'green' targets to implement the precautionary husbanding of the world's genetic resources (the conservation agenda); and 'blue' targets to protect the long-term capacity of the system to maintain the supply of valued ecosystem services (the sustainability agenda) (see Mace *et al.*, 2010). To be effective, each set of targets should be collectively agreed, be mutually consistent, be time-bound, be sensitive to changes in external conditions, and be supported by adequate measures (indicators). The central point made in this chapter is that these targets should also be defensible in terms of both the science of biodiversity and environmental change, and the science of the interactions between human society and the biosphere.

Different areas of science are relevant for each category of target. Red targets, for example, would address short-term threats to human well-being from biodiversity change. They would typically be related to the threshold effects associated with disease outbreaks, population collapse, or irreversible changes of state in ecosystems. Accordingly, they might be expected to focus on species abundance, habitat extent, concentrations of pollutants, and the like. Blue targets, on the other hand, would address the conditions needed to ensure long-term sustainability of ecosystem service supplies. The key indicators in this case would be volume and value measures of ecosystem services, including their importance for different groups of people. Since many services are marketed, such measures include indicators that are routinely monitored for other reasons – such as production and trade statistics. New indicators will need to be developed for non-marketed ecosystem services. The blue targets may be thought of as 'sustainable use' targets. But they might also include objectives for the governance of biodiversity and for 'access and benefit sharing'. The green targets would address global conservation goals as reflected in traditional conservation mechanisms, such as protected areas. However, they would also address conservation objectives outside of protected areas, including objectives for the 'restoration' of managed or converted ecosystems and for the protection of habitat in agricultural, urban, and industrial ecosystems.

The point is that target setting for the biosecurity, conservation, and sustainability agendas requires different indicators, supported by different areas of science. The legacy of the historical separation of production, conservation, and biosecurity is an institutional divide that is reflected in the supporting science. At both national and international levels, the institutions addressing agriculture, forestry, and fisheries are currently distinct from those addressing human, animal, plant, or ecosystem health or those addressing biodiversity conservation. So too are the branches of science. Following the Millennium Assessment, however, we have come to think of the biodiversity problem as the integration of these three areas. Indeed, the focus on ecosystem services suggests the need for targets that relate to all of the benefits that people obtain from ecosystems, with targets for species abundances, habitat area, and the like being derived from that. The areas of science, assessment, and monitoring described in this chapter are the necessary ingredients for the construction of an integrated set of targets.

Acknowledgements

This chapter arose from discussions facilitated by DIVERSITAS. We thank Nathalie Pettorelli, Belinda Reyers, and David Salt for helpful comments on an earlier version.

References

Alcamo, J., van Vuuren, D., Ringler, C., *et al.* (2005) Changes in nature's balance sheet: Model-based estimates of future worldwide ecosystem services. *Ecology and Society*, 10, 19.

Balmford, A., Crane, P., Dobson, A., Green, R.E., and Mace, G.M. (2005a) The 2010 challenge: Data availability, information needs and extraterrestrial insights. *Philosophical Transactions of the Royal Society B, Biological Sciences*, 360, 221–228.

Balmford, A., Bennun, L.A., ten Brink, B., *et al.* (2005b) The Convention on Biological Diversity's 2010 target. *Science*, 307, 212–213.

Barrett, S. (1994) The biodiversity supergame. *Environmental and Resource Economics*, 4, 111–122.

Biggs, R., Carpenter, S.R., and Brock, W.A. (2009) Turning back from the brink: Detecting an impending regime shift in time to avert it. *Proceedings of the National Academy of Sciences of the USA*, 106, 826–831.

Butchart, S.H.M., Walpole, M., Collen, B., *et al.* (2010) Global biodiversity: indicators of recent declines. *Science*, 328, 1164–1168.

Carpenter, S.R., DeFries, R., Dietz, T., *et al.* (2006) Millennium Ecosystem Assessment: Research needs. *Science*, 314, 257–258.

Carpenter, S.R., Mooney, H.A., Agard, J., *et al.* (2009) Science for managing ecosystem services: Beyond the Millennium Ecosystem Assessment. *Proceedings of the National Academy of Sciences of the USA*, 106, 1305–1312.

Chapin, F.S., Randerson, J.T., McGuire, A.D., Foley, J.A., and Field, C.B. (2008) Changing feedbacks in the climate-biosphere system. *Frontiers in Ecology and the Environment*, 6, 313–320.

Chapin F.S. III,, Carpenter, S.R., Kofinas, G.P., *et al.* (2010) Ecosystem stewardship: sustainability strategies for a rapidly changing planet. *Trends in Ecology and Evolution*, 25, 241–249.

Daszak, P., Cunningham, A.A., and Hyatt, A.D. (2000) Wildlife ecology – emerging infectious diseases of wildlife – threats to biodiversity and human health. *Science*, 287, 443–449.

DeFries, R., Pagiola, S., Adamowicz, W.L., *et al.* (2005) Analytical approaches for assessing ecosystem condition and human well-being. In: *Global Assessment Report, Vol. 1: Ecosystems and Human Well-being: Current State and Trends*. Millennium Ecosystem Assessment, Island Press, Washington, DC, Chapter 2.

Díaz, S., Fargione, J., Chapin, FS III,, and Tilman, D. (2006) Biodiversity loss threatens human well-being. *PLoS Biology*, 4, 1300–1305.

Díaz, S., Lavorel, S., de Bello, F., Quetier, F., Grigulis, K., and Robson, M. (2007) Incorporating plant functional diversity effects in ecosystem service assessments. *Proceedings of the National Academy of Sciences of the USA*, 104, 20684–20689.

Dobson, A. (2005) Monitoring global rates of biodiversity change: challenges that arise in meeting the Convention on Biological Diversity (CBD) 2010 goals. *Philosophical Transactions of the Royal Society B, Biological Sciences*, 360, 229–241.

Faith, D.P., Ferrier, S., and Williams, K.J. (2008) Getting biodiversity intactness indices right: ensuring that 'biodiversity' reflects 'diversity'. *Global Change Biology*, 14, 207–217.

Ferrier, S., Pressey, R.L., and Barrett, T.W. (2000) A new predictor of the irreplaceability of areas for achieving a conservation goal, its application to real-world planning, and a research agenda for further refinement. *Biological Conservation*, 93, 303–325.

Gerber, L.R., Beger, M., McCarthy, M.A., and Possingham, H.P. (2005) A theory for optimal monitoring of marine reserves. *Ecology Letters*, 8, 829–837.

Gottschalk, T.K., Diekotter, T., Ekschmitt, K., *et al.* (2007) Impact of agricultural subsidies on biodiversity at the landscape level. *Landscape Ecology*, 22, 643–656.

Green, R.E., Balmford, A., Crane, P.R., Mace, G.M., Reynolds, J.D., and Turner, R.K. (2005) A framework for improved monitoring of biodiversity: Responses to the World Summit on Sustainable Development. *Conservation Biology*, 19, 56–65.

Haberl, H., Erb, K.H., Krausmann, F., *et al.* (2007) Quantifying and mapping the human appropriation of net primary production in earth's terrestrial ecosystems. *Proceedings of the National Academy of Sciences of the USA*, 104, 12942–12945.

Hezri, A.A. and Dovers, S.R. (2006) Sustainability indicators, policy and governance: Issues for ecological economics. *Ecological Economics*, 60, 86–99.

Imhoff, M.L., Bounoua, L., Ricketts, T., Loucks, C., Harriss, R., and Lawrence, W.T. (2004) Global patterns in human consumption of net primary production. *Nature*, 429, 870–873.

Jennings, S., Melin, F., Blanchard, J.L., Forster, R.M., Dulvy, N.K., and Wilson, R.W. (2008) Global-scale predictions of community and ecosystem properties from simple ecological theory. *Proceedings of the Royal Society B, Biological Sciences*, 275, 1375–1383.

Jones, J.P.G., Collen, B., Atkinson, G., *et al.* (2011) The why, what and how of global biodiversity indicators beyond the 2010 target. *Conservation Biology*, 25, 450–457.

Klein, C., Wilson, K., Watts, M., *et al.* (2009) Incorporating ecological and evolutionary processes into continental-scale conservation planning. *Ecological Applications*, 19, 206–217.

Larigauderie, A. and Mooney, H.A. (2010) The Intergovernmental science-policy Platform on Biodiversity and Ecosystem Services: moving a step closer to an IPCC-like mechanism for biodiversity. *Current Opinion in Environmental Sustainability*, 2, 9–14.

Larigauderie, A., Mace, G., and Mooney, H. (2010) Colour-coded targets would help clarify biodiversity priorities. *Nature*, 464, 160.

Laurance, W.F. and Williamson, G.B. (2001) Positive feedbacks among forest fragmentation, drought, and climate change in the Amazon. *Conservation Biology*, 15, 1529–1535.

Laurance, W.F., Nascimento, H.E.M., Laurance, S.G., *et al.* (2006) Rapid decay of tree-community composition in Amazonian forest fragments. *Proceedings of the National Academy of Sciences of the USA*, 103, 19010–19014.

Lawler, J.J., Aukema, J.E., Grant, J.B., *et al.* (2006) Conservation science: a 20-year report card. *Frontiers in Ecology and the Environment*, 4, 473–480.

Lebel, L., Anderies, J.M., Campbell, B., *et al.* (2006) Governance and the capacity to manage resilience in regional social-ecological systems. *Ecology and Society*, 11, 19.

Mace, G.M. and Baillie, J.E.M. (2007) The 2010 Biodiversity Indicators: challenges for science and policy. *Conservation Biology*, 21, 1406–1413

Mace, G.M., Cramer, W., Díaz, S., *et al.* (2010) Biodiversity targets after 2010. *Current Opinion in Environmental Sustainability*, 2, 3–8.

Mackey, B.G., Berry, S.L., and Brown, T.(2008) Reconciling approaches to biogeographical regionalization: a systematic and generic framework examined with a case study of the Australian continent. *Journal of Biogeography*, 35, 213–229.

Margules, C.R. and Pressey, R.L. (2000) Systematic conservation planning. *Nature*, 405, 243–253.

McGeoch, M.A., Chown, S.L., and Kalwij, J.M. (2006) A global indicator for biological invasion. *Conservation Biology*, 20, 1635–1646.

Milner-Gulland, E.J. and Akcakaya, H.R. (2001) Sustainability indices for exploited populations. *Trends in Ecology and Evolution*, 16, 686–692.

Mooney, H.A. (2010) The ecosystem-service chain and the biological diversity crisis. *Philosophical Transactions of the Royal Society B, Biological Sciences*, 365, 31–39.

Mooney, H., Larigauderie, A., Cesario, M., *et al.* (2009) Biodiversity, climate change, and ecosystem services. *Current Opinion in Environmental Sustainability*, 1, 46–54.

Murdoch, J.C., Sandler, T., and Sargent, K. (1997) A tale of two collectives: sulphur versus nitrogen oxide emission reduction in Europe. *Economica*, 64, 281–301.

Nielsen, S.E., Bayne, E.M., Schieck, J., Herbers, J., and Boutin, S. (2007) A new method to estimate species and biodiversity intactness using empirically derived reference conditions. *Biological Conservation*, 137, 403–414.

Pace, M.L., Cole, J.J., Carpenter, S.R., and Kitchell, J.F. (1999) Trophic cascades revealed in diverse ecosystems. *Trends in Ecology and Evolution*, 14, 483–488.

Pereira, H.M. and Cooper, H.D. (2006) Towards the global monitoring of biodiversity change. *Trends in Ecology and Evolution*, 21,: 123–129.

Perrings, C. and Mooney, H.M.W. (2010) The problem of biological invasions. In: Perrings, C., Mooney, H., and Williamson, M. (eds), *Globalization and Bioinvasions: Ecology, Economics, Management and Policy*. Oxford University Press, pp. 1–18.

Perrings, C., Burgiel, S., Lonsdale, W.M., Mooney, H., and Williamson, M. (2010a) International cooperation in the solution to trade-related invasive species risks. *Annals of the New York Academy of Sciences*, 1195, 198–212.

Perrings, C., Naeem, S., Ahrestani, F., *et al.* (2010b) Ecosystem services for 2020. *Science*, 330, 323–324.

Pressey, R.L., Cabeza, M., Watts, M.E., Cowling, R.M., and Wilson, K.A. (2007) Conservation planning in a changing world. *Trends in Ecology and Evolution*, 22, 583–592.

Reid, W.V. and Mace, G.M. (2003) Taking conservation biology to new levels in environmental decision-making. *Conservation Biology*, 17, 943–945.

Rouget, M., Cowling, R.M., Vlok, J.A.N., Thompson, M., and Balmford, A. (2006) Getting the biodiversity intactness index right: the importance of habitat degradation data. *Global Change Biology*, 12, 2032–2036.

Sandler, T. (2004) *Global Collective Action*. Cambridge University Press, Cambridge.

Scheffer, M. and Carpenter, S.R. (2003) Catastrophic regime shifts in ecosystems: linking theory to observation. *Trends in Ecology and Evolution*, 18, 648–656.

Scheffer, M., Carpenter, S., Foley, J.A., Folke, C., and Walker, B. (2001) Catastrophic shifts in ecosystems. *Nature*, 413, 591–596.

Scholes, R.J. and Biggs, R. (2005) A biodiversity intactness index. *Nature*, 434, 45–49.

Scholes, R.J., Mace, G.M., Turner, W., *et al.* (2008) Toward a global biodiversity observing system. *Science*, 321, 1044–1045.

Thomas, C. and Ohlemüller, R. (2010) Climate change and species distributions: an alien future? In: Perrings, C., Mooney, H., and Williamson, M. (eds), *Bioinvasions and Globalization: Ecology, Economics, Management, and Policy*. Oxford University Press,

Waddel, S. and Khagram, S (2005) Multi-stakeholder global networks: emerging systems for the global common good. In: Glasbergen, P., Bierman, F., and Mol, A.P.J. (eds), *Partnerships, Governance and Sustainable Development: Reflections on Theory and Practice*. Edward Elgar Publishing, pp. 261–287.

Walpole, M., Almond, R., Besançon, C., *et al.* (2009) Tracking progress toward the 2010 biodiversity target and beyond. *Science*, 325, 1503–1504.

Walther, B.A., Larigauderie, A., Ash, N., Geller, G.N., Jürgens, N., and Lane, M.A. (2007) Toward a global biodiversity observation network. In: *GEOSS Components – Observing Systems. The Full Picture*. Tudor Rose, Geneva, pp. 79–81.

Watson, R.T. (2005) Turning science into policy: challenges and experiences from the science-policy interface. *Philosophical Transactions of the Royal Society B, Biological Sciences*, 360, 471–477.

Wätzold, F., Drechsler, M., Armstrong, C.W., *et al.* (2006) Ecological-economic modeling for biodiversity management: Potential, pitfalls, and prospects. *Conservation Biology*, 20, 1034–1041.

Wilson, K.A., Carwardine, J., and Possingham, H.P. (2009) Setting conservation priorities. *Annals of the New York Academy of Sciences*, 1162, 237–264.

Yahara, T., Donoghue, M., Zardoya, R., Faith, D.P., and Cracraft, J. (2010) Genetic diversity assessments in the century of genome science. *Current Opinion in Environmental Sustainability*, 2, 43–49.

Part IV
Biodiversity Monitoring in Practice

14

Building Sustainable National Monitoring Networks

Sarah M. Durant

Institute of Zoology, Zoological Society of London, London, UK
Wildlife Conservation Society, Bronx, NewYork, USA

Biodiversity monitoring is a necessary part of conservation management and is key to management plans of protected areas and ecosystems. Quantitative evidence of biodiversity decline or conservation success is often an essential precursor for political action, and information from biodiversity monitoring enables better management of biodiversity (du Toit *et al.*, 2004; Rands *et al.*, 2010; Watson and Novelly, 2004). Furthermore, the complexity of biological systems means they are best managed adaptively within a holistic socioecological framework, necessitating a sound understanding of resource and ecosystem dynamics (Caddy and Seij, 2005; Folke *et al.*, 2005). Biodiversity monitoring is also key to addressing national obligations to international treaties relevant to nature conservation, and provides important information for international conservation policies. Finally, information from monitoring is key to accountability, as it can be used to assess whether biodiversity conservation interventions are effective. However, biodiversity monitoring uses valuable funds that might otherwise be targeted at conservation implementation (Adams, 2010; McDonald-Madden *et al.*, 2010; Sheil, 2001; see also Chapter 15), and hence if monitoring is to be of use to conservation it needs to be clearly tied to implementation. Therefore although some degree of biodiversity monitoring is unavoidable and desirable; because of the scarcity of resources for conservation, it is also mandatory to ensure that every penny spent on monitoring counts.

Unfortunately, although there should be clear linkages between monitoring and management and policy, there are concerns that monitoring plans often fail to address such linkages (Lawrence, 2010a; Moffat *et al.*, 2008; Sheil, 2001). The definition of biodiversity is broad, ranging across ecosystem function through species and population

---

Biodiversity Monitoring and Conservation: Bridging the Gap between Global Commitment and Local Action,
First Edition. Edited by Ben Collen, Nathalie Pettorelli, Jonathan E.M. Baillie and Sarah M. Durant.
© 2013 John Wiley & Sons, Ltd. Published 2013 by John Wiley & Sons, Ltd.

abundance through to genes. This has the consequence that the very selection of the component of biodiversity for monitoring usually includes a subjective element. An incomplete understanding of the relationships between these different components of biodiversity and their threats further confounds interpretation of monitoring results. Therefore engagement of policy-makers and managers as stakeholders right at the beginning of developing a monitoring plan, before components of biodiversity to be monitored are identified, is critical to ensuring that results of monitoring are heeded and incorporated into management and policy and used to the benefit of biodiversity conservation.

In this chapter I address in detail all the aspects of monitoring that need to be considered if data are to be used for maximum effect. I argue for embedding biodiversity monitoring within a national context, to ensure that it is aligned with local and national priorities and engages managers and policy-makers from the beginning. I first outline reporting obligations incurred by international conventions and then discuss a national approach to monitoring. I go on to explore the different stages that should be considered in the development and implementation of an effective monitoring plan, grounded in good science, but focused on clear conservation goals. Finally, I place this approach in the context of developing the institutional networks and capacity necessary to ensure that information from monitoring is put to effective use.

International conventions for biodiversity conservation

There are a number of key reporting obligations for signatory countries relating to international conventions relevant to biodiversity conservation, including the Convention on Biological Diversity (CBD), Convention on International Trade in Endangered Species in Wild Fauna and Flora (CITES), Convention on Migratory Species (CMS), Convention on Wetlands of International Importance (RAMSAR), and the UNESCO Convention Concerning the Protection of the World Cultural and Natural Heritage (World Heritage Convention) (Table 14.1). The CBD provides an international framework for guiding biodiversity conservation activities with the overall objective of the conservation and sustainable use of biodiversity and the fair and equitable sharing of benefits from genetic resources (www.cbd.int). The CBD 2010 target was to reduce the rate of loss of biodiversity, and most are now agreed that this was not achieved (see Chapter 1; Butchart *et al.*, 2010; Mace *et al.*, 2010). The evidence suggests that the rate of loss of biodiversity has at best been constant, but is probably still increasing (Rands *et al.*, 2010). The most recent revised mission of the CBD is to 'take effective and urgent action to halt the loss of biodiversity in order to ensure that by 2020 ecosystems are resilient and continue to provide essential services, thereby securing the planet's variety of life, and contributing to human well-being, and poverty eradication'. This mission of halting the loss of biodiversity is even more

Table 14.1 International conventions covering different aspects of biodiversity and national reporting requirements

Convention name	Website	Year started	Number of parties	Relevant components of biodiversity	Reporting requirements	Sanctions on failure to implement
Convention on Biological Diversity (CBD)	www.cbd.int	1993	193	All biodiversity	National Biodiversity Strategies and Action Plans Biodiversity reports	None
Convention on International Trade in Endangered Species of Wild Fauna and Flora (CITES)	www.cites.org	1973	176	Species threatened by trade	Non-detriment reports for listed species Export and import permits	Loss or reduction in trade quotas
Convention on Wetlands of International Importance (RAMSAR)	www.ramsar.org	1971	163	Wetlands	Triennial National Reports to the Conference of the Contracting Parties Reports on any changes or threats to listed wetlands	Loss of RAMSAR status
Convention on Migratory Species (CMS)	www.cms.int	1979	117	Migratory species	National reports on measures taken to implement the provisions of the Convention for listed species submitted 6 months prior to each COP	None
Convention Concerning the Protection of the World Cultural and Natural Heritage	whc.unesco.org	1972	189	Natural heritage sites (cultural heritage also covered)	Regular national reports about the state of conservation and the various protection measures put in place at designated world heritage sites	Loss of World Heritage status

ambitious, but may be easier to measure, although problems associated with a lack of adequate indicators and data may still need to be addressed.

The CBD, however, remains the only convention governing overall biodiversity. The other conventions concerned with biodiversity either address particular species (e.g. CITES, which covers internationally traded species, and CMS, which covers migratory species) or are focused on sites (such as RAMSAR, which focuses on wetlands, and the World Heritage Convention, which focuses on sites of international importance). All five conventions require some form of national reporting. The CBD requires the establishment of National Biodiversity Strategies and Action Plans (NBSAPs), and national reports in line with these plans. Some form of biodiversity monitoring information is a crucial component of most NBSAPs, while the reports themselves should ideally provide monitoring and evaluation information on the progress of implementation of NBSAPs.

A summary of all five conventions and their reporting requirements is provided in Table 14.1. It is worth noting that there are no sanctions on countries not implementing the CBD. However, some other conventions have effective sanctions – a loss of trade for valuable species can be a significant deterrent against non-compliance for CITES, whilst the loss of highly regarded UNESCO World Heritage Status may be a significant potential source of political embarrassment that may help to ensure compliance and, in some areas, resulting potential declines in tourism may provide an additional deterrent.

Why national long-term sustainable monitoring?

Many of the earlier chapters outline reasons why biodiversity monitoring is important, not only to document progress in maintaining biodiversity, but also as a foundation to adaptive management, whereby information is used to assess impacts of management or policy change, and to adapt management strategies accordingly. Monitoring has traditionally been largely a decentralized activity (Marsh and Trenham, 2008), often focused on particular sites (often protected areas) or species. This can make it difficult to use biodiversity monitoring data on a larger scale, such as the national scale. However, if countries are to be accountable for biodiversity conservation within their borders it is necessary for biodiversity monitoring to be reported on a larger scale (Lawrence, 2010a).

The reality of the world in which we live is that while management is often implemented locally, policy, which may have extensive and dramatic effects on biodiversity, is implemented within a geopolitical framework, under overall control of national governments, albeit with varying degrees of regional devolution. Moreover, international reporting obligations are incurred at the national level (Table 14.1). There is, therefore, a need for a systematic national approach to biodiversity monitoring. Such an approach should not only identify areas of biodiversity loss and conservation success, but also identify areas where outside support is needed, particularly in the case of developing countries with few resources. It would also allow countries to be held to account for their progress in meeting their international obligations, both in the international arena and nationally to their citizens. However, to date, many countries do not have systematic national biodiversity monitoring programmes (e.g. Chandra and Idrisova, 2011). Instead countries tend to collate whatever data are available through a combination of different methods, containing, inevitably, various sources of variability and bias. This has two main consequences: first, it is difficult to be sure that reports of biodiversity change provide accurate results and are not a consequence of the selection of biodiversity or sites monitored; second, it is difficult to compare one country's biodiversity conservation performance against another. In this book we have seen a number of possible approaches to biodiversity monitoring, which could generate systematic indicators of biodiversity (see Chapters 2–12). Several, or all, of these could also be used to provide a national indicator.

Since monitoring is not an end result in itself, but is undertaken to address particular conservation management and policy-maker needs or international reporting obligations, it is very unlikely that each country will have the same monitoring needs. Management and policy needs are likely to vary nationally. For example, countries in Africa with high levels of wildlife tourism might be particularly concerned to monitor their charismatic wildlife, which serve to attract economic benefits from tourism, whilst countries that are rich in potential genetic resources that could be used to generate new drugs and medicines, such as in Amazonia, might be particularly concerned to monitor their plant biodiversity. Monitoring therefore needs to be tailored to different national needs (Jones *et al.*, 2011), so that measures are relevant to the national context, as well as meeting international reporting obligations. Monitoring information can then serve to inform policy-makers about impacts of different policies on those measures and consequent economic benefits, and meet national reporting needs. Whilst purists may prefer more complete measures of biodiversity, there is a need to be realistic – as it is impossible to monitor all components of biodiversity (Lawrence, 2010a; Pereira and Cooper, 2006), and any biodiversity monitoring that does take place will inevitably take away resources that could be directly used for conservation (see Chapter 15). Hence, although monitoring should remain an intrinsic component of conservation management, it should also be as cost-effective as possible.

Developing a plan for biodiversity monitoring at the national level for components of biodiversity that matter to the citizens in each country is therefore necessary for national accountability and adaptive management of biodiversity, and this information should also be used for international reporting. But how should this be done? I identify 10 key components for a sustainable national monitoring strategy:

1. an overall goal;
2. monitoring plan design;

3. data collection;
4. data storage;
5. data analysis;
6. standardization;
7. human and financial capacity;
8. institutional framework;
9. establishing trust and credibility; and
10. communication and dissemination (Table 14.2).

All these components have practical constraints; for example, there may be little point in designing a plan to cover all taxa, if resources can only stretch to monitoring a subsection of taxa, or in using highly technical methods if it is impossible to train sufficient people in these methods to ensure long-term implementation.

In the following sections these key components of a monitoring plan are each addressed in turn. Meaningful long-term monitoring will need to be sustainable. Engagement of stakeholders, including government, non-governmental organizations (NGOs), and local communities is critical to building a foundation of trust and understanding (Lebel *et al.*, 2006), which is ultimately key to long-term sustainability. Continual engagement of stakeholders in all aspects of a monitoring plan will not only ensure stakeholder needs are met and that data are useful, but also ensure that results from monitoring are used appropriately, and will help to establish capacity for self-organization and an institutional memory for ongoing and long-term implementation of a consistent and standardized monitoring methodology.

Monitoring goals

Resources for conservation are valuable, and hence the goal of biodiversity monitoring needs to be set out clearly before developing any monitoring programme. Although this may seem obvious, explicit statements of monitoring objectives are rare (Yoccoz *et al.*, 2001). The goal should be clearly tied to management and policy objectives, and should also, where possible, be tied to international reporting obligations. It should also be practical and achievable, aiming for a balance between delivering data relevant for adaptive management and national and international reporting; cost; and practicality.

Although there is an increasing realization that national and site-based monitoring goals need also to satisfy reporting obligations to international conventions (see Chapters 13 and 15), monitoring and evaluation mechanisms are absent from most countries' national biodiversity reports, including NBSAPs (Carter, 2007; Prip *et al.*, 2010). Where monitoring is a component of NBSAPs some countries choose to target a limited suite of internationally threatened species; sometimes these may

Table 14.2 The different stages of monitoring and the key questions that need to be addressed

Stage	Key questions
Monitoring goal	Is the goal relevant to local, national, and international needs?
	Is the goal practical and achievable?
Monitoring plan design	What should be monitored and what methodology should be used?
	Does the plan have on-the-ground support?
	Will the plan deliver its objectives (e.g. will monitoring have sufficient power to detect change in biodiversity)?
	Can data from existing monitoring programmes contribute to the plan?
	Is the plan sustainable or can it be made sustainable in the long term?
	Do data need external verification?
Data collection	Who will collect data?
	Are they appropriately trained?
	If not, then how will they be trained?
	How will training be maintained across personnel changeovers?
	How will data collection be coordinated between multiple sites?
Data storage	What storage capacity is needed for data and samples?
	How should data be secured?
	How should data be backed up?
Data analysis	What data analysis methods should be used?
	Are methods sufficiently accessible to stakeholders to ensure credibility of data is maintained?
	If not, how can analysis methods be made more accessible and more easy to understand, or can alternative methods be used?
	How will national capacity in the necessary data analysis skills be developed and maintained?
Standardization	Are data comparable between years?
	Is it necessary to compare data between sites or across regions?
	How should data be standardized to enable comparison?
Human and financial capacity	Do data managers have sufficient capacity to understand all aspects of monitoring?
	If not, how will sufficient capacity be established?
	Is there a sustainable financial plan to ensure long-term support for monitoring?
Institutional frameworks	How will the monitoring plan be embedded into government systems to ensure effective dissemination of monitoring results to policy-makers?
	How will the different institutions and individuals that are needed to implement monitoring coordinate?

(*continued overleaf*)

Table 14.2 (*continued*)

Stage	Key questions
Institutional frameworks (cont'd)	Are relevant government departments engaged with the monitoring plan?
	Who will be responsible for coordinating and implementing each aspect of the monitoring plan and how will monitoring be enforced?
	How will data sharing be organized? What agreements need to be put in place?
Establishing trust and credibility	Which organizations and individuals are critical to the credibility of the data?
	How should they best be engaged in monitoring?
	How should a sense of ownership of the data be best engendered?
	How can marginalized groups best be engaged in the monitoring process?
Communication and dissemination	Who should be included in the communication network for the monitoring plan?
	How will information during monitoring be coordinated and disseminated?
	Can results from monitoring be incorporated into accessible and short documents to ensure they are understood by stakeholders?
	Who will be responsible for dissemination of results?
	Is there a mechanism to ensure that information is interpreted for policy-makers?

be just vertebrates, as in Bhutan, which has identified three keystone species in its national biodiversity action plan (BAP) – tiger, snow leopard, and the white-bellied heron (National Biodiversity Centre, 2009). But monitoring can also include a full range of nationally threatened species, such as in the UK's list of biodiversity indicators (http://www.jncc.gov.uk/page-5161) and the United States' Endangered Species Program (http://www.fws.gov/endangered/). Other countries have taken a habitat approach towards monitoring, by focusing on vegetation type and land use, which has the added benefit of often being able to be monitored remotely (see Chapter 5). As well as direct measures of biodiversity (e.g. species richness, species abundance), threats and drivers of biodiversity loss are often part of the monitoring component of NBSAPs and protected area management plans (Prip *et al.*, 2010). This can include, for example, various measures of illegal offtake, for example fishing, timber extraction, bushmeat; habitat loss; perceived human-wildlife conflict; invasive species, etc., which tend to be easier to monitor than direct measures of biodiversity. These threats and drivers also tend to be the focus of governments and protected area managers in day-to-day conservation activities. There should always be a logical and mechanistic link between a reduction in drivers of biodiversity loss and a reduction in biodiversity loss, and it is important to understand these linkages, particularly the interaction between threats, target species, and overall biodiversity. In particular, if a crucial threat is left out of a list of monitored threats it may be possible to unknowingly lose an entire species. This can be a particular risk in the case of new emerging threats. For example, before the threat of chytrid fungus to amphibian populations was understood, the disappearance of populations of amphibians from many areas went completely unnoticed (Crawford *et al.*, 2010; Wake and Vredenburg, 2008).

Monitoring plan design

Once goals are established, a clear identification of what should be monitored and how monitoring should be carried out is fundamental to biodiversity monitoring plans (Yoccoz *et al.*, 2001). Monitoring needs to be designed in a way that best targets its goals with minimum cost. The power of the monitoring – its ability to detect significant trends – should be an important consideration at this stage, and is discussed in full in Chapter 15. Decisions will need to be taken on the ability of a monitoring plan to detect change – e.g. does it need to detect a 10% or a 50% change? Generally, as sensitivity in detecting trends increases, sampling effort, in terms of quantity of data and spatial coverage, and hence usually cost, will also have to increase. Moreover, particular methods will be appropriate for different levels of monitoring power. How the data are to be analysed should also be considered in the monitoring plan design, as this will also impact on how the data should be collected.

One or more appropriate biodiversity indices need to be identified. Where possible, existing ongoing monitoring programmes should be taken into account – can existing data and programmes be used to address the monitoring goals? The Living Planet Index (LPI) provides an excellent example of how existing monitoring information can be used to build an index of biodiversity (see Chapter 4). An LPI is likely to have inherent biases, as most existing monitoring programmes do not address a random cross-section of biodiversity, but tend, for example, to be focused on national priorities for biodiversity monitoring. Hence it may be necessary to develop additional monitoring plans to address key gaps. Other indices may be more objective; for example, habitat monitoring could be addressed using remote sensing data at minimal cost, providing some thorough ground truthing work has been done beforehand (see Chapter 5).

Some countries, particularly developed countries, have extensive networks of volunteer citizen scientists who are able to provide sufficiently reliable data for monitoring trends (Cunningham and Olsen, 2009; Goffredo *et al.*, 2010; see also Chapter 10). Moreover, the growing trend towards community-based natural resource management (CBNRM) has resulted in community-led monitoring programmes, particularly within developing countries (Danielsen *et al.*, 2005, 2010). Data generated through these activities can be disseminated to a centralized database, which has the potential to provide a large-scale and cost-effective monitoring programme (Danielson *et al.*, 2010). Some imaginative thinking and coordination may thus be used to design monitoring programmes using existing data that can meet national goals at minimal cost.

A good monitoring plan should stand the test of time. Tanzania prioritized its large mammal populations as a key monitoring goal, in keeping with an overall goal of maintaining these populations. This led to the design of an aerial monitoring programme, starting from 1958 in the Serengeti, that has since been used to track population trends in all large mammals across all the major protected areas in the country, a monitoring design that is still used today (Sinclair *et al.*, 2007; Stoner *et al.*, 2007). The UK national bat monitoring programme is an example of a relatively low-cost approach, as this programme has been designed to be implemented largely by volunteers, targeting eight species of bat and using information about their life history to focus data collection on call counts, maternal colony counts, and hibernation colony counts (Bat Conservation Trust, 2001; see also Chapter 10).

In some situations monitoring may also need to provide verifiable data, to enable a third party to verify results. This can be particularly important for politically sensitive species or sites. For example, arguments about whether tiger numbers in India had been artificially inflated (Ramesh, 2005) could have been addressed directly had verifiable records existed. For this species, a monitoring approach using camera traps not only produces more accurate data than the disputed pug mark method, but the photographs generated are also verifiable by an outside observer should a dispute arise (Karanth *et al.*, 2003, 2011).

Data collection

Data collection should proceed according to the monitoring plan. Crucial to this is the assignment of tasks to responsible individuals as, unless an individual is identified to implement each monitoring activity, it may well not happen. Where monitoring is ongoing and across multiple sites, there is a need to identify individuals responsible for coordinating monitoring and centralizing data. Very often data collection requires particular skills. For example, aerial monitoring in Tanzania requires trained and skilled observers who are able to rapidly estimate the size of a herd from a plane flying at over 200 km/h. The use of camera traps in surveys (see Chapter 3) requires personnel who are able to identify good sites for placing cameras to ensure maximum capture of passing wildlife, and many monitoring programmes require personnel who are skilled at species identification. Where a national monitoring programme depends on such skills, it is important that an institutional memory is established to ensure that these skills can be transferred when skilled individuals move or retire. However, data collection can sometimes be implemented by volunteers with very little training; France relies entirely on volunteers for data collection to meet its international reporting obligations to CBD on indicators of trends in abundance and distribution of selected species (Levrel *et al.*, 2010).

Data storage

Data storage is a key aspect of a monitoring programme but is surprisingly rarely considered. Whilst the increasing capacity of electronic media means that large amounts of data can be stored in a very small area, the danger is that they can also be easily lost. Where substantial electronic data are generated, such as might be the case from photographs from camera trap surveys, then a safe and secure electronic storage regime needs to be established, with a regular and frequent back-up system whereby at least one back-up is stored at a different site to the main database. Remote-sensing data with high resolution has similar storage requirements. Physical storage space may also be necessary; for example, invertebrates are often trapped and identified in hand, and very often identification is problematic and specimens need to be kept to aid future identification. If data verification is important, then samples from entire surveys may need to be stored, and steps need to be taken to ensure that they are well preserved.

Data will also need to be processed and entered into an electronic database that is easy to access and browse. Where similar data are collected across different sites, these will need to be standardized so databases can be merged. As with raw electronic data, an effective and rigorous back-up regime is essential. Care should be taken to ensure that data are kept clean. Errors can be minimized through careful design of the database, where database managers are constrained to choose between preselected categories when entering data, and by database checks, such as for valid dates or GPS points. Sophisticated software programs are nowadays available on PDAs, which can enable data to be entered in the field, but here, as with the central database, care needs to be taken about backing up. Data also need to be secured, particularly against modification by untrained individuals.

Data analysis

Once data are collected and entered into a database they will need to be analysed. Nowadays there are abundant techniques and software available for data analysis. However, it is important that the analysis regime is carefully considered in the survey design in advance of data collection, as this will influence which data are collected, and how the database is constructed. The data analysis component is often the most technically demanding component of the monitoring process. Data analysis should be appropriate to the capacity available in country, and consideration should be given to developing further capacity to enable the use of more technically demanding methods, particularly when they may enhance precision of results or reduce long-term costs.

Although continual advances are made in developing new statistical techniques and software, these are very often only understood by a handful of scientists around the world at their inception. This capability tends to be concentrated in the developed world; however, national sensitivities make it desirable that data analysis, which might uncover politically sensitive wildlife declines or increases, or changes in threats, is undertaken by national institutions. There is a real need for scientists based in developed countries to better engage with developing country nationals to transfer skills needed to analyse monitoring data. This can often be done with very little effort; for example, a few straightforward programming lines in statistical software, such as R, and a basic understanding may be all that is required for a government scientist to undertake the appropriate analysis year after year during a long-term monitoring plan.

Standardization

Some form of standardization will be necessary to ensure that data can be compared over time and/or between sites. To ensure this happens, a clear protocol should be established and documented, and this protocol should be adhered to over each monitoring cycle. Comparing data over time is often straightforward; however, comparing between sites can be more problematic, because of external factors such as different habitats and topography; hence, if comparison between sites is part of the monitoring goal then this will require particular attention. Standardization of the way data are stored and analysed is also important as there may be a need to aggregate data from different sites, for example to ensure that data can be readily combined for meta-analyses (see Chapter 15), as for the LPI (see Chapter 4).

There may also be a need to compare monitoring results between countries. This may be required by international and regional agreements, such as the European Union (EU) in Europe, the Southern African Development Community (SADC) in Africa, and the Association of Southeast Asian Nations (ASEAN) in Asia, all of which encompass aspects of biodiversity conservation. In these situations there is a need for international standardized data collection and shared databases to enable comparable analyses (see Chapter 13).

Human and financial capacity

To be sustainable and effective and to ensure that countries maintain ownership of the results, monitoring is best led and implemented by a country's nationals within national institutions. This means that the necessary capacity in terms of finances, equipment, and trained personnel needs to be established in country. While most countries have capacity gaps – for example, there is a recognized global need for capacity in taxonomic identification of lesser known groups, such as insects (Makenzie, 2010) – capacity for biodiversity monitoring is a particular concern in developing countries. Very often ministries responsible for biodiversity conservation are under-resourced, and there is little financial support available for monitoring. Where monitoring plans are in place, then there are often insufficient trained personnel, particularly for the data analysis and dissemination components of monitoring. In the Pacific, for example, capacity needs for biodiversity conservation have been identified at all levels, from individual staff members, through to institutions, and regulatory and accountability frameworks (Carter, 2007).

Where sufficient human capacity exists, it can often be limited to one or two individuals who, if they change institution or retire, leave no institutional memory behind. Hence there is also a need to ensure that institutional memory is maintained when providing any form of training. Candidates identified for training need to be well placed within appropriate institutions, and governments need to plan carefully to maintain their trained staff in positions where their skills are relevant.

Identification of financial resources that can be used to undertake monitoring in the long term is critical. Very often, these will depend to some extent on national governments; however, the time of volunteers or of communities can play a key role where there is sufficient interest or local economic incentive (Boissiere *et al.*, 2010; Lawrence, 2010b). Such schemes can save a substantial amount of money. For example, in France the economic contribution of volunteers to biodiversity monitoring has been estimated at between 680 000 and 4 420 000 euros (Levrel *et al.*, 2010). Countries can also decrease financial costs by mutual cooperation; in Europe it has been estimated that countries could save 45% of overall costs of protecting vertebrate biodiversity in the Mediterranean basin by working together (Kark *et al.*, 2009). Where countries lack resources for monitoring, then the CBD provides a framework for additional funding assistance; however, there is a need for further resources, particularly beyond short-term funding cycles. In practice, in many situations, funds for initial monitoring are secured externally, but it is important that there is a strategy to move towards in-country financing to ensure long-term sustainability of monitoring.

Institutional frameworks

Conservation resources are so valuable that it is imperative that there is a plan to embed data monitoring within an institutional framework that ensures data dissemination and use. For example, monitoring may be carried out by a university department but is there any mechanism for that data to be transferred to government? Centralized coordinators, tasked with collating data from various sources across the country, and pulling them together into the centralized database, are key to all national-level monitoring processes. For example, where data are collated across a network of volunteers or CBNRM schemes, data will need to be pulled together within a central database, which will require substantial communication and coordination. Coordinators need to be embedded within an institutional framework that has appropriate institutional linkages to conservation management agencies, usually based within government. This ensures that results from monitoring can be easily disseminated to the relevant management agencies. For example, in Tanzania, the Tanzanian Wildlife Research Institute (TAWIRI) is a government organization responsible for wildlife research activities in Tanzania; it is mandated to advise the government on wildlife policy, and hence is well placed to act as a nodal agency for storing national biodiversity data.

To facilitate communication and participation between and within institutions engaged in monitoring it is important that results are processed and disseminated in accessible formats and that agreements for data sharing are put in place. A system, whereby decentralized monitoring results need to be channelled to a centralized information node, will need to operate according to a prior agreed framework for sharing data between contributing individuals, communities, and organizations. Many such contributors may wish to ensure that issues such as data ownership, privacy, and publication rights are safeguarded. For example, CBNRM programmes in Namibia transmit only processed data to the government for use in national-level monitoring (Danielsen *et al.*, 2010). Clear guidelines for disseminating and accessing data should be established prior to any data collection (Lawrence, 2010a).

Where specific monitoring programmes are designed and planned for long-term implementation, they are likely to provide higher quality data than an aggregate of results from ongoing activities, which may be tailored to target more short-term and site-based questions. However, it will also be necessary to plan for administrative continuity through government changes and restructuring processes (Watson and Novelly, 2004). Where this is impossible due to high levels of institutional instability, then serious thought needs to be given to whether such a monitoring plan, which will only reap benefits after many years of operation, is likely to generate any useful information. Five years of data from a 20-year monitoring plan may be virtually useless and a waste of money. There are few examples of long-term monitoring programmes driven by non-governmental institutions on a broad scale; those that exist tend to be restricted to small areas and tend to be driven by individuals or academic institutions (Watson and Novelly, 2004). However the US-based Long Term Ecological Research Network (LTER) provides an example of how academically driven long-term monitoring projects can evolve into a nationally coordinated programme of monitoring (Michener *et al.*, 2011).

Establishing trust and credibility

Although trust and credibility have been identified in the Millennium Ecosystem Assessment process as one of three key principles that need to be addressed in developing indicators of biodiversity – the others being relevance and legitimacy (Raudsepp-Hearne and Capistrano, 2011) – it is particularly often neglected from monitoring (Lawrence, 2010a). Trust is critical to building and implementing a meaningful monitoring programme, since without the trust of the end users of the data – be it local communities or national and international policy-makers – there is a risk that data will be ignored. Ensuring ownership of the monitoring process by the end users of data is key to building trust and establishing credibility. Engaging local communities in data gathering can be a very effective means of establishing credibility of the data and a sense of data ownership within those communities (Lawrence, 2010a). However, political buy-in from decision-makers is also clearly important, and so a focus on the national as well as the local is critical. A lack of political buy-in was cited as a key problem in achieving national acceptance of the Biodiversity Strategy and National Action Plan in India (Bhatt and Apte, 2010). An added complication is that the technical aspects of monitoring, such as data analysis, are often driven by people outside the immediate community where monitoring takes place. In such situations particular efforts need to be made to maintain trust, which can be done by developing local capacity in monitoring techniques to ensure that monitoring can ultimately be conducted and understood by in-country and/or site-based personnel.

Trust between different stakeholders is not always easy to secure, and sometimes a scientist from outside can be useful as an intermediary between different conflicting groups. Where possible, a monitoring strategy should be explicitly designed to encompass the concerns of particular groups, helping to secure trust between different sectors. By ensuring that legitimate concerns are addressed from the start, it may be possible to develop a monitoring process that will provide data that is credible to all stakeholders. Where conflict exists between stakeholders, it is best addressed directly, because if concerns are suppressed they may ultimately undermine the credibility of the data (Lawrence, 2010a).

Unequal power relations between different stakeholders are an additional factor that can undermine trust and credibility in a monitoring process. Lawrence (2010a) argues that there is 'an ethical obligation on the researcher to ensure that the process is beneficial to the marginalized', and to ensure that the research itself does not reinforce marginalization. Scientists thus may also need to consider taking additional steps of engaging marginalized sectors and countries, where relevant. While this may often be challenging and difficult, it can ultimately result in a far more constructive response to biodiversity data and conservation.

Communication and dissemination

Establishing and maintaining effective communication networks is essential for coordinating monitoring and maintaining credibility between the different agencies involved in the monitoring process. The internet helps in this task enormously; however, it cannot replace face-to-face meetings, which are critical to the establishment of trust (see previous section), and can be achieved through small stakeholder group workshops and informal meetings. A dissemination plan of monitoring results should be budgeted within the monitoring plan and explicitly embedded within local and regional management and national reporting frameworks. This plan should incorporate mechanisms to ensure data are used to inform collective decision-making processes for adaptive management (Caddy and Seij, 2005) and to ensure that results are tied into national reporting obligations. Identifying an individual responsible for coordinating and delivering on dissemination obligations is key to effective dissemination.

A communication network for the monitoring programme needs to be explicitly established and regular communication across the network maintained by the dissemination coordinator. This network should include all stakeholders, including the researchers and the end-users of the data, such as local communities and government. Thought needs to be given to ensuring that appropriate mechanisms are put in place to enable effective communication of important monitoring results to policy-makers. In particular, information from monitoring ultimately needs to be collated in accessible formats that can be used both by managers and policy-makers, and disseminated nationally and internationally to assess progress towards national and international targets for biodiversity conservation. There is an increasing tendency for long reports that do not succinctly summarize data findings in accessible short formats for time-stretched managers and policy-makers. Where results are not summarized, then they may as well not be disseminated, as managers and policy-makers are unlikely to have sufficient time to read long and verbose reports. Data reporting systems should be streamlined and aggregated, and the data-gathering chapters in this book show how this can be approached; for example, an aggregated LPI per country (see Chapter 4), a Wildlife Picture Index (WPI) per protected area (see Chapter 3), and an index of habitat loss from remote-sensing data (see Chapter 5) could all be collated together for a single annual report, which can then feed into international reporting agendas.

Discussion

Many of the steps involved in implementing a good monitoring programme are also relevant to good conservation implementation. Bridging the gap between science and implementation, developing capacity, and embedding it within an effective institutional framework, are all essential steps for gathering information for conservation and for managing biodiversity. While these steps might superficially appear straightforward, one or more of them are often ignored in practice. In particular, monitoring, in common with conservation implementation, all too often does not take steps to embed monitoring within appropriate institutions and governance structures – the middle enabling tier in the three-tiered response to biodiversity loss (foundational; enabling; and instrumental; Rands *et al.*, 2010). A good monitoring programme is dependent on wider issues of governance, including communication, transparency, and societal engagement, and conservationists, if they wish to achieve good conservation, need to engage with these issues. To do this in areas where governance systems are not yet in place is likely to be slow and depend on sustained international support, until capacity and institutional systems are sufficiently embedded to be self-organizing (du Toit *et al.*, 2004). This, in turn, will depend on a wider network of trust – individuals from communities, from government, and from NGOs will all need to be engaged and understand monitoring if it is to become sustainable and prioritized against the plethora of competing interests within ministries responsible for biodiversity conservation today. Moreover, if a programme is ever to become long term then none of the components of a monitoring process should be dependent on a single individual and should be secured against inevitable changes in personnel and government.

Conservation plans on a regional, national, or local scale can provide important frameworks to use monitoring results to formulate and address larger management-orientated goals. For example, the recent regional conservation planning process for cheetah and wild dog across process now covers all of Africa, although final report (north west and central) is not yet out Africa has generated regional strategies leading to national action plans that embed monitoring and information needs within activities needed to address and measure progress towards overall conservation objectives and goals (IUCN/SSC, 2007a, 2007b). These, in turn, are now the focus of stakeholders tasked with implementing activities, who actively participate within constantly expanding regional and national networks, established and maintained by regionally appointed coordinators. However, although there is an abundance of examples of monitoring goals in various national documents including NBSAPs, species action plans, or general management plans for protected areas, these plans are often disconnected, even within the same country. For example, a species may be monitored within a site-based management plan but this information is not passed through to a NBSAP. In many countries information pathways between local management and national reporting need to be improved.

The importance to conservation of government, management, institutions, communities, and society necessitate their involvement in conservation activities, including monitoring. However, there are good reasons why effective engagement with national institutions and governance is rarely prioritized in monitoring. Conservation is poorly funded and hence the best conservation often tends to be opportunistic. Monitoring practitioners are often overstretched and have very little time to prioritize activities that are essential to developing the foundations of relationships of trust. These might include training a promising local scientist to undertake data collection and analysis themselves, or spending time engaging government officials even if meetings are difficult to schedule. However, if this is not done, then there is a risk that all the careful results from their monitoring will be wasted, as data that are not understood and trusted by policy-makers and communities may ultimately be ignored. Ongoing data collection and analysis additionally depends on institutional engagement, to ensure that institutional memories are maintained if the original project leader moves on. Why start a long-term monitoring plan if the data will never be used and sustainability can never be established? By contrast, where efforts have been made and foundations of trust established, this can lead to an enabling institutional environment for gathering data, an effective network of engaged policy-makers for disseminating results, and engaged communities interested in building on conservation activities and monitoring activities may snowball.

The essential role that individuals play in conservation is crucial but rarely receives formal recognition, and this is also true in monitoring. These include individuals who always keep a critical and rigorous eye on data collection methods; those who put time and effort into training young conservationists; those who are popular and engage with local communities and government officials; those who keep a close and expert eye on important and expensive equipment; and those who are always able to find new methods and gadgets to make data collection and analysis easy and quick. These individuals, their skills and their social networks, provide the social capital that is key to long-term success and sustainability (Folke *et al.*, 2005). There is also a need to understand and adapt to the sociological and political dimensions during the establishment of long-term conservation programmes, and scientists do not necessarily have all the skills for this. However, there is a growing network of anthropologists, socioeconomists, and political scientists who are increasingly eager to engage with these issues (Agrawal and Ostrom, 2006). By working together it is possible to achieve so much more than the sum of efforts working apart.

Once all countries are able to regularly provide relevant, legitimate, and credible reports on national trends in biodiversity then what? CBD has no mechanism to enforce biodiversity conservation, and so if a country shows accelerating declines in its biodiversity there is little that can be done to encourage or enforce the reversal of this process. While national reports are an important first step to accountability, there will be limited impact if there are no measures that can be taken to ensure enforcement. CITES, RAMSAR, and the World Heritage Convention have tools available that allow enforcement, centring in the economics of trade for CITES, and on the international kudos from holding RAMSAR or World Heritage sites of international importance. Compliance can be increased by enforcement and by providing international support; the latter may be particularly relevant for countries that require additional capacity. Compliance may also be increased through the voluntary embedding of CBD or NBSAPs within national or regional legislation. Without means of increasing compliance, there is a risk that international conventions may have little impact.

The analysis presented here represents an ideal approach to monitoring. However, it is naive to assume that this will be possible in all situations. Countries with poor governance and lack of transparency may wish to keep evidence of biodiversity loss hidden, while civil unrest and instability can disrupt the best laid monitoring plans. There will always be a role for the maverick scientist or conservationist who wishes to bring attention to the decline of a particular species or habitat, despite being hampered by a lack of support and sometimes even being obstructed by government. However, these situations should be exceptional; if monitoring is to help address the big conservation questions of the day, there is a need for attention to be given to the 10 components of monitoring outlined here, so that monitoring is designed to optimal effect.

Conclusions

There is a need for monitoring at a national level to enable countries to be accountable for biodiversity conservation within their borders – both successes and failures. Such monitoring needs to be pragmatic, weighing the benefits of information that can be used for adaptive management against cost, ensuring that resources used for monitoring are put to best use, and that monitoring is firmly tied to conservation implementation. If the questions in Table 14.2 can be addressed at the start of monitoring then this would be a good first step towards effective, sustainable, and relevant national biodiversity monitoring. In particular, the importance of the credibility of data is often overlooked. Ownership of data and engagement of stakeholders in biodiversity data need to take place at all levels of monitoring – from monitoring plan design to dissemination – to ensure that data collected are credible and will be used. For many developing countries this requires a concerted effort to improve technical capacity, with which international scientists should engage. The recent trend for citizen scientists to contribute to monitoring, through volunteer and community monitoring programmes, should be supported and encouraged and needs to be integrated into national monitoring plans. Finally, there is a need to remain pragmatic, and if data cease to be relevant then monitoring plans need to be updated, and goals re-evaluated. Biodiversity conservation ultimately is for people, whether for economic, social, health, or cultural benefits, and a nation's citizens are likely to have the best understanding of the ultimate benefits of biodiversity in their country.

Acknowledgements

I am grateful to Sultana Bashir, Ben Collen, Nathalie Pettorelli, and two anonymous reviewers for their comments on earlier drafts of this chapter. I am also grateful for numerous discussions with my colleagues in Tanzania, particularly Charles Foley, Alex

Lobora, Simon Mduma, and Maurus Msuha, which have contributed to the ideas underlying this chapter.

References

Adams, W.M. (2010) Conservation plc. *Oryx*, 44, 482–484.

Agrawal, A. and Ostrom, E. (2006) Political science and conservation biology: a dialog of the deaf. *Conservation Biology*, 20, 681–682.

Bat Conservation Trust (2001) *The UK's National Bat Monitoring Programme – Final report 2001*. Bat Conservation Trust, London.

Bhatt, S. and Apte, T. (2010) How thousands planned for a billion: lessons from India on decentralised participatory planning. In: Lawence, A. (ed.), *Taking Stock of Nature: Participatory Biodiversity Assessment for Policy, Planning and Practice*. Cambridge University Press, Cambridge, pp. 211–231.

Boissiere, M., Sassen, M., Sheil, D., *et al.* (2010) Researching local perspectives on biodiversity in tropical landscapes: lessons from ten case studies. In: Lawence, A. (ed.), *Taking Stock of Nature: Participatory Biodiversity Assessment for Policy, Planning and Practice*. Cambridge University Press, Cambridge, pp. 113–141.

Butchart, S.H.M., Walpole, M., Collen, B., *et al.* (2010) Global biodiversity: indicators of recent declines. *Science*, 328, 1164–1168.

Caddy, J.F. and Seij, J.C. (2005) This is more difficult than we thought! The responsibility of scientists, managers and stakeholders to mitigate the unsustainability of marine fisheries. *Philosophical Transactions of the Royal Society B, Biological Sciences*, 360, 59–75.

Carter, E. (2007) *National Biodiversity Strategies and Action Plans. Pacific Regional Review*. SPREP/COMSEC.

Chandra, A. and Idrisova, A. (2011) Convention on Biological Diversity: a review of national challenges and opportunities for implementation. *Biodiversity and Conservation*, 20, 3295–3316.

Crawford, A.J., Lips, K.R., and Bermingham, E. (2010) Epidemic disease decimates amphibian abundance, species diversity, and evolutionary history in the highlands of central Panama. *Proceedings of the National Academy of Sciences of the USA*, 107, 13777–13782.

Cunningham, R. and Olsen, P. (2009) A statistical methodology for tracking long-term change in reporting rates of birds from volunteer-collected presence-absence data. *Biodiversity and Conservation*, 18, 1305–1327.

Danielsen, F., Burgess, N.D., and Balmford, A. (2005) Monitoring matters: examining the potential of locally-based approaches. *Biodiversity and Conservation*, 14, 2507–2542.

Danielsen, F., Burgess, N., Funder, M., *et al.* (2010) Taking stock of nature in species-rich but economically poor areas: an emerging discipline of locally based monitoring. In: Lawence, A. (ed.), *Taking Stock of Nature: Participatory Biodiversity Assessment for Policy*, Planning and Practice. Cambridge University Press, Cambridge, pp. 88–112.

du Toit, J.T., Walker, B.H., and Campbell, B.M. (2004) Conserving tropical nature: current challenges for ecologists. *Trends in Ecology and Evolution*, 19, 12–17.

Folke, C., Hahn, T., Olsson, P., and Norberg, J. (2005) Adaptive governance of social-ecological systems. *Annual Review of Environment and Resources*, 30, 441–473.

Goffredo, S., Pensa, F., Neri, P., *et al.* (2010) Unite research with what citizens do for fun: "recreational monitoring" of marine biodiversity. *Ecological Applications*, 20, 2170–2187.

IUCN/SSC (2007a) *Regional Conservation Strategy for the Cheetah and African Wild Dog in Eastern Africa*. IUCN/SSC, Gland, Switzerland.

IUCN/SSC (2007b) *Regional Conservation Strategy for the Cheetah and African Wild Dog in Southern Africa*. IUCN/SSC, Gland, Switzerland.

Jones, J.P.G., Collen, B., Atkinson, G., *et al.* (2011) The why, what, and how of global biodiversity indicators beyond the 2010 target. *Conservation Biology*, 25, 450–457.

Karanth, K.U., Nichols, J.D., Seidensticker, J., *et al.* (2003) Science deficiency in conservation practice: the monitoring of tiger populations in India. *Animal Conservation*, 6, 141–146.

Karanth, K.U., Gopalaswamy, A.M., Kumar, N.S., *et al.* (2011) Counting India's wild tigers reliably. *Science*, 332, 791.

Kark, S., Levin, N., Grantham, H.S., and Possingham, H.P. (2009) Between-country collaboration and consideration of costs increase conservation planning efficiency in the Mediterranean Basin. *Proceedings of the National Academy of Sciences of the USA*, 106, 15368–15373.

Lawrence, A. (2010a) Introduction: learning from experiences of participatory biodiversity assessment. In: Lawence, A. (ed.), *Taking Stock of Nature: Participatory Biodiversity Assessment for Policy*, Planning and Practice. Cambridge University Press, Cambridge, pp. 1–29.

Lawrence, A. (2010b) The personal and political of volunteers' data: towards a national biodiversity database for the UK. In: Lawence, A. (ed.), *Taking Stock of Nature: Participatory Biodiversity Assessment for Policy*, Planning and Practice. Cambridge University Press, Cambridge, pp. 251–265.

Lebel, L., Anderies, J.M., Campbell, B., *et al.* (2006) Governance and the capacity to manage resilience in regional social-ecological systems. *Ecology and Society*, 11 [online]: http://www.ecologyandsociety.org/vol11/iss1/art19/.

Levrel, H., Fontaine, B., Henry, P.Y., *et al.* (2010) Balancing state and volunteer investment in biodiversity monitoring for the implementation of CBD indicators: A French example. *Ecological Economics*, 69, 1580–1586.

Mace, G., Cramer, W., Dıaz, S., *et al.* (2010) Biodiversity targets after 2010. *Current Opinion in Environmental Sustainability*, 2, 1–6.

Makenzie, R. (2010) Monitoring and assessment of biodiversity under the Convention on Biological Diversity and other international agreements. In: Lawence, A. (ed.), *Taking Stock of Nature: Participatory Biodiversity Assessment for Policy*, Planning and Practice. Cambridge University Press, Cambridge, pp. 30–48.

Marsh, D.M. and Trenham, P.C. (2008) Current trends in plant and animal population monitoring. *Conservation Biology*, 22, 647–655.

McDonald-Madden, E., Baxter, P.W.J., Fuller, R.A., *et al.* (2010) Monitoring does not always count. *Trends in Ecology and Evolution*, 25, 547–550.

Michener, W.K., Porter, J., Servilla, M., and Vanderbilt, K. (2011) Long term ecological research and information management. *Ecological Informatics*, 6, 13–24.

Moffat, A.J., Davies, S., and Finer, L. (2008) Reporting the results of forest monitoring – an evaluation of the European forest monitoring programme. *Forestry*, 81, 75–90.

National Biodiversity Centre (2009) *Bhutan National Biodiversity Action Plan*. NBC, Thimphu.

Pereira, H.M. and Cooper, H.D. (2006) Towards the global monitoring of biodiversity change. *Trends in Ecology and Evolution*, 21, 123–129.

Prip, C., Gross, T., Johnston, S., and Vierros, M. (2010) *Biodiversity Planning: an Assessment of National Biodiversity Strategies and Action Plans*. Institute of Advanced Studies, United Nations University.

Ramesh, R. (2005) Indian reserve emptied of tigers. *The Guardian*, 31 March.

Rands, M.R.W., Adams, W.M., Bennun, L., *et al.* (2010) Biodiversity conservation: challenges beyond 2010. *Science*, 329, 1298–1303.

Raudsepp-Hearne, C. and Capistrano, D. (2011) The Millennium Ecosystem Assessment: a multi-scale assessment for global stakeholders. In: Lawence, A. (ed.), *Taking Stock of Nature: Participatory Biodiversity Assessment for Policy*, Planning and Practice. Cambridge University Press, Cambridge, pp. 49–68.

Sheil, D. (2001) Conservation and biodiversity monitoring in the tropics: Realities, priorities, and distractions. *Conservation Biology*, 15, 1179–1182.

Sinclair, A.R.E., Mduma, S.A.R., Hopcraft, J.G.C., Fryxell, J.M., Hilborn, R., and Thirgood, S. (2007) Long-term ecosystem dynamics in the Serengeti: Lessons for conservation. *Conservation Biology*, 21, 580–590.

Stoner, C., Caro, T., Mduma, S., Mlingwa, C., Sabuni, G., Borner, M., and Schelten, C. (2007) Changes in large herbivore populations across large areas of Tanzania. *African Journal of Ecology*, 45, 202–215.

Wake, D.B. and Vredenburg, V.T. (2008) Are we in the midst of the sixth mass extinction? A view from the world of amphibians. *Proceedings of the National Academy of Sciences of the USA*, 105, 11466–11473.

Watson, I. and Novelly, P. (2004) Making the biodiversity monitoring system sustainable: Design issues for large-scale monitoring systems. *Austral Ecology*, 29, 16–30.

Yoccoz, N.G., Nichols, J.D. and Boulinier, T. (2001) Monitoring of biological diversity in space and time. *Trends in Ecology and Evolution*, 16, 446–453.

15

Monitoring in the Real World

Julia P.G. Jones

School of the Environment, Natural Resources and Geography,
Bangor University, Bangor, UK

Introduction

In the early 1990s the UK government became concerned about the status of the long-horned zebaroo, a species endemic to the uplands of northern England and thought to be threatened by habitat loss and poaching. Advisors decided that information on population trends was needed and an annual monitoring scheme was set up. After 10 years the government asked an independent reviewer to look at the amassed data. The reviewer reported that despite many hundreds of thousands of pounds being spent it was impossible to conclude whether the zebaroo population was increasing, decreasing, or remaining stable.

I have been asked to focus this chapter on monitoring in the real world. I hope the story of the zebaroo, while clearly a fairy tale, can remind us just how easy it is to get monitoring wrong. Other chapters in this book (see Chapters 1–9) have demonstrated why we need monitoring. Here, I want to emphasize that the real world is one constrained by limited resources and that this must influence how we think about conservation monitoring. I will focus on programmes aiming to monitor ecological trends (such as species abundance, distribution, habitat extent, or quality). I will start by considering the importance of doing monitoring powerful enough to be informative, then introduce some specific challenges to monitoring in the real world, and finally discuss what would make an ideal real-world monitoring scheme.

The importance of power in monitoring

In 2005 Legg and Nagy published an influential paper with the provocative title 'Why most conservation monitoring is, but need not be, a waste of time'. They noted

---

Biodiversity Monitoring and Conservation: Bridging the Gap between Global Commitment and Local Action, First Edition. Edited by Ben Collen, Nathalie Pettorelli, Jonathan E.M. Baillie and Sarah M. Durant.
© 2013 John Wiley & Sons, Ltd. Published 2013 by John Wiley & Sons, Ltd.

that despite many papers in the ecological literature laying out the criteria of good, statistically powerful monitoring, this information was not making it into textbooks, undergraduate training courses, or tenders for government or commercially funded conservation monitoring work. Statistical power is the probability that an analysis will reject a null hypothesis that is indeed false. In the context of monitoring change over time, it is the probability that an analysis will correctly identify a true trend (Gerrodette, 1987). While this may sound rather dull and something only of interest to statisticians, in fact the concept is of central importance to anyone concerned with conservation monitoring. If you have ever thrown a quadrat, walked a transect, or got up at dawn to extract birds from mist nets and care that you are not wasting your time, then you need to care about power. Unless you understand the power of your study, you will not know whether failing to detect a trend is because of limitations in your study design or because of the lack of a trend – a rather fundamental distinction.

Lack of proper power analysis is not specific to conservation or ecological studies; it has long been overlooked in many fields that depend on statistics. For example, a review of psychology papers published in 1960 found the chance of recognizing a significant result when it was present was about the same as would be found by tossing a coin (Cohen, 1962). The problem continues. In 2009 the UK National Audit Office published a report criticizing a nationwide chlamydia screening programme because the lack of proper data collection made it impossible to say whether the programme was successful at reducing infection rates among the target population (NAO, 2009).

What factors affect the power of monitoring?

A number of factors affect the power of a monitoring study. Power is positively related to the sample size, the effect size, and the risk of a false positive, and inversely related to the variability in the system (Di Stephano, 2001). This can be put more simply. The more data points you have, the more likely you are to detect a trend. The larger the trend, the more likely you are to detect it. The more willing you are to accept that you may sometimes conclude there is a trend when there is in fact no trend, the more likely you are to detect a true trend. The less variability there is in your data (due to natural variability, measurement error, or even cyclical patterns in the dynamics of the target species), the more likely you are to detect a trend.

The relationship between power and sample size is probably the most intuitive. If you have only a small sample size, a non-significant result cannot be interpreted as meaning that there is no trend. As you can see from Figure 15.1, it could also be simply because you don't have enough samples to detect the trend that is there in the population.

It is interesting to consider the way in which the risk of a false positive affects the power of a monitoring study. A monitoring programme can make two types of

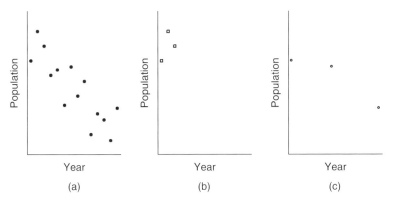

Figure 15.1 The power to detect a trend depends on the number of samples. That this population is declining is clear from the full dataset shown in (a) ($P < 0.001$), but would not be detected ($P = 0.67$) had data only been collected over a shorter period of time (b), or with less frequent samples (c) ($P = 0.232$).

mistake. A type I error (more simply called a false positive) is when the monitoring wrongly concludes there is a change when in fact there is no change. A type II error (or false negative) is when monitoring wrongly concludes there is no change when there is in fact a change. For many years it has been standard in much of ecology to take the risk of a false positive as 5% (Di Stephano, 2003), that is, to assume that if the probability of a given result occurring if there is no trend is less than 5%, a trend must be present. However, by setting the acceptable level of risk for a false positive so low, we increase the risk of a false negative (i.e. missing a true trend) and of course both types of errors have costs. Mapstone (1995) described this beautifully in the context of monitoring in environmental impact studies. He pointed out that should monitoring conclude that there is no impact when in fact a development is having deleterious impact (false negative), development that should be stopped won't be and irreversible damage may be done (resulting in the collapse of a fishery, catastrophic pollution, or extinctions of species). Conversely, by concluding that a development is having an impact when in fact it is not (a false positive), development may be halted, causing unnecessary economic hardship. If statistical significance (risk of a false positive) levels are set too low, power to detect change will inevitably be poor, undermining a fundamental aim of ecological monitoring, which must be to detect real change soon enough to act (Field et al., 2007). However, if significance levels are set too high, then non-existent trends will be detected and unnecessary, and perhaps costly, conservation interventions may be made. Therefore anyone involved in the design of conservation monitoring should think carefully about the relative costs of getting it wrong in both directions (Field et al., 2004).

Power analysis in the real world

Power analysis asks the question 'with the resources available is it possible to gain the information I want?' Hockley *et al.* (2005) asked whether it was practical to monitor the population size of harvested endemic freshwater crayfish in eastern Madagascar with a realistic investment in resources. They compared two methods that required very different levels of field effort and technical capacity: a simple method (1-day index of abundance) and a scientific method (5-days mark and recapture). By looking at the variability in each method and by carrying out some simple simulations, they estimated the power of each type of monitoring to detect various declines in the crayfish population. They found that the scientific method detected large declines with reasonable power *but* would need a very high effort (100 days of survey to detect a 20% decline with 80% power) as well as requiring unrealistic capacity. The simple method (attractive to policy-makers as it required only the capacity available locally and no computers or complex mathematical skills) had such low power that it could not possibly provide useful information (Figure 15.2).

Similarly, power analysis can be used to identify future investment required to give a certain amount of information. Scott Field and colleagues (Field *et al.*, 2007) looked at monitoring of rare birds in Australia and found very low power to detect changes in population size. However by using power analysis they were able to identify the species

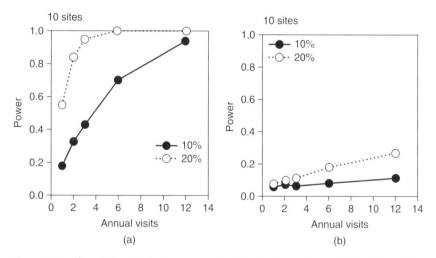

Figure 15.2 **The relationship between power to detect a change in population size and the number of visits to survey crayfish at 10 sites for two effect sizes (a 10% and a 20% decline) using (a) 5-days mark and recapture or (b) 1-day index of abundance. The significance level (risk of a false positive) was set at** $\alpha = 0.05$.

for which a small further investment would result in greatly increased power to detect trends and those where further monitoring would yield little. For a technical treatment of how to carry out power analysis I suggest you refer to Murphy and Myors (2004).

Is monitoring that lacks power a waste of time?

Because monitoring is costly, powerless monitoring can be worse than simply a waste of time. Firstly, it creates the illusion that something useful has been done when it has not (Peterman, 1990). Secondly it diverts resources from potentially effective interventions (Field *et al.*, 2004; Chades *et al.*, 2008). Revisiting our zebaroo example can illustrate this nicely. The government could use their zebaroo monitoring programme to argue that they were addressing zebaroo conservation – removing any pressure there may otherwise have been to directly address the known threats. In addition, the money that was spent on powerless monitoring could have been spent on anti-poaching patrols or habitat conservation. Monitoring can therefore make direct conservation interventions less likely and therefore be harmful to conservation.

Yoccoz *et al.* (2001) defines monitoring as 'gathering information about state variables ... for the purpose of ... drawing inferences about changes in state over time'. If this is the purpose of monitoring then considerations of power have to be central to the design of the scheme. However, conservation monitoring may have other objectives including engaging the public and/or policy-makers or detecting unexpected events (Jones *et al.*, 2011). An *a priori* power analysis may not be possible or appropriate in all these situations.

Real-world challenges

There are many conservation monitoring programmes around the world targeted at collecting and using data at very different scales, including monitoring of a species at a site (Marino *et al.*, 2006), national monitoring of a range of biodiversity (Firbank *et al.*, 2003), or global efforts to collate and synthesize monitoring data to uncover global trends (Collen *et al.*, 2009; Butchart *et al.*, 2010). And yet many of the challenges are the same whatever scale work is carried out at (Jones *et al.*, 2011).

Lack of resources and capacity to collect data

The ability to identify target species or habitats is essential for effective monitoring. Unfortunately, traditional taxonomic skills are in decline and there are very few individuals able to identify some poorly known taxa (Hopkins and Freckleton, 2002;

Wheeler *et al.*, 2004). At the same time, however, the increased availability of field guides has democratized species identification of certain groups, and increased participation in field-based monitoring in some parts of the world (Hidayanto *et al.*, 2008).

Resources for conservation will always be limited so we must always be on the lookout for ways of reducing the costs of monitoring. Citizen science, where volunteers collect data according to formal protocols, is common in many developed countries and can reduce the cost of collecting data over a large area (Silvertown, 2009). In North American and Europe there are a number of well-established monitoring schemes that provide robust trend data for popular taxonomic groups such as birds and butterflies (Link and Sauer, 1998, Gregory *et al.*, 2005; Thomas, 2005). A recent review of 395 species-based monitoring projects in Europe found that of the 46 000 people involved, more than 85% were volunteers (Schmeller *et al.*, 2009). In less developed countries there may be a smaller pool of potential monitoring volunteers due to limited leisure time (Danielsen *et al.*, 2009) but by providing specialist training for local people with limited formal education, projects can massively increase the sampling effort compared to what would be possible were only university-trained staff employed (Janzen, 2004). Such parataxonomists and paraecologists have made an invaluable contribution to rainforest inventories and monitoring programmes, particularly of lesser known taxa (Basset *et al.*, 2000).

There has been a surge in interest over the last decade in the potential of participatory monitoring to reduce the cost of biodiversity monitoring (Danielsen *et al.*, 2005; Reed *et al.*, 2008) in developing country settings. The term 'participatory monitoring' means different things to different people but in principle it means assessing change by involving people who would be affected by that change (Guijt 1999). In the context of biodiversity monitoring this often means the people who directly use the species being monitored, or who live and work in the target habitat (Danielsen *et al.*, 2005).

Participatory monitoring methods might include asking local people about trends in their resource rather than carrying out direct monitoring. Above we reported on a study that found that powerful monitoring of harvested crayfish populations in Madagascar would be too costly to be practical (Hockley *et al.*, 2005). With colleagues, we followed up on this study by investigating whether robust information on trends could be obtained from interviews with local crayfish collectors (Jones *et al.*, 2008). We compared data collected in a set of one-off interviews where we asked harvesters about their activity over the previous year, with data collected from the same harvesters in daily resource use interviews where the crayfish caught were counted and measured. Then we used the accuracy with which informants were able to report their activity in the annual one-off interviews to simulate the power with which researchers would be able to detect declines of various magnitudes given a simple annual interview with informants. The study was encouraging as we found that interviews could detect changes in catches and harvesting effort with reasonable power, especially when the

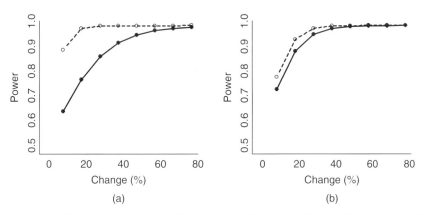

Figure 15.3 **The power to detect a change of variable magnitude (*x*-axis) using one-off interviews in (a) the number of crayfish caught at a site and (b) the proportion of days per month spent catching crayfish. Open symbols represent a paired analysis with the same people visited in subsequent years, filled symbols represent a new set of people sampled (Redrawn from Jones *et al.*, 2008, With permission from Elsevier).**

same informants were interviewed each year (Figure 15.3). Of course the details would vary from case to case but in general interviews with harvesters might provide useable information at low cost where individuals are interviewed independently and do not have any reason to hide their harvests.

Lack of resources and capacity to store data

Monitoring does not end with data collection. In 2007, the Durrell Wildlife Conservation Trust in Madagascar organized a workshop for government and NGOs to meet and discuss how biodiversity data could be better used in policy. The minister with responsibility for national parks mentioned in her address that despite the volume of research being carried out in Madagascar by academics and NGOs, she had difficulty accessing the biological information she needed to make informed decisions about park management (Durrell Wildlife Conservation Trust, 2007). This was because the data collected were held by a myriad of institutions and individuals in ways not easily accessed by decision-makers. Madagascar in fact has better systems in place for biodiversity data sharing than many countries: the REBIOMA project (led by the Wildlife Conservation Society) has been successful at creating a meta-database of a large number of important biodiversity datasets that have been used for national conservation planning (Kremen *et al.*, 2008). However, the point that this minister made, that monitoring data are often not available to those who need them, is very pertinent and its relevance is far from exclusive to low-income countries.

During a Welsh Assembly Government-funded project to develop a biodiversity indicator for Wales, we spent 3 months contacting holders of biodiversity datasets to try to obtain data disaggregated to the Welsh scale. We initially identified 101 possible datasets, and contacted the 35 that best met our criteria, but were only able to obtain five within the 3-month time frame. Our inability to access these other datasets wasn't because of any unwillingness to share data (in most cases the data holders were delighted for their data to contribute to the indicator) but the organizations lacked the resources, particularly staff time, to extract the data in a form useful to our project. There have been repeated calls for ecologists to pay more attention to ensuring that their data are as useful as possible to others (Michener *et al.*, 1997; McIntosh *et al.*, 2007; see also Chapter 14). It is a common problem that funders have not historically been willing to fund the costly work of maintaining and updating databases, which is so essential to make monitoring data as useful as they can be. Some recent positive steps have been made; for example, the UK Natural Environment Research Council (NERC) now requires that data collected on NERC-funded projects must be submitted to a recognized data archive.

Verification of data quality

Data quality is of central importance to any monitoring programme. Noisy data will mask true trends (reduce the power) and systematic biases can result in totally misleading conclusions. Biases can arise from a number of sources. Most are accidental, for example systematic misidentification of certain species, but the incentives faced by data collectors to bias data towards a preferred result should also be considered (Keane *et al.*, 2011).

Intentional biasing of data is a risk wherever data collectors are stakeholders in the conclusions drawn by the survey; for instance, managers of a reserve who fear criticism if negative trends of target species are detected. For example, from 1997 to 2003, the number of tigers in Sariska Tiger Reserve, India, reported by the local authorities to the national government was relatively stable at between 24 and 27 individuals. This fell slightly in 2004 to 17 individuals. However, later investigation revealed that the number of tigers in the reserve was much lower throughout the period and by 2004 the species had disappeared from this reserve altogether (Tiger Task Force, 2005; Figure 15.4). A mixture of poor methodology and unwillingness of local authorities to report bad news meant that serious poaching and decline of tigers at Sariska went unnoticed until it was too late. The Indian government is now in the process of implementing a much improved national tiger monitoring programme (Jhala *et al.*, 2011).

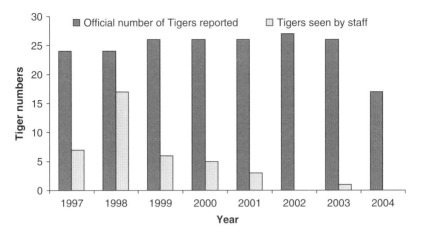

Figure 15.4 **The official number of tigers reported from Sariska Tiger reserve (Rajasthan, India), compared with the number of individuals seen by staff and tourists. Redrawn data from Project Tiger 2005.**

An ideal monitoring programme?

The British Trust for Ornithology coordinates a number of UK bird monitoring schemes (including the Breeding Birds Survey, the Big Garden BirdWatch, and the national bird ringing scheme). These schemes rely on volunteers to collect data, while professional ecologists design the protocols and carry out the bulk of the analysis. Some, such as the national bird ringing scheme, require that volunteers are qualified to a very high level in bird identification and handling, while others, such as the garden bird watch, are open to anyone (Greenwood, 2003). The large volume of data in such volunteer schemes compensates, to some extent, for the noise introduced due to misidentification (Schmeller *et al.*, 2009, although biases are more difficult to deal with (Boakes *et al.*, 2010). BTO data have been used to answer many fundamental and applied questions about bird ecology (e.g. Prendergast *et al.*, 2003; Baillie *et al.*, 2000; Atkinson *et al.*, 2003; McCleod *et al.*, 2005) and since 1997 the UK government has included the Wild Bird Indicator (based on BTO data) as one of their 15 official quality of life indicators (Gregory *et al.*, 2004). The BTO schemes have therefore been enormously successful at providing good quality data at high spatial and temporal resolution, while engaging the public and influencing policy.

Should we be looking to replicate the sort of scheme run by the BTO for other elements of biodiversity in other parts of the world? Certainly, those designing monitoring for other taxa can learn from the BTO bird monitoring schemes. Like the BTO, the UK Tracking Mammal partnership relies on volunteers collecting data according to specific protocols, which they then enter into a centralized database (Battersby and Greenwood, 2004). Using the BTO breeding bird survey squares for part of the scheme adds value as the mammal data can be directly compared with the bird data. The BTO are also contributing their expertise to the design of a UK national amphibian and reptile monitoring scheme (Baker and Gleed-Owen, 2007). Many other countries, particularly in the developed world, also have effective volunteer-based monitoring (Silvertown, 2009). Like the BTO scheme these tend to focus on larger, charismatic and easy to identify groups (Dobson, 2005).

We have much less information about smaller, less charismatic species. Many look at this imbalance in our knowledge of the world's biodiversity and conclude that it must be put right. For example, Sean Nee (2004) calls on biologists to be more like Star Trek's notoriously unemotional Mr Spock – avoiding sentimentality and treating all of biodiversity equally, regardless of size. Of course research to tackle our woeful lack of understanding of the many small and poorly known taxa that underpin much ecosystem functioning is extremely valuable (Dobson, 2005), but we have to accept that we can't expect to know everything about everything. Replicating the coverage that is possible in monitoring relatively large and charismatic species for other, less well known, elements of biodiversity would be prohibitively expensive and in some cases simply impossible. There can be no such thing as an ideal monitoring scheme, because what is sensible and desirable will depend on the cost of monitoring, the available capacity, and the need for the data (see Chapter 14). Sometimes we have to accept that less detailed monitoring, or even no monitoring, will be best for conservation.

Conclusions

We will never have enough money to spend on all the possible conservation interventions that we would like to make. Therefore the decision of how to carry out a monitoring programme, and even whether monitoring is justified, must be taken seriously by everyone involved in conservation practice. Much conservation monitoring is not as valuable as it could be (or may even have no value) as too little attention has been paid to the design of the study, and especially to the power of the study to detect trends. Other valuable and well-designed monitoring has been carried out but the data produced are not stored in a way that makes it possible for others to make the best use of them. In their excellent paper, Legg and Nagy (2005) question our attachment to monitoring: 'Monitoring seems to be the automatic response of conservationists to any change or development that is seen as a potential threat to the environment, whether or not it is appropriate.' I hope that after reading this chapter you will avoid the trap of assuming that monitoring is always good and take care to ensure that any monitoring you do in the real world makes a real contribution to conservation.

Acknowledgements

Thanks to Neal Hockley, Sue Hearn, James Gibbons, Aidan Keane, E.J. Milner-Gulland, and Hugh Possingham for useful discussion, and to the Leverhulme Trust for funding the research project 'optimizing monitoring as a conservation tool'.

References

Atkinson, P.W., Clark, N.A., Bell, M.C., Dare, P.J., Clark, J.A., and Ireland, P.L. (2003) Changes in commercially fished shellfish stocks and shorebird populations in the Wash, England. *Biological Conservation*, 114, 127–141.

Baillie, S.R., Sutherland, W.J., Freeman, S.N., Gregory, R.D., and Paradis, E. (2000) Consequences of large-scale processes for the conservation of bird populations. *Journal of Applied Ecology*, 37, 88–102.

Baker, J. and Gleed-Owen, C. (2007) *National Amphibian and Reptile Recording Scheme*. The Herpetological Conservation Trust, Bournemouth, UK.

Basset, Y., Novotny, V., Miller, S.E., and Pyle, R. (2000) Quantifying biodiversity: Experience with parataxonomists and digital photography in Papua New Guinea and Guyana. *Bioscience*, 50, 899–908.

Battersby, J.E. and Greenwood, J.J.D. (2004) Monitoring terrestrial mammals in the UK: past, present and future, using lessons from the bird world. *Mammal Review*, 34, 3–29.

Boakes, E.H., McGowan, P.J.K., Fuller, R.A., et al. (2010) Distorted views of biodiversity: spatial and temporal bias in species occurrence data. *PLoS Biology*, 8(6), e1000385.

Butchart, S.H.M., Walpole, M., Collen, B., et al. (2010) Global biodiversity: indicators of recent declines. *Science*, 328, 1164–1168.

Chades, I., McDonald-Madden, E., McCarthy, M.A., Wintle, B., Linkie, M., and Possingham, H.P. (2008) When to stop managing or surveying cryptic threatened species. *Proceedings of the National Academy of Sciences of the USA*, 105, 13936–13940.

Cohen, J. (1962) The statistical power of abnormal social psychological research: A review. *Journal of Abnormal and Social Psychology*, 65, 145–153.

Collen, B., Loh, J., Whitmee, S., McRae, L., Amin, R., and Baillie, J.E.M. (2009) Monitoring change in vertebrate abundance: the Living Planet Index. *Conservation Biology*, 23, 317–327.

Danielsen, F., Burgess, N., and Balmford, A. (2005) Monitoring matters: examining the potential of locally-based approaches. *Biodiversity and Conservation*, 14, 2507–2542.

Danielsen, F., Burgess, N.D., Balmford, A., et al. (2009) Local participation in natural resource monitoring: a characterization of approaches. *Conservation Biology*, 23, 31–42.

Di Stefano, J. (2001) Power analysis and sustainable forest management. *Forest Ecology and Management*, 154, 141–153.

Di Stefano, J. (2003) How much power is enough? Against the development of an arbitrary convention for statistical power calculations. *Functional Ecology*, 17, 707–709.

Dobson, A. (2005) Monitoring global rates of biodiversity change: challenges that arise in meeting the Convention on Biological Diversity (CBD) 2010 goals. *Philosophical Transactions of the Royal Society B, Biological Sciences*, 360, 229–241.

Durrell Wildlife Conservation Trust (2007) *Process verbaux du symposium national dur le suivi ecologique "Monitoring Matters": analysis comparatives des approaches innovatrices.* DWCT, Antananarivo, Madagascar.

Field, S.A., Tyre, A.J., Jonzen, N., Rhodes, J.R., and Possingham, H.P. (2004) Minimizing the cost of environmental management decisions by optimizing statistical thresholds. *Ecology Letters*, 7, 669–675.

Field, S.A., O'Connor, P.J., Tyre, A.J., and Possingham, H.P. (2007) Making monitoring meaningful. *Austral Ecology*, 32, 485–491.

Firbank, L.G., Barr, C.J., Bunce, R.G.H., *et al.* (2003) Assessing stock and change in land cover and biodiversity in GB: an introduction to Countryside Survey 2000. *Journal of Environmental Management*, 67, 207–218.

Gerrodette, T. (1987) A power analysis for detecting trends. *Ecology*, 68, 1364–1372.

Greenwood, J.J.D. (2003) The monitoring of British breeding birds: a success story for conservation science? *Science of the Total Environment*, 310, 221–230.

Gregory, R.D., Noble, D.G., and Custance, J. (2004) The state of play of farmland birds: population trends and conservation status of lowland farmland birds in the United Kingdom. *Ibis*, 146, 1–13.

Gregory, R.D., van Strien, A., Vorisek, P., *et al.* (2005) Developing indicators for European birds. *Philosophical Transactions of the Royal Society B, Biological Sciences*, 360, 269–288.

Guijt, I. (1999) *Participatory Monitoring and Evaluation for Natural Resource Management and Research*. Natural Resources Institute, Greenwich, London.

Hidayanto, Y., Saryanthi, R., Amama, F.P., and Utomo, A.B. (2008) *Impact Study on Field Guides: Assessing the Usefulness of Bird of Java, Bali, Sumatra and Kalimantan Field Guide*. Bank-Netherlands Partnership Programme.

Hockley, N.J., Jones, J.P.G., Andriahajaina, F., Manica, A., and Randriamboahary, J.A. (2005) When should communities and conservationists monitor exploited resources? *Biodiversity and Conservation*, 14, 2795–2806.

Hopkins, G.W. and Freckleton, R.P. (2002) Declines in the numbers of amateur and professional taxonomists: implications for conservation. *Animal Conservation*, 5, 245–249.

Janzen, D.H. (2004) Setting up tropical biodiversity for conservation through non-damaging use: participation by parataxonomists. *Journal of Applied Ecology*, 41, 181–187.

Jhala, Y., Qureshi, Q., and Gopal, R. (2011) Can the abundance of tigers be assessed from their signs? *Journal of Applied Ecology*, 48, 14–24.

Jones, J.P.G. (2011) Monitoring species distribution and abundance at the landscape scale. *Journal of Applied Ecology*, 48, 9–13.

Jones, J.P.G., Andriamarovolona, M.M., Hockley, N., Gibbons, J.M., and Milner-Gulland, E.J. (2008) Testing the use of interviews as a tool for monitoring trends in the harvesting of wild species. *Journal of Applied Ecology*, 45, 1205–1212.

Jones, J.P.G., Collen, B., Atkinson, G., *et al.* (2011) The why, what and how of global biodiversity indicators beyond the 2010 Target. *Conservation Biology*, 25, 450–457.

Keane, A., Jones, J.P.G., and Milner-Gulland, E. (2011) Encounter data in resource management and ecology: Pitfalls and possibilities. *Journal of Applied Ecology*, 48, 1164–1173.

Kremen, C., Cameron, A., Moilanen, A., *et al.* (2008) Aligning conservation priorities across taxa in Madagascar with high-resolution planning tools. *Science*, 320, 222–226.

Legg, C.J. and Nagy, L. (2005) Why most conservation monitoring is, but need not be, a waste of time. *Journal of Environmental Management*, 78, 194–199.

Link, W.A. and Sauer, J.R. (1998) Estimating population change from count data: application to the North American Breeding Bird Survey. *Ecological Applications*, 8, 258–268.

Mapstone, B.D. (1995) Scalable decision rules for environmental-impact studies – effect size, type-I, and type-II errors. *Ecological Applications*, 5, 401–410.

Marino, J., Sillero-Zubiri, C., and Macdonald, D.W. (2006) Trends, dynamics and resilience of an Ethiopian wolf population. *Animal Conservation*, 9, 49–58.

McCleod, R., Barnett, P., Clark, J.A., and Cresswell, W. (2005) Body mass change strategies in blackbirds *Turdus merula*: the starvation-predation risk trade-off. *Journal of Animal Ecology*, 74, 292–302.

McIntosh, A.C.S., Cushing, J.B., Nadkarni, N.M., and Zeman, L. (2007) Database design for ecologists: Composing core entities with observations. *Ecological Informatics*, 2, 224–236.

Michener, W.K., Brunt, J.W., Helly, J.J., Kirchner, T.B., and Stafford, S.G. (1997) Nongeospatial metadata for the ecological sciences. *Ecological Applications*, 7, 330–342.

Murphy, K.R. and Myors, B. (2004) *Power Analysis: a Simple and General Model for Traditional and Model Hypothesis Tests*. Lawrence Erlbaum Associates, Mahwah, NJ.

NAO (2009) *Young People's Sexual Health: the National Chlamydia Screening Programme*. National Audit Office, London.

Nee, S. (2004) More than meets the eye – Earth's real biodiversity is invisible, whether we like it or not. *Nature*, 429, 804–805.

Peterman, R.M. (1990) Statistical power analysis can improve fisheries research and management. *Canadian Journal of Fisheries and Aquatic Sciences*, 47, 2–15.

Prendergast, J.R., Quinn, R.M., Lawton, J.H., Eversham, B.C. and Gibbons, D.W. (1993) Rare species, the coincidence of diversity hotspots and conservation strategies. *Nature*, 365, 335–337.

Reed, M.S., Dougill, A.J. and, Baker, T.R. (2008) Participatory indicator development: What can ecologists and local communities learn from each other? *Ecological Applications*, 18, 1253–1269.

Schmeller, D.S., Henry, P.Y., Julliard, R., *et al.* (2009) Advantages of volunteer-based biodiversity monitoring in Europe. *Conservation Biology*, 23, 307–316.

Silvertown, J. (2009) A new dawn for citizen science. *Trends in Ecology and Evolution*, 24, 467–471.

Thomas, J.A. (2005) Monitoring change in the abundance and distribution of insects using butterflies and other indicator groups. *Philosophical Transactions of the Royal Society B, Biological Sciences*, 360, 339–357.

Tiger Task Force (2005) *Joining the Dots*. Union Ministry of Environment and Forests (Project Tiger), New Delhi.

Wheeler, Q.D., Raven, P.H., and Wilson, E.O. (2004) Taxonomy: impediment or expedient? *Science*, 303, 285–285.

Yoccoz, N.G., Nichols, J.D. and Boulinier, T. (2001) Monitoring of biological diversity in space and time. *Trends in Ecology and Evolution*, 16, 446–453.

16
Monitoring in UNDP-GEF Biodiversity Projects: Balancing Conservation Priorities, Financial Realities, and Scientific Rigour

Sultana Bashir

Institute of Zoology, Zoological Society of London, London, UK

Introduction

There are many good reasons to monitor biodiversity (Jones *et al.*, 2010). One important reason is to understand the status and trends in biodiversity, given the intimate relationship between biodiversity and the delivery of ecosystem services that are critical to human well-being and economic development (MA, 2005; SCBD, 2010; TEEB, 2010). Such monitoring data can help us identify and evaluate critical threats to biodiversity, which in turn can be used to guide conservation policy and action (see Chapter 1). Armed with such information, decision-makers, planners, conservation managers, and other concerned actors and stakeholders can prioritize how best to allocate limited financial, technical, and other resources to different biodiversity conservation needs. For those implementing conservation on the ground, biodiversity monitoring is often essential to guide the design of specific interventions, measure the impact of actions on biodiversity targets, and adapt interventions as needed based on monitoring feedback (Kinnaird and O'Brien, 2001).

There are many challenges to implementing effective biodiversity monitoring systems, however, particularly at national and global scales (see Chapters 1 and 15). Different aspects of biodiversity[1] are perceived and valued in many different ways

---

[1] The United Nations (UN) Convention on Biological Diversity (CBD) defines biodiversity as 'the variability among living organisms from all sources, including, inter alia, terrestrial, marine and other aquatic ecosystems and the ecological complexes of which they are part; this includes diversity within species, between species, and of ecosystems' (Article 2).

by different stakeholders, and achieving agreement on what should be measured, how, and at what scale is often difficult (Bubb *et al.*, 2005; Mace *et al.*, 2010). Other obstacles include identifying scientifically valid indicators that are also practical and cost-effective and obtaining reliable and consistent datasets (Walpole *et al.*, 2009; see also Chapter 13). Many countries, including most developing countries, lack the technical, human, and financial capacity needed to develop and implement systematic biodiversity monitoring systems. There is thus considerable variation in the quality and scope of existing global and national biodiversity monitoring efforts, particularly in developing countries (CBD, 2007a, 2007b; UNEP-WCMC, 2009).

Existing gaps in capacity and other resources for biodiversity monitoring are unlikely to be redressed any time soon, given the plethora of competing demands on government budgets; and despite recent international pledges to provide additional funding, dedicated funding for biodiversity remains small relative to estimated need.[2] Greater support for biodiversity monitoring per se from traditional donors in the developed countries, whether from aid budgets or private or public sources, seems unlikely in the current political and economic climate. Meanwhile, global biodiversity continues to decline rapidly (Butchart *et al.*, 2010; SCBD, 2010), and without urgent action the outlook for biodiversity, and the environment generally, is grim, with potentially very serious implications for our collective future well-being (OECD, 2012; SCBD, 2010; UNEP, 2007).

Ideally conservation policy, planning, and management should be informed by adequate data (OECD, 2012). However, where multiple priorities exist and resources are limited, trade-offs are unavoidable: for example, investing resources in biodiversity monitoring may mean fewer resources available for other forms of conservation action (see Chapter 14). Furthermore, the conservation benefits arising from the additional information gained may not outweigh the costs of obtaining it (McDonald-Madden *et al.*, 2010). Thus, decisions must be made in any given situation about how much to invest in biodiversity monitoring, including the level of scientific rigour that is feasible, vis-à-vis other equally or more pressing conservation management activities (McDonald-Madden *et al.*, 2010; Salzer and Salafsky, 2006; Yoccoz *et al.*, 2010). This means, amongst other things: ensuring that the data collected address agreed policy and management priorities; tailoring monitoring methods and systems to the resources

---

[2] Additional funds to support the implementation of the CBD's Strategic Plan for 2011–2020 and the Aichi Targets, were announced at the meeting of the Tenth Conference of the Parties (COP) to the CBD in Nagoya in 2010, notably by Japan, which announced the establishment of a US$2 billion Japan Biodiversity Fund as well as by France, the European Union, and Norway. While this is promising, the total funds available globally are calculated to be at least an order of magnitude lower than most estimates of conservation financing needs, particularly in developing countries (Parker and Cranford, 2010). For example, IUCN 2010b (cited in Parker and Cranford, 2010) propose a target of US$300 billion/year to finance the conservation and sustainable use of biodiversity (also see CBD, 2007c). The CBD is currently preparing a more systematic assessment of the financing required to implement its Strategic Plan in order to develop a Resource Mobilization Strategy (see http://www.cbd.int/financial/assessment/ and related links).

available and the local context; and ensuring that mechanisms are in place to maximize the long-term utility of such data for conservation planning and management (see Chapter 14). How such decisions are made in practice, and who is involved in making them, is of course key, but beyond the scope of the present chapter.

If we agree that the resources invested in monitoring must be used as efficiently and effectively as possible, then there is also a need to explore opportunities for making better use of existing ad hoc efforts to monitor biodiversity. For example, many government agencies, conservation organizations, and research institutions undertake biodiversity assessments and monitoring schemes in specific sites and landscapes, often through government or donor-assisted projects and programmes (CBD, 2007a). However, the data generated through such initiatives are frequently difficult to access for use in broader or long-term biodiversity monitoring because of lack of standardization, variable data quality, gaps in coverage, and the relatively short timeframes involved (CBD, 2007a; Dobson, 2005; see also Chapter 13).

This chapter considers the particular case of Global Environment Facility (GEF)-supported biodiversity projects implemented through the United Nations Development Programme (UNDP) and their existing and potential contribution to broader-scale biodiversity monitoring efforts. Over the past 20 years, the GEF has disbursed over US$155 million per year on average for biodiversity projects and programmes in developing countries and countries with economies in transition (GEF, 2010a).[3] Although biodiversity monitoring is rarely the primary focus of UNDP or GEF projects for reasons discussed later, many projects already provide data on selected indirect measures of biodiversity for global monitoring purposes, such as information on protected area numbers, coverage, and status (GEF EO, 2010; see later). Additionally, various monitoring reports and studies show that a significant number of GEF biodiversity projects undertake some direct biodiversity assessment and monitoring, often through, or in collaboration with, various partners, although the data collected are often problematic for the reasons mentioned earlier (Dublin and Volonte, 2004).

The information presented in this chapter is based mainly on the knowledge and experiences gained by the author as a result of managing a portfolio of GEF-supported biodiversity projects implemented through UNDP in the Asia-Pacific region between 2004 and 2009.[4] Key policies, studies, and reports relating to GEF and UNDP biodiversity projects have also been consulted. I begin with some general background on the GEF and UNDP and their respective work on biodiversity conservation.

---

[3] Based on dividing the figures given in GEF (2010a), i.e. US$3.1 billion of GEF funds by 20 years. This excludes co-financing, which was on a scale of over US$400 million/year over the same period or (US$8.3 billion in total).

[4] GEF and UNDP-supported biodiversity projects are referred to as UNDP-GEF biodiversity projects hereafter for convenience. However, these projects are identified, owned, and ultimately delivered by recipient country governments and other national and local stakeholders. Between 2004 and 2009, I managed 30 to 35 UNDP-GEF biodiversity projects under varying stages of implementation in the Asia-Pacific region.

This is followed by a review of the role of biodiversity monitoring in the overall monitoring and evaluation of UNDP-GEF[5] biodiversity projects. I then go on to discuss examples of direct biodiversity assessment and monitoring in UNDP-GEF projects in Asia-Pacific. I conclude with a discussion of how the GEF and its partners such as UNDP could help to improve the quality, broader relevance, and sustainability of biodiversity monitoring efforts associated with GEF projects by better leveraging the substantial collective technical, scientific, and capacity development expertise and resources contained within the GEF network.

The policies and practices of both the GEF and UNDP are continually evolving, particularly between the four-yearly donor replenishments of the GEF. This chapter is based mainly on GEF-3 (2002–06) and GEF-4 (2006–10) projects and GEF-4 policies and practices. Relevant changes in policy and practice in GEF-5 are also reviewed. The views put forward here are solely those of the author and not of the GEF, UNDP, or any other agency mentioned unless otherwise stated.

The GEF and biodiversity conservation

The GEF is a multilateral trust fund and the officially designated financial mechanism of the Convention for Biological Diversity (CBD) and several other environmental conventions.[6] It is the largest source of public grant financing for environmental projects in developing countries (Clémençon, 2006). Originally set up in 1991 as a US$1 billion pilot programme to support the protection of the global environment and promote sustainable development, it was restructured and formally established in 1994 (GEF, 2011f).[7] The GEF has a unique institutional and governance structure, which includes: an Assembly, with representatives from 182 member countries; a Council with 32 Members representing 32 constituencies (16 from developing countries, 14 from developed countries, and two from economies in transition); a Secretariat tasked with managing the day-to-day business of the GEF; three Implementing Agencies and seven Executing Agencies, mandated to help countries develop and manage GEF

---

[5] Hereafter, unless otherwise stated, 'UNDP-GEF biodiversity projects(s)' refers to medium-sized (up to US$1 million in GEF-4) and full-sized biodiversity projects (more than US$ 1 million in GEF-4) under the GEF Biodiversity Focal Area.

[6] It is also the financial mechanism of the and the United Nations Framework Convention on Climate Change (UNFCCC), United Nations Convention to Combat Desertification (UNCCD), and the Stockholm Convention on Persistent Organic Pollutants. The major focal areas of the GEF are: biodiversity, climate change (including mitigation and adaptation), international waters, land degradation, persistent organic pollutants, and ozone layer depletion. Sustainable forest management, which generally cross-cuts more than one focal area, became an important area of work under GEF-4. See http://www.thegef.org/gef/Areas_work.

[7] The GEF was officially launched at the first 'Rio Conference', the 1992 UN Conference on Environment and Development (UNCED). See Clémençon (2006), GEF (2011f), and Porter et al. (2008) for more background and additional references on the history of the GEF.

projects;[8] an independent Scientific and Technical Advisory Panel (STAP); and an independent Evaluation Office. The Assembly is the highest political body of the GEF, while the Council is its main governing body: together these bodies determine the operational policies, strategies, and programmes of the GEF, while the GEF Agencies have mainly an advisory role (Perez del Castillo, 2009; GEF, 2011f).[9]

A central feature of the GEF is that it provides funds that are additional to Official Development Assistance (ODA) to cover the extra costs associated with generating broader environmental benefits that would not be produced otherwise by existing baseline or proposed non-GEF interventions, that is, the GEF will provide 'new and additional grant and concessional funding to meet the agreed incremental costs of measures to achieve agreed global environmental benefits' (GEF, 1996, p. 6). However, it was recognized from the outset that the additional funds that were likely to be made available through the GEF by donor countries could not possibly address the full gamut of global environmental problems (Clémençon, 2006; Porter et al., 2008). Thus, the GEF's role was regarded as being primarily catalytic in nature, to leverage additional resources for sustainable environmental management and to test new approaches and solutions with potential for replication and scale-up beyond the life of a project to achieve long-term conservation impacts (GEF, 1996; GEF EO, 2010). As replication and scale-up of successful project results and strategies rely on the continued support and active engagement of primarily national actors and stakeholders, various criteria are used during project selection and design (and subsequent management) to increase the likelihood of this happening (GEF EO, 2010).

In the biodiversity arena, the GEF derives its objectives from the CBD, and GEF support is intended to help recipient countries meet their commitments under the CBD. Given the GEF's incremental cost rule, all GEF biodiversity projects must ultimately generate global environmental benefits, for example by helping to conserve globally significant biodiversity, and must not focus solely on local or national benefits. Since GEF-3 (2002–2006), the GEF has had a dedicated biodiversity strategy that focuses on a subset of priorities under the CBD and provides a framework for identifying and developing GEF projects. This strategy has been further refined and expanded in GEF-4 (2006–10) and GEF-5 (2010–14) (GEF, 2011a). Both GEF-4 and GEF-5 goals, objectives, and priorities are presented in Tables 16.1 and 16.2, which show that despite some restructuring and minor reformulation, most of the key elements of the GEF Biodiversity strategy in GEF-5 are essentially the same as in GEF-4. Thus,

---

[8] The Implementing Agencies are: the World Bank, UNDP, and the UN Environment Programme (UNEP). The Executing Agencies are: the UN Food and Agricultural Organization (FAO); the International Fund for Agricultural Development (IFAD); the UN Industrial Development Organization (UNIDO); the Inter-American Development Bank (IADB); the African Development Bank (AfDB), the Asian Development Bank (ADB), and the European Bank for Reconstruction and Development (EBRD). Collectively, these organizations are often just referred to as the GEF Agencies. In GEF-5, the GEF has put in place mechanisms for accreditation of additional GEF agencies.

[9] Further details of the institutional structure and governance of the GEF can be found in 'The Instrument for the Establishment of the Restructured Global Environment Facility' (GEF, 2011f), the GEF website http://www.thegef.org/gef/home, and http://www.gefcountrysupport.org/report_detail.cfm?projectId=140.

Table 16.1   **Goal, objectives, and programs of the GEF-4 Biodiversity Strategy (2006–10)**

GOAL: *The conservation and sustainable use of biodiversity, the maintenance of the ecosystem goods and services that biodiversity provides to society, and the fair and equitable sharing of the benefits arising out of the utilization of genetic resources*

Strategic Objectives:	Strategic Programs (SP)
1. To catalyze the sustainability of Protected Area Systems	1.1 Sustainable financing of Protected Area Systems at the national level (SP1) 1.2 Increasing representation of effectively managed marine protected areas in Protected Area Systems (SP2) 1.3 Strengthening terrestrial Protected Area Networks (SP3)
2. To mainstream biodiversity conservation in production landscapes/seascapes and sectors	2.1 Strengthening the policy and regulatory framework for mainstreaming biodiversity (SP4) 2.2 Fostering markets for biodiversity goods and services (SP5)
3. To safeguard biodiversity	3.1 Building capacity for the implementation of the Cartagena Protocol on Biosafety (SP6) 3.2 Prevention, control and management of Invasive Alien Species (SP7)
4. To build capacity on Access and Benefit Sharing	4.1 Building capacity on Access and Benefit Sharing (SP8)

Source: GEF (2007a).

the overall goal of the GEF biodiversity focal area in GEF-4 was 'the conservation and sustainable use of biodiversity, the maintenance of the ecosystem goods and services that biodiversity provides to society, and the fair and equitable sharing of the benefits arising out of the utilization of genetic resources'. In GEF-5, this goal has been simplified to 'the conservation and sustainable use of biodiversity and the maintenance of ecosystem goods and services'. One notable change, however, is the inclusion of an explicit fifth objective on integrating CBD obligations into national planning processes through Enabling Activities (EAs),[10] although the GEF has in fact supported a range

---

[10] Enabling Activities (EAs) and their scope were originally defined in the first GEF Operational Strategy (GEF, 1996). These are activities that help recipient countries prepare the foundation to design and implement effective measures to achieve the objectives of the CBD, including the development of national strategies, plans, or programmes (CBD Article 6), and the identification of biodiversity components and threats (CBD Article 7).

Table 16.2 Goal, objectives, and priority areas of the GEF-5 Biodiversity Strategy

GOAL: *Conservation and sustainable use of biodiversity and the maintenance of ecosystem goods and services*

Objective	Priorities for project support
1. Improve sustainability of Protected Area (PA) Systems	• Improving sustainable financing of PA Systems • Expanding marine and terrestrial ecosystem representation within PA Systems • Expanding threatened species representation within PA Systems • Improving management effectiveness of existing PAs
2. Mainstream biodiversity conservation and sustainable use into production landscapes, seascapes and sectors	• Strengthening policy and regulatory frameworks for mainstreaming biodiversity • Implementing Invasive Alien Species management frameworks • Strengthening capacities to produce biodiversity-friendly goods and services
3. Build capacity for the implementation of the Cartagena Protocol on Biosafety	• Implementation of activities identified in country stock-taking analyses and in Conference of the Parties (COP) guidance to the GEF
4. Build capacity on access to genetic resources and benefit sharing (ABS)	• Priority areas for project support to be identified after completion of negotiations on an international regime for ABS
5. Integrate Convention on Biological Diversity (CBD) obligations into national planning processes through enabling activities	• Revising National Biodiversity Strategic Action Plans (NBSAPs) in line with CBD's new strategic plan • Integrating biodiversity into sectoral planning • National reporting to the CBD • Implementation of guidance related to the Clearing House Mechanism

Source: GEF (2009, 2011a).

of EAs since its inception, such as the development of National Biodiversity Strategies and Action Plans (NBSAPs) and national reports to the CBD.[11]

---

[11] Parties to the CBD are called upon to develop an NBSAP and to report on progress in its implementation every 4 years through national reports to the Conference of the Parties to the CBD.

Since 1991, the GEF has provided US$3.1 billion in grants and mobilized a further US$8.3 billion in cofinancing,[12] including leveraged resources, to support the implementation of over one thousand biodiversity projects in more than 155 developing countries and countries with economies in transition (GEF, 2010a). GEF support to other focal areas such as Land Degradation, International Waters, and Climate Change may also contribute to biodiversity objectives, but these focal areas are not covered here. To date, the vast majority of GEF biodiversity funding has been allocated towards improving the sustainability of protected areas (PA) systems (Objective 1), followed by mainstreaming biodiversity conservation and sustainable use into production landscapes, seascapes, and sectors (Objective 2) (GEF EO, 2010). The GEF also has a number of global projects, some of which are specific to the biodiversity focal area, such as the support given to biodiversity indicator development for national use (Bubb *et al.*, 2005) and on various CBD 2010 Target-related activities, notably the development of the Biodiversity Indicators Partnership (BIP, 2010). Other global corporate programmes include the long-running GEF Small Grants Programme (SGP) and the capacity development programme.[13]

UNDP and biodiversity conservation

As the United Nations' global development programme, UNDP has long recognized the importance of biodiversity and well-functioning natural ecosystems for global poverty reduction and sustainable development. UNDP's Human Development Report for 2007/2008 identified ecological collapse as a key threat to the achievement of the UN Millennium Development Goals (MDGs), and also as a potential driver of reversals in human development, with disproportionate impact on the poorest and most vulnerable sections of society.[14] Additionally, from a human development perspective, ecosystems and biodiversity are considered to have intrinsic value (UNDP, 2007a). Environment and sustainable human development thus comprise one of UNDP's major areas of focus, alongside poverty reduction, democratic governance, and crisis

---

[12] Co-financing comprises financial resources from non-GEF Trust Fund sources that will be managed with the GEF allocation as part of the initial financing package of a GEF project at the time of approval and without which the GEF objectives could not be met. It is considered distinct from 'leveraged resources', which are defined as the additional resources, beyond those committed to the project itself at the time of approval, that are mobilized later as a direct result of the project. Leveraged resources are viewed as an important indicator of the catalytic effect of the GEF (GEF, 2002). However, recent GEF publications and the GEF website often describe all co-financing as leveraged resources (e.g. GEF, 2010a).
[13] For further details see http://www.thegef.org/gef/Areas_work.
[14] Human development has many dimensions and is defined broadly by UNDP as a process of enlarging people's choices and building human capabilities. It is distinct from economic development and does not assume that economic growth automatically leads to human progress. For further details see http://hdr.undp.org/en/humandev/.

prevention and recovery.[15] UNDP's work on environment, including its work with the GEF, is embedded within its broader development agenda and aligned with its Strategic Plan, specifically with the strategic priority on 'Environment and Sustainable Development for the Millennium Development Goals'.[16]

Environmental sustainability, biodiversity conservation, and sustainable development are also long-standing goals of the UN system as a whole. 'Ensuring environmental sustainability' is one of the UN MDGs (MDG 7) and includes the CBD's 2010 Target as one of its four targets (UNDP, 2008b, 2010a). While this target was not met by 2010 (Butchart et al., 2010), it remains relevant as the UN General Assembly has declared 2011–2020 to be the UN Decade on Biodiversity in support of the CBD's Strategic Plan for 2011–2020 and the Aichi Targets.[17]

UNDP played a key role in the pilot phase of the GEF along with UNEP and the World Bank, and in 1994 the three agencies were formally designated as GEF Implementing Agencies (IAs) under the GEF Instrument[18] (GEF, 2011f). Thus, UNDP has been actively engaged in helping recipient countries identify, develop, and implement GEF projects since 1991 with a strong focus on national stakeholder engagement, policy advice, capacity development, and institutional strengthening (UNDP, 2008a). In doing so, the agency has capitalized on its experience of undertaking inter-country and national programming, and of working closely with governments, non-governmental organizations (NGOs), and communities through its global network of field offices (GEF, 2011a, Annex D). All UNDP projects, including GEF projects, are developed in close collaboration with national governments and other partners under country-specific UN Development Assistance Frameworks (UNDAFs) and are thus aligned with national, as well as GEF, policies and priorities (UNDP, 2011a). Since 1991, UNDP has supported the development and implementation of over 570 major projects in all the GEF focal areas in over

---

[15] For more details see http://www.undp.org/content/undp/en/home/ourwork/overview.html.

[16] UNDP's Strategic Plan 2008–2013 includes four Key Results under the Strategic Priority on Environment and Sustainable Development, focused on supporting the achievement of the MDGs. These are: (i) mainstreaming environment and energy in MDG-based national policy and planning frameworks; (ii) generating new environment-based sources of finance to significantly scale up investment in environment and energy to achieve the MDGs; (iii) promoting adaptation to climate change in order to lower risks to the poor in developing countries and enable the attainment of the MDGs; and (iv) expanding access to environmental and energy services for the poor for poverty reduction and economic growth (UNDP, 2008c). See http://www.undp.org/content/undp/en/home/ourwork/environmentandenergy/overview.html and associated pages.

[17] See http://www.cbd.int/2011-2020/goals/.

[18] The GEF Instrument is the document used to establish the GEF after its pilot phase and essentially constitutes the statutes and bylaws of the GEF. The GEF Instrument includes provisions for the governance, participation, replenishment, and fiduciary and administrative operations of the GEF (see GEF, 2011f). It also outlines the roles and responsibilities of different actors in the GEF. Thus, each IA was expected to contribute and engage in the GEF based on their special areas of competence, or comparative advantage, as outlined in Annex D of the GEF Instrument on Principles of Cooperation among the Implementing Agencies. Also see GEF (2006). The World Bank is also the Trustee of Fund.

140 countries.[19] UNDP has also implemented the well-regarded GEF Small Grants Programme (SGP) for many years.[20]

Some 36% of all (non-SGP) GEF grants made through UNDP to date have been for biodiversity projects (UNDP, 2011b). The objective of UNDP's biodiversity work (which also includes non-GEF-financed projects and programmes[21]), is to maintain and enhance natural ecosystem services that are of critical importance to continued human well-being and development (UNDP, 2010b, 2010c, 2011c). Specifically, UNDP helps countries to address biodiversity loss and degradation through two signature programmes, which build on the agency's core strengths, are aligned with the GEF Biodiversity Strategy, and leverage the specific expertise developed by the agency over more than two decades of implementing GEF biodiversity projects. These two programmes are:

1. To improve the sustainability of protected areas (PAs) systems by tapping into their full economic potential such that PAs are effectively managed, sustainably financed, and contributing towards sustainable development objectives.
2. To integrate biodiversity management objectives into economic sector activities to ensure that production processes do not compromise biodiversity and the essential ecosystem functions that underpin human well-being (UNDP, 2011f).

Together, these two approaches, which relate to GEF Strategic Objectives 1 and 2, respectively (see Tables 16.1 and 16.2), target a broad swathe of global biodiversity. While PAs are widely agreed to be an important element of any global conservation strategy, they cover only some 22% of the Earth's surface area. Not only does most biodiversity reside outside PAs, but the existing global PA network also has many gaps in biogeographical coverage and is vulnerable to climate change impacts. It is therefore vitally important to maintain biodiversity both within and outside PAs, especially in a changing climate.

In 2010, UNDP had 116 biodiversity projects under implementation, funded mainly by the GEF, with a total value of nearly US$1.6 billion, including nearly US$434 million in GEF grants and the rest as co-financing from UNDP, recipient governments, and other sources (GEF, 2010c). These included 67 PA projects and 45 biodiversity mainstreaming projects in some 50 countries (UNDP, 2010a). Earlier UNDP-GEF biodiversity projects (and GEF biodiversity projects generally) focused more on site-based project activities and addressing the immediate threats to biodiversity,

---

[19] UNDP has over 140 country offices and several regional centres; the agency works in over 170 countries and territories.
[20] See http://sgp.undp.org/index.cfm?module=ActiveWebandpage=WebPageands=AboutSGP
[21] See UNDP Ecosystems and Biodiversity (http://www.undp.org/content/undp/en/home/ourwork/environmentand energy/focus_areas/ecosystems_and_biodiversity.html) and UNDP-GEF Biodiversity (http://web.undp.org/gef /do_biodiversity.shtml).

although the importance of addressing the underlying drivers of biodiversity loss and degradation was recognized early on (GEF, 1996, paragraph 1.12(2)). Over the years, there has been growing emphasis on achieving wider and more lasting impacts by tackling the underlying barriers that prevent countries from effectively addressing the root causes of biodiversity degradation and loss, such as policy, capacity, and financial constraints (UNDP, 2010a, 2011f). UNDP is currently developing a biodiversity strategy to further clarify its framework for engagement on biodiversity, specifically to establish how best to help countries scale-up their responses to meet the CBD's Aichi Biodiversity Targets (N. Sekhran, personal communication).

The role of biodiversity monitoring in the monitoring and evaluation of UNDP-GEF biodiversity projects

Monitoring and evaluation (M&E) of UNDP-GEF projects and programmes is guided by numerous policies and requirements.[22] M&E is considered essential by both the GEF and UNDP for assessing progress towards planned results (i.e. outputs, outcomes, and impacts) from project to corporate level and for adapting targets, strategies, and policies as needed at each level to strengthen their relevance, efficiency, effectiveness, and sustainability. M&E also serves a broader function in terms of generating lessons and new knowledge that may have wider application in the design and implementation of other interventions that support sustainable development and the generation of global environmental benefits (GEF, 2010b; GEF EO, 2006). Since GEF-4, the GEF has been placing increasing emphasis on Results-Based Management, which stresses the importance of moving beyond reporting results to better use of results for accountability, internal management, learning and knowledge management (GEF, 2007b, 2010d, pp. 70–74).

M&E takes place at different levels and times, serving different functions. Thus, monitoring focuses on performance and results delivered during the life of a project, while evaluation seeks also to determine the sustainability of project results and long-term impacts (GEF EO, 2006, 2011). GEF Agencies, such as UNDP, are responsible for monitoring and reporting results at the project level based on the indicators and

---

[22] The overarching objectives of M&E in the GEF are to: '1) Promote accountability for the achievement of GEF objectives through the assessment of results, effectiveness, processes, and performance of the partners involved in GEF activities. GEF results are monitored and evaluated for their contribution to global environmental benefits; 2) Promote learning, feedback and knowledge sharing on results and lessons learned among the GEF and its partners as a basis for decision-making on policies, strategies, program management, projects and programs, and to improve knowledge and performance' (GEF, 2010b, p. 8; GEF EO, 2011). UNDP's M&E policies have similar objectives and are laid out in UNDP (2002) and the organization's internal Programmes and Operational Procedures and Policies (POPP). M&E in UNDP-GEF projects is further guided by UNDP's Evaluation Office as well as GEF policies and guidelines.

targets included in individual project logframes or results frameworks,[23] while the GEF Secretariat is responsible for monitoring and reporting progress at the project portfolio and corporate levels, based on the indicators and targets in the results framework for each focal area and the overall GEF Strategic Results Framework. These indicators seek to capture the GEF's contribution to higher-level results, such as its contributions to generating global environmental benefits, the environmental conventions, and the MDGs (GEF, 2010d).

Independent evaluations take place at both the project and corporate levels in line with various GEF and UNDP policies. All GEF projects must undergo an independent terminal (or final) evaluation. A key feature of the terminal evaluation is an assessment of the sustainability of project results and the likelihood of achieving desired long-term impacts.[24] Full-size projects also undergo a mid-term review (also referred to as a mid-term evaluation by UNDP). Mid-term reviews are not mandatory for medium-size projects, but are often undertaken for adaptive management purposes, especially for longer or more complex projects.

To date, direct biodiversity monitoring, such as assessments of species populations and trends or habitat change, has not been a major focus of M&E in UNDP-GEF biodiversity projects for several reasons, including the time and costs involved in primary data collection and analysis; the difficulties of standardizing data collection methods and ensuring data quality across a vast portfolio of projects; and concerns about the long-term usefulness and sustainability of such monitoring without a clear institutional base for continued monitoring beyond the life of the project.[25] This last is of particular concern given that effecting real change, particularly in relation to natural ecosystems, generally takes longer than the average life of a project, and it may take even longer for reductions in pressures to be reflected in changes in biodiversity status and trends.

Given the many challenges of biodiversity monitoring and also the GEF and UNDP's emphasis on removing barriers to biodiversity conservation, M&E of UNDP-GEF biodiversity projects has focused more on assessing the key factors that are likely to yield successful and sustainable project results and ultimately deliver conservation impacts, such as extent of stakeholder engagement, delivery of co-financing, risk management, financial and institutional sustainability, and implementation of strategies for replication and scale-up. This focus is in line with the findings of the most recent overall

---

[23] All UNDP-GEF projects develop a results framework during their design phase (RF, previously known as the logframe) and a budgeted M&E plan that includes measurable indicators for different planned outcomes and the overall objective stated in the project RF. All indicators must have baselines and clear time-bound targets, including end-of-project targets and the means through which achievement of these targets will be verified, together with the risks and assumptions associated with achieving each project outcome and proposed risk monitoring and mitigation strategies.

[24] See http://www.thegef.org/gef/Guidelines%20Terminal%20Evaluations.

[25] For the same reasons, the GEF does not generally fund research at the project level.

performance study of the GEF (OPS4), which suggests that the key determinants of the long-term conservation impact of GEF biodiversity projects are stakeholder ownership and support, effective financial mechanisms, adequate information flows, partly derived from monitoring, and adequate replication and scale-up (GEF EO, 2010).[26]

Nevertheless, both the GEF and UNDP have always recognized the importance of being able to assess the impact of their interventions on globally significant biodiversity. To this end, a number of indicators, albeit mostly indirect measures of biodiversity, have been developed to track progress in the GEF biodiversity focal area (GEF and UNEP-WCMC, 2003). Tables 16.3 and 16.4 show the results framework for Objectives 1 (PA systems) and 2 (biodiversity mainstreaming) of the GEF-4 and GEF-5 biodiversity strategies, respectively (GEF, 2011a). The indicators for the expected outcomes of these two objectives essentially measure changes in pressures on biodiversity and/or in societal responses to biodiversity problems and are based on certain assumptions about the relationship between a particular management action and changes in biodiversity status and trends (GEF and UNEP-WCMC, 2003).

Two biodiversity tracking tools for use at the individual project level were introduced in GEF-3 (and further revised during GEF-4) to measure progress in achieving portfolio-level expected outcomes under the strategic objectives on PAs and on biodiversity mainstreaming.[27] Each tracking tool includes a set of general questions on project background, coverage, and the specific indicators included in the GEF biodiversity strategy (GEF, 2011b, 2011c). The remainder of the PA tracking tool consists of the Management Effectiveness Tracking Tool (METT) and a Financial Sustainability Scorecard. The former scores answers to a set of 30 questions based on a framework developed by the World Commission on Protected Areas (WCPA) for assessing PA management effectiveness (Hockings *et al.*, 2006; World Bank and WWF International, 2007). The latter is only completed by projects working on PA system financing (and not individual PAs) and is also based on scoring answers to a set of questions. The tracking tool for mainstreaming projects is less detailed than the one on PAs and includes questions on the area of project interventions, as well as in relation to specific types of management practices such as certification and payment for ecosystem services. The tool also includes questions on the policy and regulatory framework of the economic sectors targeted by the project.

Tracking tools are completed three times during a project's lifetime: at the time of project submission for final approval by the GEF (to establish a baseline), and at

---

[26] OPS4 covers GEF-4, i.e. the 2006–2010 cycle of the GEF.

[27] A tracking tool has also been developed for biosafety projects, which is not covered here as UNDP does not implement biosafety projects, and a tracking tool is under development for GEF Access and Benefit Sharing projects (GEF, 2010c). In GEF-5, an additional short tracking tool has been introduced for projects working on Invasive Alien Species under Objective 2. These are not reviewed here but for further information see http://www.thegef.org/gef/tracking_tools.

Table 16.3 Results framework for Strategic Objectives 1 and 2 of the GEF-4 Biodiversity Strategy

Strategic Objective	Expected long-term impacts	Indicators
Objective 1 To Catalyze Sustainability of Protected Area Systems	Biodiversity conserved and sustainably used in protected area systems	Extent of habitat cover (hectares) by biome type maintained as measured by cover and fragmentation in protected area systems Extent and percentage increase of new habitat protected (hectares) by biome type in protected area systems that enhances ecosystem representation Protected area management effectiveness as measured by protected area scorecards that assess site management, financial sustainability, and capacity
Objective 2 To Mainstream Biodiversity Conservation in Production Landscapes/Seascapes and Sectors	Conservation and sustainable use of biodiversity incorporated in the productive landscape and seascape	Number of hectares in production landscapes/seascapes under sustainable management but not yet certified Number of hectares/production systems under certified production practices that meet sustainability and biodiversity standards Extent (coverage: hectares, payments generated) of payment for environmental service schemes

Source: GEF (2007a).

Note: In GEF-4, outcomes and indicators were also specified for each strategic priority under each objective (see GEF, 2007a).

Table 16.4 Results framework for Objectives 1 and 2 of the GEF-5 Biodiversity Strategy

Objective	Expected outcome	Indicator	Outcome target*	Core outputs
1. Improve sustainability of protected area (PA) systems	1.1 Improved management effectiveness of existing and new protected areas	1.1 PA Management Effectiveness score as recorded by Management Effectiveness Tracking Tool (METT)	80% of projects meet or exceed their PA management effectiveness targets covering 170 million ha of existing or new PAs	1. Number of new PAs and coverage (ha) of unprotected ecosystems 2. Number of new PAs and coverage (ha) of unprotected threatened species (number) 3. Number of sustainable financing plans
	1.2 Increased revenue for PA systems to meet total expenditures required for management	1.2 Funding gap for management of PA systems as recorded by PA financing scorecard	80% of projects meet or exceed their target for reducing the PA management funding gap in PA systems that develop and implement financing plans	
2. Mainstream biodiversity conservation and sustainable use into production landscapes, seascapes, and sectors	2.1 Increase in sustainably managed landscapes and seascapes that integrate biodiversity conservation	2.1 Landscapes and seascapes certified by international or nationally recognized environmental standards that incorporate biodiversity considerations (e.g. Forest Stewardship Council, Marine Stewardship Council) measured in hectares and recorded in GEF tracking tool	Sustainable use and management of biodiversity in 60 million ha of production landscapes and seascapes	1. Number of policies and regulatory frameworks for production sectors 2. Number of national and sub-national land-use plans that incorporate biodiversity and ecosystem services valuation 3. Area of certified production landscapes and seascapes (ha)

2.2 Measures to conserve and sustainably use biodiversity incorporated into policy and regulatory frameworks	2.2 Policies and regulations governing sectoral activities that integrate biodiversity conservation as recorded by GEF tracking tool as a score	50% of projects achieve a score of 6 (i.e. biodiversity conservation and sustainable use is mentioned in sector policy through specific legislation, regulations are in place to implement the legislation, regulations are under implementation, implementation of regulations is enforced, and enforcement is monitored)
2.3 Improved management frameworks to prevent, control, and manage invasive alien species (IAS)	2.3 IAS management framework operational score as recorded by the GEF tracking tool	80% of projects meet or exceed their target for a fully operational and effective IAS management framework

Source: GEF (2011a).

*Outcome target based on a US$4.2 billion replenishment scenario for the entire GEF fund in GEF-5.

the time of the project's mid-term and terminal evaluations. Data from the tracking tools are mostly qualitative apart from basic information on numbers, area, and coverage of PAs and of mainstreaming interventions. These allow a broad brush analysis of directional trends and patterns at the portfolio level and enable the GEF to report, for example, that GEF funding has led to the creation of more than 1600 PAs covering 360 million hectares and improved sustainable use and management of biodiversity in production landscapes by mainstreaming biodiversity in over 100 million hectares of productive landscapes and seascapes. Some data from the tracking tools are analysed by the GEF Secretariat and reported to the GEF Council through Annual Monitoring Reports. Data from both tracking tools also contribute to global monitoring processes, notably to measuring progress under some of the focal areas of the CBD's 2010 Biodiversity Target, such as 'Status and trends of the components of biological diversity' and 'Sustainable Use'.[28] The METT in particular has been widely adopted by many agencies and is one of the tools used to track global progress on various WCPA targets (GEF EO, 2010).

Such figures, however, tell us very little about the actual condition of biodiversity within these PAs or production landscapes (Walpole *et al.*, 2009). Furthermore, the underlying assumptions about the relationship between METT indicators and scores and biodiversity status and trends – notably that a high management effectiveness score corresponds to effectively conserved biodiversity – are yet to be tested empirically (Dublin and Volonte, 2004; GEF EO, 2010). The limitations of the tracking tool data are fully acknowledged by the GEF, which is seeking to address some of these (see GEF, 2009, Annex 2). For example, in GEF-5 the accuracy and quality of habit cover data at the portfolio level are to be improved by using remote-sensing data to measure indicators such as extent of intact vegetative cover and degree of fragmentation in both terrestrial and coastal/marine systems, supported where possible in the latter case by visual or other verification methods (GEF, 2011a, Annex Three; see also Chapter 5). Additionally, some of the underlying assumptions of the METT are being tested to establish 'a solid evidence basis' to better correlate a PA METT score with desired biodiversity outcomes and impacts within the PA (GEF, 2009; Zimsky *et al.*, 2010). This will be done through a series of country case studies and field visits to selected countries that have been applying the METT to their PA system over a long period of time and that also have quantitative data on the status of biodiversity and pressures on biodiversity within their PAs (GEF, 2009).

Finally, since the 2008 reporting cycle of GEF projects, additional information on biodiversity is being collected by UNDP through the annual project implementation

---

[28] See CBD 2010 Biodiversity Target 'Indicators' (http://www.cbd.int/2010-target/framework/indicators.shtml) and Biodiversity Indicators Partnership 'The Indicators' (http://www.bipindicators.net/indicators).

reviews (PIRs), which are also submitted to the GEF.[29] The PIR form includes a series of questions designed to help track progress against each project's results framework, the GEF tracking tools, and selected CBD areas of work to assess how many projects contribute to the conservation of globally threatened species, biodiversity hotspots and ecoregions, management of invasive species and climate change impacts, and working with indigenous people. Considerable time and effort is invested in this annual exercise by project teams, government counterparts, UNDP, and the GEF, which uses the aggregated data to monitor portfolio performance and report to the GEF Council. However, like the data from the tracking tools, only limited analyses have been possible to date because of the nature of the data, which are mostly nominal, ordinal, or in narrative form, and often based on subjective assessments (e.g. see GEF, 2010c; UNDP, 2009, 2011d). Furthermore, until recently data were recorded in Word documents further limiting analysis. Data are now recorded in Excel spreadsheets, but meaningful analysis will remain a challenge without greater standardization and quality assurance.

Direct biodiversity monitoring in UNDP-GEF projects

Despite it not being a requirement of either the GEF or UNDP, a significant number of UNDP-GEF biodiversity projects nevertheless undertake some amount of direct assessment and monitoring of biodiversity. An earlier comprehensive review of the GEF Biodiversity Program in 2004 found that 40% of all reviewed projects reported establishing baseline data for biodiversity components, while 33% reported some form of biodiversity monitoring (Dublin and Volonte, 2004).[30] The study, however, also found that many projects were unclear about the purpose of such monitoring, failed to distinguish clearly between research and monitoring, and that linkages between management and monitoring were weak (Dublin and Volonte, 2004, p. 91).

Almost all the biodiversity projects managed by the author in Asia-Pacific between 2004 and 2009 also included some direct assessment and monitoring of selected biodiversity components. Exchanges with colleagues managing UNDP-GEF biodiversity

---

[29] Also known as Annual Performance Reviews (APRs) in UNDP. The PIR also asks project teams to report on progress towards achievement of planned project outputs, outcomes, and objectives using the indicators specified in each project's M&E plan. Additionally project teams, government, and UNDP are asked to rate progress and impacts subjectively based on a qualitative scale from 'highly satisfactory' to 'highly unsatisfactory'. Between 2004 and 2009, I was involved in four cycles of annual project reporting and analysis of the performance of the UNDP-GEF regional biodiversity portfolio.

[30] The study team conducted detailed standardized reviews of 99 full- and medium-sized projects that were under implementation and more than halfway through their planned duration by 30 June 2003, and 42 projects that had been completed in the previous three fiscal years (Dublin and Volonte, 2004).

projects in other regions indicated this was also the case elsewhere. Independent evaluations of individual UNDP-GEF projects also confirm that biodiversity surveys and some direct monitoring of biodiversity are undertaken in many UNDP-GEF projects, either under the auspices of the project, and/or through other partners. This is not entirely surprising given that all UNDP-GEF projects address globally significant biodiversity, including many unique, rare, and threatened species and habitats, which are often already priorities for national and international action and support. Furthermore, as co-financing is a prerequisite for obtaining GEF financing, GEF projects, particularly protected area projects, are invariably partly co-financed by others with an interest in the target biodiversity, such as the government departments tasked with the management of the area, all or some of whom may also be doing some direct biodiversity monitoring, such as periodic habitat and wildlife surveys, particularly of threatened species (e.g. see CBD, 2007a).

While it was not possible to undertake a comprehensive analysis of past evaluations of GEF biodiversity projects, findings from the independent evaluations of the author's projects between 2004 and 2009 (13 in total) indicated that all but one collected some data on selected components of globally significant biodiversity to establish project baselines for monitoring purposes (see Table 16.5).[31] Furthermore, while such monitoring often had positive features, such as generating baseline data where none existed previously, or serving as inputs for the preparation of species management plans or for promoting policy change, most attempts to monitor biodiversity in the evaluated projects suffered from one or more of the following problems:

- Lack of standardization.
- Lack of adequate scientific rigour.
- Weak data analysis and knowledge management.
- Lack of institutionalization.

In the Bhutan LINKPA Project, for example, detailed biodiversity surveys were conducted in 2001 and 2007–08, but the data could not be compared because of differences in techniques, incomplete documentation of methodology, and lack of access to the earlier data (Table 16.5, Project No. 2). Similarly in the Bangladesh Wetlands Project, although ecological data had been collected for some project sites over nearly 10 years, lack of standardization had limited the usefulness of the data (Table 16.5, Project No. 1). The evaluation of the Pakistan Baluchistan Project stressed the importance of establishing the biological sustainability of harvesting wild populations of markhor, urial, and reptile species (Table 16.5, Project No. 10).

---

[31] All GEF project Terminal Evaluations reports are publicly available through the GEF online Project Database (GEF, 2011e). All Mid-Term and Terminal Evaluation reports of UNDP-GEF projects are available through UNDP's global website through the Evaluation Resource Centre (UNDP, 2011d) as well as often through the concerned UNDP Country Office website.

Table 16.5 Main findings and recommendations on biodiversity monitoring from the Independent Evaluation Reports of UNDP-GEF biodiversity projects that were evaluated between 2005 and 2008 in South Asia, West Asia and the Pacific

Project	Types of biodiversity surveyed/monitored	Evaluation observations relating to biodiversity assessments and monitoring	Recommendations/lessons[†] learned relating to biodiversity assessments and monitoring	Source (MTE/TE)[‡]
1. Bangladesh: *Coastal and Wetland Biodiversity Management at Cox's Bazar and Hakaluki Haor (FSP: 2002–2011)[§]	• Birds, including migratory species • Fish • Gangetic dolphin • Wild rice species	• No systematic ecological monitoring and data management programme although this was a planned output for both project sites • Useful ecological information collected for project sites over a nearly 10-year period. Some trends beginning to emerge, notably for migratory birds. Access to and usefulness of data limited by lack of standardized data collection or any kind of data management system • Project staff need greater guidance and support for knowledge acquisition and management through involvement of others with complementary specialist skills, e.g. through networking exchanges with research organizations and universities	No. 4: Identify the genetic biodiversity (e.g. rice crop biodiversity) of project sites (Ecologically Critical Areas) and address their management requirements No. 5: Make the governments of Australia and Japan aware of project measures to increase the populations and improve the management of migratory bird species protected through the Japan-Australia Migratory Birds Treaty and seek further support from those countries No. 12: Establish a suitable project database as soon as possible No. 13: Seek further guidance to determine which of all the biodiversity factors are best to measure, how, and when ... to establish meaningful baselines for project monitoring and evaluation (M&E)	MTE Report (Oct 2008)

(continued overleaf)

Table 16.5 (continued)

Project	Types of biodiversity surveyed/monitored	Evaluation observations relating to biodiversity assessments and monitoring	Recommendations/lessons† learned relating to biodiversity assessments and monitoring	Source (MTE/TE)‡
		• Poor coordination with government-funded marine research lab in project area	No. 14: Make more effort to capture and document local knowledge of biodiversity and reflect in conservation management plans No. 15: Establish a technical advisory group drawn from universities and/or other research organizations to provide technical backstopping and support to project team through some form of quid pro quo arrangement so there is some incentive for research institutions to participate, such as teaching or research opportunities although not necessarily a monetary incentive	

2. **Bhutan:** *Linking and Enhancing Protected Areas in the Temperate Broad-leaved Forest Eco-Region of Bhutan – "LINKPA"* (MSP:[3] 2003–2008)	• Tiger • Leopard • Musk deer • Sambar deer • Forest cover	• No monitoring framework or schedule including what should be monitored, by whom, or for what purpose • No baseline on biodiversity values or other indicators used although potentially available • Detailed biodiversity surveys conducted in 2001 and 2007–08 but using different techniques. 2001 survey data effectively baseline but not left in Bhutan and documentation of methods incomplete • Some analysis undertaken using both datasets to establish trends in some species, but data not as useful as they might have been • M&E protocol and system developed June 2007 with quarterly monitoring by park staff, mainly of pressures on biodiversity such as grazing • M&E system proving extremely useful and being shared with other parks and park headquarters in capital	• The Royal Government of Bhutan's Nature Conservation Division should seek greater mentorship and support in research design for park staff • Prioritizing research on rare species such as red panda is not the best approach. Focus should be instead on threats such as grazing and road impacts, corridor bottlenecks, and overall forest regeneration, i.e. the habitat of the rare species • Donors need greater understanding that sustainable conservation of resources in complex ecosystems takes time and that projects cannot change centuries old resource use patterns in 3–5 years • Greater technical backstopping would have helped logframe revision and development of new indicators and integration of these into M&E system	TE Report (Oct 2008)

(continued overleaf)

Table 16.5 (*continued*)

Project	Types of biodiversity surveyed/monitored	Evaluation observations relating to biodiversity assessments and monitoring	Recommendations/lessons[†] learned relating to biodiversity assessments and monitoring	Source (MTE/TE)[‡]
		• Database for flora and fauna developed in June 2007 but agreement still to be reached on who owns, maintains, and uses the database • Ecological studies started on some focal and charismatic species with support from Critical Ecosystems Partnership Fund (CEPF) • METT scores had improved between start and end of project • Threat Reduction Assessments conducted during terminal evaluation (for 5-year period) suggested significant reduction in some threats, but a few such as human-wildlife conflict and grazing remained problematic		

| 3. **Federated States of Micronesia (FSM)** *Community Conservation and Compatible Enterprise Development on Pohnpei* (MSP: 2000–2005) | • Forest monitoring and inventory
• Snail survey planned for 2005 (after final evaluation) | • Forest monitoring conducted on a quarterly basis in 14 sites from 2001 to 2005
• The Watershed Forest Reserve (WFR) programme has also influenced and helped to develop the community-based monitoring methodology for the marine environment and the marine PA programmes
• The data from the monitoring of fixed plots and stations were not available to the TE. There were also no technical reports available for assessment. CSP is working with the US Forest Service and Pohnpei Department of Lands and Natural Resources to fix permanent plots and implement a long-term forest-monitoring programme. The current monitoring includes the number and sizes of forest clearings and the eradication of invasive species. There is no monitoring of the forest health and biodiversity | Recommendation 11: Long-term Monitoring of Health and Biodiversity within WFR areas
The Conservation Society of Pohnpei (CSP – the project implementers in the latter part of the project) should work with the US Forest Service and academic institutions in the region to implement a long-term monitoring of health and biodiversity within the WFR areas. Local Pohnpeians studying for higher degrees can also be encouraged to take up such projects as their thesis research topics | TE Report (Nov 2005) |

(continued overleaf)

Table 16.5 (continued)

Project	Types of biodiversity surveyed/monitored	Evaluation observations relating to biodiversity assessments and monitoring	Recommendations/lessons[†] learned relating to biodiversity assessments and monitoring	Source (MTE/TE)[‡]
4. India: *Conservation and Sustainable Use of the Gulf of Mannar's Biosphere Reserve's Coastal Biodiversity* (FSP: 2002–ongoing)	• Marine biodiversity but not clear from evaluation which aspects of biodiversity surveyed although some individual species management plans prepared, such as for globally endangered dugong	• Component 4 on developing a 'Biodiversity Overlay' and establishing a marine biosphere reserve (BR), envisaged biological research and inventories to establish a baseline, conservation planning and biodiversity and pollution monitoring but lacked a clear strategy linking research and the establishment and management of a marine BR. Some surveys undertaken but a marine environment or biodiversity monitoring system was not yet in place although a biodiversity management plan was prepared for the integrated national park and BR • Draft Management Plan is focused primarily on the natural resources and biodiversity features of the area	• No specific recommendations under component 4 on biodiversity assessment and monitoring. Instead recommendations focus on finalizing the legal establishment and management of the BR, and especially on taking action on issues where sufficient information is already available, e.g. Recommendation 12: strengthening natural regeneration by: removal of island weeds, litter, remains of fires; preventing direct benthic damage, by anchors, feet, destructive fishing gears, and construction; and reducing air and water pollution from human activities in and around the reserve. Recommendation 15: Development of Gulf of Mannar BR Management Information System (MIS):	MTE Report (April 2008)

The BR management needs adequate capacity to receive, organize, disseminate, assimilate, and apply the range of results and datasets being generated by research commissioned by the project. There is a risk that too much unrefined data will come in at once, with no system in place to handle it, and no capacity for it to be used to achieve the underlying objectives of devising and establishing an effective marine resource management and conservation regime

An integrated MIS system is recommended to organize all aspects of BR management and development.

Recommendation 18: Establish an effective management planning function

The BR MIS should focus more dynamically on monitoring anthropogenic influences or threats to the whole BR.

Conservation area management has to focus on people, i.e. resource users, their actions, and impacts, rather than on the natural site or the biodiversity present

(continued overleaf)

Table 16.5 (continued)

Project	Types of biodiversity surveyed/monitored	Evaluation observations relating to biodiversity assessments and monitoring	Recommendations/lessons† learned relating to biodiversity assessments and monitoring	Source (MTE/TE)‡
5. Iran: *Conservation of the Asiatic Cheetah, its Natural Habitat and Associated Biota in the I.R. of Iran (CACP) (MSP: 2001–2008)	• Cheetah • Cheetah prey • Cheetah habitats	• Relatively little known about Asiatic cheetah compared to African cheetah. Project generated valuable information on distribution of cheetah in Iran and good cheetah habitats and important prey populations • Initial rapid surveys resulted in a number of recommendations, protocols, and training workshops for protected area (PA) guards aimed at establishing continuous monitoring of cheetah and its prey in project sites but this was not consistently implemented • Knowledge of cheetah population dynamics, behaviour, and survival factors in Iran still rudimentary, however. Although a lot of new information collected over 7 years, much of the information hypothetical or anecdotal	• Greater support was needed on technical and scientific aspects from the project design stage through to its implementation – in particular there was need to reach agreement between all partners involved in project implementation, research, and PA management on what needed to be monitored, why, when, and how • Also greater realism was needed in setting targets for a short-term project with limited funds working across five project sites with a combined area of 38 000 km² scattered within a huge, remote area of nearly 200 000 km²	TE Report (Jan 2009)

		• Very few biological/ecological data collected through statistically robust methods and analysed and reported according to accepted scientific standards • No reliable monitoring system for cheetah population or prey populations in place • Potential for stronger engagement of PA guards in systematic wildlife monitoring was not realized despite promising start	• The conservation and recovery of a top predator such as the Asiatic cheetah in an environment like the Dasht-e-Kavir in Iran is a complex and long-term endeavour. A good start has been made by CACP, and the Asiatic cheetah's chances of survival are better today than they were in 2001 when the project started • Field survey and monitoring of cheetah and prey populations must be improved. Steps needed for this include: (i) further testing and adaptation of scientifically robust methods; (ii) continuous training of field staff; (iii) education of data analysts; and (iv) consistent reporting	
6. Maldives: *Atoll Ecosystem Conservation (AEC) Project, Baa Atoll (FSP: 2005– ongoing)*	• Coral reefs • Marine and island biodiversity • Sharks	• Project design includes major emphasis on establishing a comprehensive baseline on biodiversity and ecosystem health with regular monitoring as a basis for developing a conservation plan and a network of protected areas. However, this is not a realistic strategy as	Recommendation 9: The strong recommendation is for the AEC project to adopt a more pragmatic and direct strategy for introducing ecosystem-based conservation on Baa Atoll, by focusing on specific conservation issues and implementing management	MTE Report (Sep 2008)

(continued overleaf)

Table 16.5 (continued)

Project	Types of biodiversity surveyed/monitored	Evaluation observations relating to biodiversity assessments and monitoring	Recommendations/lessons† learned relating to biodiversity assessments and monitoring	Source (MTE/TE)‡
		conservation problems on Baa and elsewhere in the Maldives are accumulating faster than solutions are being found. Pressures on biodiversity from tourism, reef fishing, and island development have increased since the project was originally designed. The project should focus on priority conservation actions that could be planned and implemented without waiting for the results of further surveys and studies	actions to manage these issues. Under this approach, the conservation plan for Baa Atoll would consist of a number of resource use management plans Recommendation 11: Revise and simplify existing M&E system and greatly reduce focus on biological, ecological, and socioeconomic surveys other than targeted information collection to address specific management information needs	
7. **Nepal:** *Upper Mustang Biodiversity Conservation Project* (MSP: 2000–2006)	• Birds including Himalayan griffon vulture, lammergeier, Tibetan snowcock, Tibetan partridge, Tibetan sandgrouse, Tibetan snowfinch	• Biodiversity monitoring and project implementation extremely tough given location, terrain, and climate. The study area is a 9-day trek from nearest motorable road and a 3–5-day trek from nearest airport. Villages are scattered. Temperatures fall to −26°C in	• A duplicate of the MIS in the local conservation office should be set up as soon as possible • Three-dimensional models of the project areas were particularly effective compared with better 2-dimensional contour maps in enabling local communities with high levels of	TE Report (Sep 2006)

- Mammals including blue sheep and globally endangered Tibetan wild ass (kiang), Tibetan gazelle, Tibetan bighorn sheep (argali), snow leopard
- Butterflies
- Plants including Non-Timber Forest Products (NTFPs)

winter and snow blocks movement – computers often stop working
- Virtually no baseline at the start of the project
- Baseline surveys and repeat surveys undertaken including seasonal coverage allowing detection of rough trends
- Surveys of human-wildlife conflicts and livestock depredation undertaken as indirect measure of biodiversity
- Survey data for four globally significant mammals and one bird species obtained but only two data points, in some case only for two consecutive years. However, very tough to repeat surveys and conduct them systematically in this environment.
- Survey data entered into a Geographical Information System (GIS) and biodiversity hotspots in Mustang District identified. GIS analysis used to establish a zoning system and as inputs to the development of an integrated management plan for Upper Mustang for 2006–10

illiteracy and low levels of education to better visualize their surroundings and to participate in project planning and decision-making

(continued overleaf)

Table 16.5 (continued)

Project	Types of biodiversity surveyed/monitored	Evaluation observations relating to biodiversity assessments and monitoring	Recommendations/lessons[†] learned relating to biodiversity assessments and monitoring	Source (MTE/TE)[‡]
		• The MIS developed through the project is located in Pokhara, several days by horse and foot from project implementation area • Biodiversity surveys to be continued at 2-year intervals by the Annapurna Conservation Area Programme (ACAP)		
8. Nepal: *Landscape-scale Conservation of the Endangered Tiger and Rhino Populations in and around Chitwan National Park (Tiger-Rhino Conservation Project) (MSP: 2001–2006)	• Forest vegetation • Rhino • Tiger and prey species • Birds	• Large amounts of information generated from specific sector and topic-based research on proposed wildlife corridor and regular biodiversity monitoring activities • Biodiversity monitoring included: annual 2-month-long camera trapping surveys of tiger; rhino censuses on elephant back in seven consecutive blocks pre-monsoon, during the monsoon, and post-monsoon; monthly tiger prey counts using	• Methodology for monitoring ungulates needs modification to enhance robustness of density estimates and allow more detailed analyses • Institutionalization of community-based forest and wildlife monitoring cannot be achieved without adequate training and resourcing, long-term supervision, and feedback of the collated and analysed results	TE Report (May 2007)

| | | line transects by foot and vehicle; bird surveys using line transects in winter, summer, and autumn. Additionally vegetation in most of proposed corridor mapped using satellite imagery and 32 permanent plots established for monitoring future changes
• Research reports available electronically and in hard copy at the Biodiversity Conservation Centre, and biodiversity survey data stored electronically in a Management Information System (MIS), but not widely disseminated or readily accessible
• Evaluation team could not review the MIS as none of the staff present knew how to access it. Previous GIS/database officer had left more than a year ago (when project ended) | • Management interventions should be informed by rigorous science with monitoring as appropriate
• Undertake a census of vehicle and animal movement along section of highway that bisects the corridor to inform development and enforcement of appropriate measures to mitigate impacts of vehicle movement through a wildlife corridor
• Vegetation studies needed of ongoing changes to grasslands in the floodplains of Chitwan Valley over past 20–30 years to better understand potential impacts of different management interventions and guide future management of grasslands within the corridor, particularly as a habitat for ungulates | |
| **9. Pakistan:** *Mountain Areas Conservancy Project (MACP)* (FSP: 1999–2006) | • Markhor
• Ibex
• Some habitat monitoring
• Medicinal plants | • Many surveys and assessments undertaken of different biodiversity components including globally threatened mammals, birds, and medicinal plants, but regular monitoring only instituted for markhor and | • The project has published an impressive number of publications, information packages, and awareness material, but production and publication have not been undertaken systematically and | TE Report (Aug 2006) |

(continued overleaf)

Table 16.5 (*continued*)

Project	Types of biodiversity surveyed/monitored	Evaluation observations relating to biodiversity assessments and monitoring	Recommendations/lessons[†] learned relating to biodiversity assessments and monitoring	Source (MTE/TE)[‡]
		ibex as part of establishing trophy hunting schemes • MTE had concerns about the methods used in species and ecosystem monitoring but accepted that every effort had been made to standardize the surveys to reduce potential for variability and that direct count method may be the only workable method in the MACP context. However, while monitoring results were adequate for determining potential for trophy hunting, results were not sufficiently reliable to be used as proof of the project's success in protecting particular species • Knowledge base of MACP is satisfactory but knowledge management is unsatisfactory although this could easily be remedied	products may not reach their target audience or realize their full potential impact. The MTE recommends that MACP catalogues all printed and electronic material that was produced or acquired by the project and bring to the attention of, and make accessible to, potential users. In doing this MACP should ensure harmonization with similar, parallel, or related initiatives. This work needs to be done as part of the exit strategy to ensure that the impressive publications records of MACP are not lost out of circulation in some archive	

10. Pakistan: *Conservation of Habitats and Species of Global Significance in arid and semi-arid ecosystems in Baluchistan (CHAS) (MSP: 2004– ongoing)	• Straight-horned Suleiman markhor • Afghan urial • Endemic reptile species	• Socio-ecological baseline studies conducted including vegetation baseline assessments in the two target project areas; range management and ungulates survey in Torghar; reptile and small mammal survey in Nushki-Chagai; ethnobotanical study; studies of human activities • Biodiversity surveys between 2005 and 2008 focused on strengthening the scientific basis for the main sustainable use initiatives piloted through the project, i.e. trophy hunting of markhor and urial in Torghar; reptile ranching and/or harvesting in Nushki-Chagai; medicinal plant harvesting; and habitat protection and restoration through improved livestock management • Surveillance patrols by local community rangers and game guards contributed to continued cessation of hunting in Torghar and apprehension and prosecution of illegal reptile trappers in Nushki-Chagai	• Further strengthen activities under Component 4, which includes the biodiversity assessment and monitoring work, by focusing on a clear set of outputs needed for natural resource management in the two proposed conservancies of the project, covering protection, rehabilitation, recovery, and monitoring of land, water, vegetation, forests, and wildlife • Under Component 5 there is a need to establish the biological sustainability of harvesting local wildlife populations amongst other issues	MTE Report (Jul 2008)

(continued overleaf)

Table 16.5 (continued)

Project	Types of biodiversity surveyed/monitored	Evaluation observations relating to biodiversity assessments and monitoring	Recommendations/lessons[†] learned relating to biodiversity assessments and monitoring	Source (MTE/TE)[‡]
11. **Papua New Guinea:** *Milne Bay Community-based Coastal and Marine Conservation Project* (FSP: 2002–2006)	• Dugongs • Marine turtles	• Marine PA (MPA) establishment in Milne Bay cannot be disconnected from issues of resource ownership and usage. Considerable data acquired for preparing options for sustainable development activities that could reduce negative impacts on marine biodiversity. The idea is sound, but the papers are too long and written in a technical style unsuited for the communities for which they are written. Much information in the papers would be helpful for communities if presented in a more meaningful style and format • A good set of biodiversity data has been obtained and applied to management zoning of several community 'owned' marine areas. A number of areas might be regarded as being close to 'ready' for marine protected area status, depending on how that is defined	• Pay more attention to first determining community subsistence needs from marine resources and use that as a base for moving towards notions of biodiversity conservation Data: • Carefully assess what level of data needed, for what purpose, to fit the ecological and social circumstances of the Project area, and with particular attention to developing methodology that is meaningful to local communities and can be adopted (and adapted) by them • Consciously and consistently seek out traditional knowledge of biodiversity, resources, and environment through a methodology (to be developed) that results in the documentation and application of such knowledge as can be used in resource management and conservation plans, and do	TE Report (Jul 2006)

- Turtle and dugong surveys conducted and provincial strategic action plan drafted. Sufficient data on turtles to develop detailed management plans for turtles in two areas. More survey-based scientific data needed to develop dugong management plans, notably to establish their local feeding ranges and migratory pathways before proposing MPA arrangements for protecting dugong
- Strong sense of ownership of the project concept at all levels, but weak ownership of actual implementation and results. Insufficient capacity development or institutionalization of project within local communities or provincial and local governments making it difficult to sustain its results or potential conservation impacts

this in ways that empower the communities concerned
- Make effective use of the data in hand, identifying gaps and proposing measures to deal with those gaps
- Reach out to other conservation groups, both within Milne Bay and beyond, to drawn on their experience, lessons learned, and ideas regarding community-based resource and biodiversity conservation and to share data

Biodiversity in context:
- Refocus approaches at community level so that a foundation of traditional biodiversity knowledge is first established – and external information, skills, and technological interventions are then introduced as appropriate in the context of, and building on, local knowledge
- Bring dugong and turtle surveys to a satisfactory conclusion with the minimal dataset needed to achieve a useful level of management and awareness of these species; an emphasis here could be on understanding the level of exploitation to determine a useful level of management

(*continued overleaf*)

Table 16.5 (continued)

Project	Types of biodiversity surveyed/monitored	Evaluation observations relating to biodiversity assessments and monitoring	Recommendations/lessons[†] learned relating to biodiversity assessments and monitoring	Source (MTE/TE)[‡]
12. Sri Lanka: *Conservation of Biodiversity through Integrated Collaborative Management in the Rekawa, Usangoda, and Kalametiya (RUK) Coastal Ecosystems, Sri Lanka (MSP: 2000–2005)	• Inland and marine subtidal biodiversity • Coral reefs • Five species of globally threatened marine turtles	• A number of biodiversity assessments and surveys conducted by IUCN with some follow-up surveys to examine trends, but no continuous monitoring or regular follow-up; surveys useful for planning but not for establishing monitoring programmes, i.e. no institutionalization of biodiversity monitoring • Turtle Conservation component of the project proposed monitoring of turtle nesting and ecological research on different turtle species but was only partly completed because of lack of funds and possibly insufficient technical backstopping • Turtle database developed by IUCN and handed over to Department of Wildlife Conservation (DWC) but never used as community-based monitoring programme did not materialize	• None relating specifically to biodiversity monitoring as TE was more concerned about more fundamental problems relating to overall project implementation progress, achievement of planned outcomes, and sustainability issues	TE Report (Mar 2007)

| 13. **Sri Lanka:** *Contributing to the Conservation of Unique Biodiversity in the Threatened Rain Forests of South-west Sri Lanka* (MSP: 2000–2006) | • No systematic biodiversity monitoring | • Despite a number of earlier biodiversity studies, no impact monitoring, so no hard data on biodiversity status
• Anecdotal information, evaluation inspection of forest boundaries, and data on offences indicate overall biodiversity status improving
• Studies by university students suggest number of endemic plant species present has increased during 1994–2003 and forest is maturing (i.e. greater proportion of larger trees)
• Ad hoc monitoring by forest department staff on patrols but no systematic recording of observations; records of charismatic megafauna sighted from time to time
• Kanneliya-Dediyagala-Nakiyadeniya (KDN) forest complexes and Sinharaja, a Man and Biosphere Reserve, are also national Important Bird Areas visited frequently by amateur ornithologists who keep sighting records, but no mechanism for making use of these for systematic monitoring | • Consider a rapid biodiversity survey building on earlier surveys before formally ending the project to enable longer-term monitoring
• Training of forest staff originally as a precursor for working with villagers and implementing co-management had added benefits of upgrading their biodiversity skills and general awareness about local and national forest issues
• Re global environmental benefits generated by the project: '…we can state that the forest is intact, the forest is recovering from past logging, and that there are high proportions of endemic plant taxa in plot surveys. We cannot say if all the endangered endemic species have viable populations, or are still present. We can state that charismatic larger fauna are seen. We cannot give information on population abundance or presence of many endemic animal taxa.' | TE Report (Dec 2007) |

(continued overleaf)

Table 16.5 (continued)

Project	Types of biodiversity surveyed/monitored	Evaluation observations relating to biodiversity assessments and monitoring	Recommendations/lessons[†] learned relating to biodiversity assessments and monitoring	Source (MTE/TE)[‡]
		• Sinharaja has long-term Smithsonian Institute 50-ha forest plot; although not on the project site, it means national capacity for forest plant monitoring exists • METT scores showed progress in PA management over course of project	• The Forest Department should convene a monitoring programme to look at a select group of taxa such as birds or trees to assess population sustainability and recruitment and regeneration levels	

*Projects marked with an asterisk included biodiversity assessment and monitoring as a specific project component.
[†]Recommendations are not always numbered in the evaluation reports.
[‡]MTE, Mid-Term Evaluation; TE, Terminal Evaluation. Reports are available from either UNDP's Evaluation Centre website (http://erc.undp.org/index.html;jsessionid=9571607B55F4FFB99752EE3D700A7286), the concerned UNDP Country Office website, or the GEF online Project Database (http://www.gefonline.org/).
[§]FSP, full-sized project (>US$1 million in GEF-4); MSP, medium-sized project (≤US$1 million in GEF-4). Dates refer to actual implementation period as stated in the MTE or TER, that is, once project team is on board and activities have started rather than date of final approval by GEF or first disbursement of funds by UNDP.

Although ibex and markhor populations were successfully monitored in the Pakistan Mountain Areas Conservation Project to establish trophy hunting quotas, the terminal evaluation expressed deep concern about the weak management of the considerable knowledge generated by the project, through numerous biodiversity assessments and surveys (Table 16.5, Project No. 9). The evaluation of the Milne Bay Project in Papua New Guinea noted that while sufficient data had been obtained for developing detailed turtle management plans, and draft provincial strategic action plans had been developed for both species, more survey-based scientific data were needed to develop the same for dugongs (Table 16.5, Project No. 11).

The many complexities and challenges of undertaking effective biodiversity monitoring are also clearly reflected in the remarks and recommendations of the evaluators, with several highlighting the need for greater technical and scientific support and capacity development to improve the quality and usefulness of project-level biodiversity monitoring (e.g. the Bhutan LINKPA project, the Bangladesh Wetlands Project, and the Iranian Cheetah Project). Despite the resources and effort put into biodiversity surveys and monitoring by most of the evaluated projects, institutionalization of such processes seemed likely in only one project – the Nepal Upper Mustang Biodiversity Conservation project, where a good information management system had been developed and periodic survey work was to be continued by the long-running Annapurna Conservation Programme (Table 16.5, Project No. 7).

As within the wider conservation community, there was considerable difference of opinion among evaluators about how much to invest in biodiversity assessment and monitoring versus other activities, depending partly on the project's overall objectives, and also on whether to focus on direct or indirect measures of biodiversity. Thus, the evaluators of the Maldives Atoll Ecosystem Conservation Project, a biodiversity mainstreaming project, felt that biodiversity surveys and monitoring should be kept to a minimum, and instead greater attention given to understanding and managing human activities, stating that: 'There is no need to describe in comprehensive detail the condition of biodiversity on Baa Atoll in order to find a solution to a problem of overfishing, inappropriate shore-line construction or conflict between dive tourists and wildlife' (Hunnam and Moosa, 2008, paragraph 106; see also Table 16.5, Project No. 4).

In the Bhutan LINKPA project, a protected areas corridors project, the evaluators did not agree with prioritizing research on charismatic rare species such as red panda, and recommended instead that conservation management and therefore monitoring, focus on information needs to address grazing pressures, corridor bottlenecks, and ensuring overall forest regeneration (Rodgers and Dorji, 2008; see also Table 16.5, Project No. 2). On the other hand, the evaluators of the Iranian Cheetah Project, a charismatic species-focused PA project,[32] noted that scientifically rigorous data on cheetah and its

---

[32] Although GEF projects focused on single species are relatively rare, the GEF has also recently increased its support for tiger conservation (see http://www.thegef.org/gef/news/tiger_grant_2010).

prey were essential for the long-term conservation of the critically endangered Asiatic cheetah and associated threatened species; however, they also stressed the importance of investing in strengthened law enforcement, increased capacity of PA management staff, development of PA co-management with local communities, and greater public outreach (Breitenmoser *et al.*, 2009; Table 16.5, Project No. 5).

Finally, it should be noted that while the evaluators may have differed in their views about the precise role of biodiversity monitoring in individual projects and therefore how much to invest in such monitoring relative to other activities, no evaluator suggested that biodiversity assessment and monitoring should be completely dispensed with, or that it should be limited to only completing the GEF tracking tools and the PIRs.

Mechanisms for increasing the contribution of GEF biodiversity projects to broader biodiversity monitoring efforts

It is clear from the 2004 Biodiversity Program Study and from the UNDP-GEF biodiversity project portfolio that a significant number of GEF projects invest in some form of biodiversity assessment and monitoring, whether with GEF resources and/or other co-financing. As discussed in the previous section, there is also evidence that the data collected may not be effectively used for improving project performance, conservation management, or delivering long-term conservation impacts.

Yet GEF biodiversity projects have considerable potential to contribute more effectively to broader systematic biodiversity monitoring efforts. First, the GEF biodiversity portfolio has good biogeographical coverage and by definition targets globally significant biodiversity, which generally is also nationally significant, making GEF projects potentially relevant to both national and global biodiversity monitoring efforts. Second, as national ownership and co-financing are key criteria for GEF financing, all GEF biodiversity projects are closely associated with government departments, including environment ministries and forest and wildlife departments, and often also with national and international research institutions and conservation organizations. This offers options for both institutionalization of monitoring and for providing greater technical support for improving the quality, management, analysis, and application of monitoring data. Finally, the GEF is a major global partnership for international cooperation, with representation from 182 member governments, the CBD, and other multilateral environmental conventions, several key UN agencies and multilateral banks, numerous international and national NGOs,[33] research institutions, the private

---

[33] The GEF has established a GEF-NGO network to increase the engagement of national and international NGOs in the work of the GEF. See http://www.gefngo.org/ and http://www.gefcountrysupport.org/report_detail.cfm?projectId=140.

sector, and other sections of civil society. Through this vast and diverse network, the GEF and its partners such as UNDP, have access to a wealth of knowledge and skills, including scientific and technical expertise. However, these resources are currently dispersed across the network with insufficient exchange and coordination between relevant actors and stakeholders.

Although the precise number of projects that undertake direct biodiversity monitoring was not established in this chapter, even if only 50% of all GEF biodiversity projects are doing so (and this is probably a conservative estimate), this still represents a significant number of projects and a sizeable investment of conservation resources, given that the GEF had over 230 biodiversity projects under implementation in 2010–11 (GEF, 2011a) and also given that GEF projects are typically larger in scope and better resourced than many other conservation projects.[34]

Given that resources for conservation are limited relative to need (Parker and Cranford, 2010), the GEF and its partner Agencies must do all they can to ensure that any resources invested in biodiversity monitoring are used as effectively and efficiently as possible, regardless of whether such monitoring is actually financed by the GEF. A major reason cited by the GEF, UNDP, and the 2004 Biodiversity Program Study for not investing significantly in biodiversity monitoring, particularly direct monitoring, is that this would not be a cost-effective use of limited GEF resources given the many shortcomings of short-term site-based monitoring. This was reiterated recently by the first GEF study on the underlying assumptions of the METT, which found that monitoring data on wildlife populations in Zambia that had been collected over the relatively short timeframe of GEF projects were too variable for measuring the impact of project interventions or for adaptive management purposes (Zimsky *et al.*, 2010). Proxy indicators of biodiversity status such as reduction in threat were found to be more reliable over a short timeframe. The study therefore recommends that where direct monitoring of species populations and trends is considered a priority, GEF support should only be provided where such monitoring will be sustained beyond the life of the GEF project.

These two provisos – that GEF resources should be used cost-effectively and that direct biodiversity monitoring should only be supported by the GEF in situations where it will be institutionalized and implemented over a meaningful time period – are fully supported by this author and by many others cited in the book. However, with some creative thinking and action, there are a number of things that could be done at relatively low cost by the GEF and its partners such as UNDP, to:

1. help GEF biodiversity projects identify when and what to monitor and with what level of scientific rigour; and

---

[34] An analysis of past evaluations of GEF biodiversity projects or a survey of projects currently under implementation could help confirm exactly how many projects undertake direct biodiversity monitoring. Analysis of METT data could also provide an indication of the extent of biodiversity monitoring in GEF PA projects as although the METT does not include specific questions on biodiversity monitoring per se, it does include questions on baseline ecological data ('Resource inventory') and about what management-oriented survey and research work is taking place.

2. where it is agreed that direct biodiversity monitoring is a priority, to improve the quality, usefulness, and sustainability of biodiversity monitoring undertaken by GEF projects.

As discussed earlier, guidance on most project M&E, including technical matters, is largely left to the GEF Agencies, such as UNDP, and other key partners, such as government departments and conservation NGOs. UNDP provides technical support to its GEF projects through its regional technical advisors and its Country Office environment teams, but again mainly on a project-by-project basis, with a focus on ensuring delivery of planned results and compliance with GEF and UNDP requirements. While some internal guidance notes on various aspects of GEF project design and management have been developed over the years, much of the guidance is of a fairly general nature, infrequently revised, and not always accessible. For example, in the past such notes used to be available only through internal workspaces. Many UNDP-GEF projects also establish technical advisory groups or have technical advisors, but the extent of their engagement and effectiveness varies greatly. As a result, whether or not project-level biodiversity data are useful, useable, or used often depends on the initiative, capacity, and professional networks of project team members.

The GEF itself has no major mechanism for providing substantive technical guidance on the more applied aspects of conservation management to its biodiversity projects since this function has been delegated to the GEF Agencies. The main technical guidance provided by the GEF is on a project-by-project basis at the time of initial development and approval, when GEF focal area task managers review project proposals and when the GEF's Scientific and Technical Advisory Panel (STAP) screens their technical and scientific quality and assesses whether the global environmental benefits to be generated by the project are measurable and whether the proposed M&E system is adequate and appropriate (Carugi and MacKinnon, 2010; GEF EO, 2010).

There are, however, at least two major ways in which the GEF, UNDP, and other GEF agencies could lead and support the development of more substantive, targeted technical guidance to strengthen the quality of biodiversity monitoring in GEF projects where appropriate and potentially increase their contribution to broader monitoring efforts. These are discussed below.

Review and codification of tools and techniques for biodiversity monitoring

Promising new methods and tools are emerging as scientists begin to respond to demands from conservation managers and policy-makers for practical and cost-effective approaches to biodiversity monitoring (e.g. Gardner *et al.*, 2008; Soberon and Townsend Peterson, 2009). Simple frameworks exist to guide decision-making and resource allocation in relation to biodiversity monitoring that could be further adapted or refined for different country contexts and GEF projects (e.g. McDonald-Madden *et al.*, 2010; Salzer and Salafsky, 2006; Royal Society, 2003). However, much of this growing body of knowledge is scattered across the published and grey literature and is often difficult to locate and/or too expensive to access and therefore inaccessible to many conservation practitioners, particularly in developing countries.

Furthermore, it is extremely inefficient for every project team to individually research these matters. Even useful initiatives such as the ConserveOnline website (http://conserveonline.org/), which makes conservation tools, techniques, and experiences freely available to conservation practitioners, requires users to have a reasonable idea of what they are looking for, or to be ready to invest considerable time searching through databases and workspaces to locate potentially useful documents and tools. UNEP-WCMC, for example, has a dedicated webpage for 'Specialist online tools' (see http://www.unep-wcmc.org/), but the existing content is of limited use at the project level. What is really needed is some critical review of the available options, easy-to-read summaries of the reviews, and some specific guidance on individual options to enable conservation practitioners and project managers to easily locate the most suitable tools and techniques for their specific needs.

Thus, a key area in which the GEF and its partners such as UNDP could add value, is by deploying their vast network and collective resources to spearhead an initiative to systematically locate, assemble, review, and codify the tools and methodologies that are likely to be most useful to biodiversity projects of GEF (and others) and to have these readily available to potential users, both through web-based platforms and other means. Such an exercise would include amongst other things identification and review of planning tools to help project developers and managers determine when, where, and how to undertake different kinds of biodiversity monitoring, and also to easily identify such things as:

- the appropriate biodiversity measure(s) to monitor given project objectives, conservation management information needs, ecological setting, country context, available budget, and other competing priorities;
- practical cost-effective, monitoring protocols;
- options for data standardization;
- opportunities for linking with other monitoring efforts, including contributing to larger-scale monitoring;
- options for institutionalization of monitoring, including effective data management systems that ensure data are properly stored, analysed, applied for furthering conservation objectives, and disseminated.

A key objective of such an exercise would be to make it as efficient, user-friendly, and state-of-the-art (and knowledge) as possible. Additional user-friendly toolkits and guidelines to meet the specific needs of those designing and implementing GEF projects may also be required and could be developed with the help of research institutions, thereby further strengthening science-policy-conservation action linkages.

A good example of a web-based system for access to technical tools, support, knowledge, and exchange (at different levels) is the GEF IW:Learn website of the International Waters Focal Area (http://iwlearn.net/abt_iwlearn).[35] Developing a similar web-based platform for the Biodiversity focal area, with a dedicated section for specialized technical support, could also help to address a perceived need among some stakeholders for greater support from the GEF and the STAP on technical and scientific matters, especially in other focal areas (Dublin and Volonte, 2004; GEF EO, 2010; ICF International, 2009).

Development of a project mentoring system

The development of a project mentoring system for coordinated technical support to projects from beginning to end (or 'cradle to grave'), rather than through periodic GEF Agency or short-term consultancy inputs, is another way in which the GEF, UNDP, and other GEF Agencies could not only improve the quality, relevance, and sustainability of biodiversity monitoring (where appropriate) but also potentially many other aspects of project implementation and results delivery.

Mentoring services could be organized in different ways, for example, thematically, such as by ecosystem type, project objectives, or geographically, and could be provided through partnerships and collaborations with relevant national and international organizations and individuals drawn from different disciplines and skill sets of relevance to the GEF biodiversity portfolio. If designed properly, such a system could ensure a certain continuity of support over a project's lifetime, helping to buffer against turnover in key project and GEF Agency personnel and government counterparts, as well as contributing to the institutionalization of the considerable knowledge and lessons generated by GEF projects, a key priority in GEF-5.

Additionally, just as biodiversity monitoring should be embedded within broader monitoring processes (see Chapter 14), a project mentoring system is likely to be more cost-effective and sustainable if it is (i) integrated into a broader and more holistic system of providing technical support to biodiversity projects by the GEF and its partners; and (ii) is designed to meet user needs as well as donor requirements. Although periodic mentoring of projects takes place all the time informally, the author is not aware of any formal long-term mentoring systems that have been developed to support conservation projects and programmes from initial design to final completion.

---

[35] IW:Learn's stated objective is 'to strengthen global portfolio experience sharing and learning, dialogue facilitation, targeted knowledge sharing and replication in order to enhance the efficiency and effectiveness of GEF IW projects to deliver tangible results in partnership with other IW initiatives'.

Financing for increased technical support to GEF biodiversity projects and other considerations

Implementing the two mechanisms proposed above for increasing technical support to GEF biodiversity projects would entail some costs, but there are many ways in which these could be kept to a minimum. Furthermore, the potential conservation benefits of doing so are very likely to outweigh the costs in the medium to long term if such mechanisms are developed and implemented in a way that genuinely meets end-user needs first, that is, at the project and country level.

Coordination in itself is a relatively low-cost activity and the assembly of tools and development of toolkits for dissemination through a website is also likely to be less expensive than designing and implementing a project mentoring system but would nevertheless involve some costs. With regard to creating a web-based platform, it would be important to learn from the experiences of IW:Learn (and any other comparable platforms), which offers a very wide-ranging suite of information and tools, to better understand what resources are needed to establish and maintain such a site over the long term.

One way of keeping additional costs to a minimum is to identify potential synergies; for example, many partners within the GEF network already have dedicated technical and scientific staff who may be working in relevant areas and who may be able to reorient their work a little to accommodate any additional tasks associated with the review and codification of tools and techniques. Some organizations may even wish to proactively support such work through their own resources, if it is in line with their strategic objectives or if they see added value in doing so. Better coordination among GEF partners would allow the identification of such opportunities. Cost savings may also be achieved as a result of reduced duplication of effort.

Conservation organizations and research institutions with experience in applied research are likely to have considerable in-house expertise and capacity for the development of such tools, although they would probably benefit from working with development practitioners and others to ensure that proposed tools and techniques are indeed practical, locally relevant, and cost-effective. For example, the 2004 Biodiversity Program Study reported that ' . . . a number of projects found that monitoring systems developed by scientists and external researchers were overly complex, did not provide the necessary information, were inappropriate to local circumstances, or were difficult to maintain consistently over time. Other projects emphasized the need for and greater sustainability of "home-grown [monitoring] approaches" that can be carried out by local communities or untrained park staff' (Dublin and Volonte, 2004, p. 91).

The GEF has direct access to many relevant organizations and institutions through its own network and the numerous networks of key GEF partners, including notably the CBD Secretariat. The CBD, for example, has a Memorandum of Understanding (MoU) with several leading scientific institutions, specifically to 'leverage the expertise and experience of these institutions in order to implement education and training activities to support developing countries that are building scientific, technical and policy skills in the area of biodiversity.'[36] The International Union for the Conservation of Nature (IUCN) also has many important global networks of experts, particularly through its Specialist Commissions.[37] Such partners, along with appropriate development NGOs and social science research institutions, could also have a role in the development and implementation of a GEF project mentoring system. Capacity development is an important cross-cutting element of many of the CBD's areas of work, and also a specific component of its Programme of Work on Technology Transfer and Cooperation.[38] It is also integral to the work of the GEF and its Agencies.

Depending on how mentoring services were organized, there could be multiple mini-mentoring systems implemented by different groups of GEF partners with the appropriate mix of technical skills, knowledge, and experience instead of a single monolithic system. Again costs may be minimized if there are opportunities for forming long-term partnerships and institutionalization of mentoring services, with perhaps a quid pro quo arrangement rather than direct financial transfers for mentoring services – for example, opportunities for institutions providing mentoring services to undertake research jointly with recipient country partners (e.g. see Table 16.5, Project No. 1, Recommendation 15). It is beyond the scope of this chapter to assess the likely costs of developing a mentoring system, but this is something that the GEF and its partners could establish relatively easily. One possibility is for the GEF or one of its partner agencies to pilot a demonstration mentoring system with a small number of projects. UNDP, with its great experience of capacity development and its network of regional and field offices, would be ideal for developing and testing such a pilot.

A slightly more radical suggestion for financing additional technical support to GEF biodiversity projects is that the GEF, UNDP, and other GEF Agencies consider reallocating some portion of the resources currently invested in costly M&E activities that focus on monitoring and reporting progress at higher levels to providing greater and more coordinated technical support to individual projects. There has been an explosion of M&E-related activity within the GEF and UNDP in recent years, resulting in a dizzying array of M&E policies, reports, guidance, and requirements (with an equally dizzying number of acronyms).[39] A significant proportion of the GEF's recent work on M&E has been targeted at facilitating portfolio- and corporate-level monitoring, reporting, and learning as the GEF must demonstrate to its participant countries that its funds are both well managed and well spent. While it is the GEF Secretariat's responsibility to undertake such 'higher-level' monitoring, GEF Agencies such as UNDP still have to supply some of the project-level data needed for this

---

[36] The Consortium of Scientific Partners on Biodiversity. See http://www.cbd.int/csp/.
[37] See http://www.iucn.org/about/union/commissions/.
[38] See http://www.cbd.int/tech-transfer/ and related webpages.
[39] For example, see GEF (2011d), UNDP (2011e).

exercise, and in such a way that the data can be easily aggregated and analysed by the GEF for reporting to the GEF Council and others.

In UNDP's case, many staff members and project teams invest many weeks of their time annually to finalize the PIRs and prepare regional reports for both UNDP and the GEF in addition to meeting a range of other monitoring requirements. As noted earlier, much of the data collected are difficult to analyse or use and have little application beyond meeting the performance monitoring requirements of donors, particularly at the project level or even country level, where GEF projects are implemented and must deliver tangible results and ultimately conservation impacts. Similarly, although tracking tools are only completed three times in a project's life, according to OPS4, substantial resources would be needed to collate, provide quality assurance, and analyse all the data collected if these are to be truly useful for monitoring purposes (GEF EO, 2010, p. 106). However, current levels of investment in these types of project monitoring exercises limit the resources available for other forms of monitoring and technical guidance at the corporate level.

Finally, it would be important to ensure that any new systems of providing additional technical or mentoring support to projects are responsive to user needs, flexible, and largely voluntary. Therefore, the GEF (and the GEF Agencies) should avoid being overly prescriptive and seek to balance corporate needs with project-level realities.[40] In particular, investment in biodiversity monitoring, or in supporting capacity development for monitoring, should not be at the expense of other equally or more pressing conservation management actions. It should also not increase the existing monitoring and reporting burden of projects, which is reported to have increased since GEF-3 (ICF International, 2009). Additionally, any attempts to make project-level biodiversity monitoring more widely relevant, should also not reduce its relevance or application at local scales in any way (Soberon and Townsend Peterson, 2009).

Conclusions

Systematic large-scale biodiversity monitoring is complex and resource intensive. Funds and capacity for conservation are limited relative to need, especially in developing countries. While biodiversity monitoring has rarely been a primary focus of GEF biodiversity projects, these projects nevertheless have considerable potential to contribute to broader monitoring efforts. Most GEF biodiversity projects already

---

[40] For example, the GEF-5 Programming Document states that the corporate approach to Knowledge Management would require amongst other things that all relevant projects would develop a GIS map of the project area using tools and technical input developed at the corporate level and that all projects would be required to develop a project-specific website for posting and transferring lessons (GEF, 2010d, paras 209–210).

contribute some data to global monitoring efforts, notably in relation to CBD and WCPA indicators and targets.

The GEF and its partners have the capacity to greatly strengthen the quality, usefulness, and sustainability of biodiversity monitoring in GEF projects by helping to create high-quality technical support systems for project implementation and management. This could be delivered through both web-based means and project mentoring services, and could potentially be achieved at relatively low cost by drawing upon the substantial technical, financial, and human resources available to the GEF through its global network, which in addition to 182 member countries, includes UN organizations, multilateral banks, numerous international and national NGOs, research institutions, and the private sector. Providing such support would also be in line with GEF-5 polices on Results-Based Management and Learning and Knowledge Management, through which the GEF seeks to improve the delivery and quality of project results (GEF, 2010d, paragraphs 195–213).

Systematic and sustained local capacity development and technical support along the lines discussed here would have numerous benefits, including likely improvement in the overall quality of project implementation and the delivery of both short-term results and long-term conservation impacts of a globally significant nature. It would also contribute indirectly to the GEF's objective of creating a stronger evidence base for specific project interventions, monitoring indicators, and conservation planning and investment generally (GEF, 2008). With greater access to the best available knowledge and tools for field-level conservation planning and management that is appropriate to local circumstances, project teams will be better equipped to assess their biodiversity management information needs and make decisions about how to allocate finite resources between different priorities, including how much and what kinds of biodiversity monitoring are necessary to achieve conservation management objectives. If such support is provided systematically to the entire GEF biodiversity portfolio, it could also lead to better integration of standardized data collection protocols where appropriate, improved data quality across the GEF portfolio, more effective institutionalization of site-based monitoring processes, and strengthened networks for data sharing. This in turn would help increase the overall contribution of GEF projects to national and global biodiversity monitoring schemes.

Acknowledgements

I am grateful to Nik Sekhran, Sameer Karki, Peter Hunnam, Nathalie Pettorelli, Sarah Durant, Ben Collen, and two anonymous reviewers for their feedback on early drafts of this chapter. Jessie Mee and Nancy Bennett are also thanked for providing important clarifications and locating key documents.

References

BIP (2010) Biodiversity Indicators Partnership. URL: http://www.twentyten.net/. Accessed 15 January 2011.

Breitenmoser, U., Alizadeh, A., and Breitenmoser-Würsten, C. (2009) Conservation of the Asiatic Cheetah, its natural habitat and associated biota in the I.R. of Iran. Project number IRA/00/G35. *Terminal Evaluation Report. GEF Project ID 865.* URL: http://www.gefonline.org/projectDetailsSQL.cfm?projID=865. Accessed 6 March 2011.

Bubb, P., Jenkins, M., and Kapos, V. (2005) *Biodiversity Indicators for National Use: Experience and Guidance.* UNEP-WCMC, Cambridge, UK.

Butchart, S.H.M., Walpole, M., Collen, B., *et al.* (2010) Global biodiversity decline continues. *Science*, 328, 1164–1168.

Carugi, C. and MacKinnon, M.S. (2010) *Biodiversity and the GEF: Findings and Recommendations from the Fourth Overall Performance Study of the GEF.* OPS4 Learning Product #1. Global Environment Facility Evaluation Office, Washington, DC. URL: www.gefeo.org. Accessed 16 February 2011.

CBD (2007a) *Updated Synthesis of Information Contained in the Third National reports.* UNEP/CBD/WG-RI/2/INF/1. May 2007. URL: http://www.cbd.int/reports/syntheses.shtml. Accessed 19 April 2012.

CBD (2007b) *Updated Synthesis of Information Contained in Third National Reports: Implementation of the Articles and Provisions of the Convention.* UNEP/CBD/WG-RI/2/INF/1/.Add.2. May 2007. URL: http://www.cbd.int/reports/syntheses.shtml. Accessed 19 April 2012.

CBD (2007c) *Review of the Implementation of Articles 20 and 21: Review of the Availability of Financial Resources.* June 2007. UNEP/CBD/WG-R1/2/INF/4. URL: http://www.cbd.int/financial/policy.shtml. Accessed 20 April 2012.

CBD (2010) Decision adopted by the Conference of the Parties to the Convention on Biological Diversity at its Tenth Meeting. X/2. The Strategic Plan for Biodiversity 2011–2020 and the Aichi Biodiversity Targets. UNEP/CBD/COP/DEC/X/2 October 2010. URL: http://www.cbd.int/sp. Accessed 16 February 2011.

Clémençon, R. (2006) What future for the global environment facility? *Journal of Environment and Development*, 15, 50–74.

Dobson, A. (2005) Monitoring global rates of biodiversity change: challenges that arise in meeting the Convention on Biological Diversity (CBD) 2010 goals. *Philosophical Transactions of the Royal Society, Series B*, 360, 229–241.

Dublin, H. and Volonte, C. (2004) *Biodiversity Program Study 2004.* GEF Office of Monitoring and Evaluation, Washington, DC. Available at: http://www.thegef.org/gef/node/2132. Accessed 7 March 2011.

Gardner, T.A., Barlow, J., Araujo, I.S., *et al.* (2008) The cost-effectiveness of biodiversity surveys in tropical forests. *Ecology Letters*, 11, 139–150.

GEF (1994) *Incremental Costs and Financing Modalities.* URL: http://www.thegef.org/gef/node/1361. Accessed 7 March 2011.

GEF (1996) *Operational Strategy of the Global Environmental Facility.* URL: http://www.thegef.org/gef/node/1238. Accessed 7 March 2011.

GEF (2002) *Cofinancing.* GEF/C.20/6. http://www.thegef.org/gef/sites/thegef.org/files/documents/C.20.6.pdf. Accessed 2 April 2012.

GEF (2006) *Roles and Comparative Advantages of the GEF Agencies.* GEF/C.30/9. November, 2006. URL: http://www.thegef.org/gef/sites/thegef.org/files/documents/C.30.9%20Roles%20and%20Comparative%20Advantages%20of%20the%20GEF%20Agencies.pdf. Accessed 20 April 2012.

GEF (2007a) *Focal Area Strategies and Strategic Programming for GEF-4.* GEF Policy Paper, October 2007. URL: http://www.thegef.org/gef/node/1783 Accessed 19 April 2012.

GEF (2007b) *Results-Based Management Framework.* GEF/C.31/11. May 2007. URL: http://www.thegef.org/gef/C31_11_RBM. Accessed 19 April 2012.

GEF (2008) *A Science Vision for GEF-5. Proposals from the Scientific and Technical Advisory Panel.* GEF/C.34/Inf.14. October 2008. URL: http://www.thegef.org/gef/node/258. Accessed 20 April 2012.

GEF (2009) *GEF-5 Focal Area Strategies.* GEF/R.5/Inf.14. GEF Secretariat, Washington, DC. URL: http://www.thegef.org/gef/GEF5_Biodiversity_Strategy. Accessed 8 February 2011.

GEF (2010a) *Behind the Numbers: A Closer Look at GEF Achievements.* Global Environment Facility, Washington, DC. URL: http://www.thegef.org/gef/pubs/Behind_the_Numbers_2010. Accessed 8 February 2011.

GEF (2010b) *Revision of the GEF Monitoring and Evaluation Policy.* GEF/ME/c.39/6/Rev.1. 17 November 2010. URL: https://www.thegef.org/gef/GEF/ME/C.39/6/Rev.1_Revision_of_the_GEF_Monitoring_and_Evaluation_Policy. Accessed 8 February 2011.

GEF (2010c) *Annual Monitoring Report (AMR) FY 2009.* GEF/C.38/4. 4 June 2010. URL: https://www.thegef.org/gef/node/3223. Accessed 8 February 2011.

GEF (2010d) *GEF-5 Programming Document.* GEF/R.5/31/CRP.1 May 2010. URL: http://www.thegef.org/gef/fifth_replenishment. Accessed 10 March 2011.

GEF (2011a) *GEF 5 Focal Area Strategies.* Global Environment Facility, Washington, DC. URL: http://www.thegef.org/gef/sites/thegef.org/files/documents/document/GEF-5_FOCAL_AREA_STRATEGIES.pdf. Accessed 19 February 2012.

GEF (2011b) *GEF-4 Tracking Tool for Biodiversity Focal Area Strategic Objective One: Catalyzing Sustainability of Protected Areas Systems.* URL: http://www.thegef.org/gef/content/BIO-portfolio-management-tracking-tool. Accessed 1 March 2011.

GEF (2011c) *GEF-4 Tracking Tool for Biodiversity Focal Area Strategic Objective Two: Mainstreaming Biodiversity Conservation in Production Landscapes/Seascapes and Sectors.* URL: http://www.thegef.org/gef/content/BIO-portfolio-management-tracking-tool. Accessed 1 March 2011.

GEF (2011d) *Evaluations and Studies.* URL: http://www.thegef.org/gef/EvaluationsStudies. Accessed 3 March 2012.

GEF (2011e) The GEF Project Database. URL: http://www.gefonline.org/. Accessed 28 February 2011.

GEF (2011f) *Instrument for the Establishment of the Restructured Global Environment Facility.* GEF, Washington, DC. URL: http://www.thegef.org/gef/instrument. Accessed 1 April 2012.

GEF (2011g) *Annual Monitoring Review FY11: Part 1.* GEF/C.41/04/Rev.02, December 2011. http://www.thegef.org/gef/content/amr-2011. Accessed 15 April 2012.

GEF EO (2006) *The GEF Monitoring and Evaluation Policy.* Evaluation Document 2006, No. 1. GEF Evaluation Office, Washington, DC. URL: http://www.thegef.org/gef/Evaluation%20Policy%202006. Accessed 19 April 2012.

GEF EO (2010) *OPS4: Progress towards Impact*. Fourth Overall Performance Study of the GEF. Full Report. Global Environment Facility Evaluation Office, Washington, DC. URL: http://www.thegef.org/gef/OPS4 and associated links. Accessed 19 April 2012.

GEF EO (2011) *The GEF Monitoring and Evaluation Policy* 2010. URL: http://www.thegef.org/gef/Evaluation%20Policy%202010. Accessed 19 April 2012.

GEF & UNEP-WCMC (2003) *Measuring Results of the GEF Biodiversity Program*. Monitoring and Evaluation Working Paper 12. GEF, Washington, DC. URL: http://www.thegef.org/gef/node/2229. Accessed 16 February 2011.

Hockings, M., Stolton, S., Leverington, F., Dudley, N., and Courrau, J. (2006) *Assessing Effectiveness – A Framework for Assessing Management Effectiveness of Protected Areas*, 2nd edn. IUCN, Gland, Switzerland.

Hunnam, P. and Moosa, L. (2008) *Atoll Ecosystem Conservation Project, Baa Atoll, Maldives. Mid-Term Evaluation Report*. URL: http://erc.undp.org/evaluationadmin/manageevaluation/viewevaluationdetail.html?evalid=3976. Accessed 19 April 2012.

ICF International (2009) *Fourth Overall Performance Study of the Global Environment Facility: Independent Monitoring and Evaluation Review*. ICF International, Washington, DC. URL: http://www.thegef.org/gef/node/2090

Jones, J.P.G., Collen, B., Atkinson, G., et al. (2010) The why, what, and how of global biodiversity indicators beyond the 2010 Target. *Conservation Biology*, 25, 450–457.

Kinnaird, M.F. and O'Brien, T.G. (2001) Who's scratching whom? Reply to Whitten et al. *Conservation Biology*, 15, 1459–1460.

Mace, G.M., Cramer, W., Díaz, S., et al. (2010) Biodiversity targets after 2010. *Current Opinion in Environmental Sustainability*, 2, 3–8.

MA (2005) *Ecosystems and Human Well-being: Biodiversity Synthesis*. Millennium Ecosystem Assessment. World Resources Institute, Washington, DC.

McDonald-Madden, E., Baxter, P.W.J., Fuller, R.A., Martin, T.G., Game, E.T., Montambault, J., and Possingham, H.P. (2010) Monitoring does not always count. *Trends in Ecology and Evolution*, 25, 547–550.

OECD (2012) Key findings on biodiversity. *Environmental Outlook To 2050. The Consequences of Inaction*. URL: http://www.oecd.org/dataoecd/51/38/49897175.pdf. Accessed 1 April 2012.

Parker, C. and Cranford, M. (2010) *The Little Biodiversity Finance Book*. http://www.globalcanopy.org/materials/little-biodiversity-finance-book. Accessed 15 April 2012.

Perez del Castillo, C. (2009) *Governance of the GEF*. OPS4 Technical Document No. 5. URL: http://www.thegef.org/gef/node/2088. Accessed 15 April 2012.

Porter, G., Bird, N., Kaur, N., and Peskett, L. (2008) *New Finance for Climate Change and the Environment*. WWF and Heinrich Böll Foundation.

Rodgers, W.A. and Dorji, Y. (2008) *Terminal Evaluation of the Project Linking and Enhancing Protected Areas in the Temperate Broad-leaved Forest Eco-Region of Bhutan – "LINKPA"*. URL: http://erc.undp.org/evaluationadmin/manageevaluation/viewevaluationdetail.html?evalid=3443. Accessed 8 March 2011.

Royal Society (2003) *Measuring Biodiversity for Conservation*. Policy document 11/03. The Royal Society, London.

Salzer, D. and Salafsky, N. (2006) Allocating resources between taking action, assessing status, and measuring effectiveness of conservation actions. *Natural Areas Journal*, 26, 310–316.

SCBD (2010) *Global Biodiversity Outlook 3*. Secretariat of the Convention on Biological Diversity, Montreal.

Soberon, J. and Townsend Peterson, A. (2009) Monitoring biodiversity loss with primary species-occurrence data: toward national-level indicators for the 2010 target of the Convention on Biological Diversity. *Ambio*, 38, 29–34.

TEEB (2010) Biodiversity, ecosystems and ecosystem services. Draft Chapter 2. In: *The Economics of Ecosystems and Biodiversity*. The Economics of Ecosystems and Biodiversity (TEEB) study report of the Ecological and Economic Foundation. URL: http://www.teebweb.org/EcologicalandEconomicFoundationDraftChapters/tabid/29426/Default.aspx. Accessed 9 February 2011.

UNDP (2002) *Handbook on Monitoring and Evaluating for Results*. UNDP Evaluation Office, New York. URL: http://www.undp.org/gef/monitoring/policies.html. Accessed March 2011.

UNDP (2007a) *Human Development Report 2007/2008*. Fighting Climate Change: *Human Solidarity in a Sivided World*. UNDP, New York.

UNDP (2007b) *Annual Monitoring Report Fiscal Year 2007*. UNDP-GEF. 21 December 2007. URL: http://www.thegef.org/gef/sites/thegef.org/files/documents/UNDP%20AMR%202007.pdf. Accessed 19 April 2012.

UNDP (2008a) *Evaluation of the Role and Contribution of UNDP in Environment and Energy*. UNDP Evaluation Office, New York. URL: http://web.undp.org/evaluation/documents/thematic/ee/EE-Full-Report.pdf. Accessed 19 April 2012.

UNDP (2008b) *Biodiversity: Delivering Results*. UNDP Global Environment Facility Unit, New York. URL: http://www.undp.org/content/undp/en/home/librarypage/environment-energy/ecosystems_and_biodiversity/biodiversity-delivering-results.html. Accessed 19 April 2012.

UNDP (2008c) *UNDP Strategic Plan, 2008–2011: Accelerating Global Progress on Human Development*. Executive Board of the United Nations Development Programme and of the United Nations Population Fund. DP/2007/43/Rev.1. URL: http://www.undp.org/execbrd/. Accessed 1 April 2012.

UNDP (2009) *2008–2009 Annual Monitoring Report of UNDP supported GEF funded Projects. Fiscal Year 2009*. UNDP EEG. URL: http://web.undp.org/gef/document/GEF_Annual PerformanceReport_web.pdf. Accessed 19 April 2012.

UNDP (2010a) *Key Results and Lessons from the UNDP-GEF Biodiversity Portfolio*. UNDP Environment and Energy Group. URL: http://www.undp.org/biodiversity/biodiversity_library.shtml. Accessed 15 February 2011.

UNDP (2010b) *UNDP's Work on Biodiversity Management*. UNDP Environment and Energy Group. UNDP, New York. URL: http://www.undp.org/content/undp/en/home/librarypage/environment-energy/ecosystems_and_biodiversity/undps-work-on-biodiversity-management.html. Accessed 19 April 2012.

UNDP (2010c) *2010 Global Annual Portfolio Review: Biodiversity, Land Degradation and Integrated Ecosystems Management*. UNDP-GEF. URL: http://web.undp.org/gef/document/2010_PPER_FINAL_DRAFT_081201.pdf. Accessed March 2011.

UNDP (2011a) *About UNDP-GEF*. URL: http://web.undp.org/gef. Accessed 23 February 2011.

UNDP (2011b) *UNDP-GEF Biodiversity Focal Area*. URL: http://web.undp.org/gef/do_biodiversity.shtml. Accessed 23 February 2011.

UNDP (2011c) *UNDP Environment and Energy: Biodiversity*. URL: http://www.undp.org/biodiversity/. Accessed 28 February 2011.

UNDP (2011d) *GEF Evaluation Status Report*. UNDP Evaluation Resource Centre. URL: http://erc.undp.org/evaluationadmin/gefEvaluationStatusReport.html. Accessed 28 February 2011.

UNDP (2011e) *UNDP Evaluation*. URL: http://www.undp.org/evaluation/. Accessed 28 February 2011.

UNDP (2011f) *2011 UNDP-GEF Global Annual Portfolio Review: Ecosystems and Biodiversity Programme*. URL: http://web.undp.org/gef/document/2011%20UNDP-GEF%20EBD%20Global.pdf. Accessed 2 April 2012.

UNDP (2012) *2010–2011 Annual Performance Report of UNDP Supported GEF Financed Projects*. UNDP, New York. URL: http://web.undp.org/gef/document/APR-10-11_WEB.pdf. Accessed 19 April 2012.

UNEP (2007). *GEO-4: Global Environment Outlook*. UNEP, Nairobi.

UNEP-WCMC (2009) *International Expert Workshop on the 2010 Biodiversity Indicators and Post-2010 Indicator Development*. UNEP-WCMC, Cambridge, UK.

Walpole, M., Almond, R., Besançon, C., et al. (2009) Tracking progress towards the 2010 biodiversity target and beyond. *Science*, 325, 1503–1504.

World Bank and WWF International (2007) *Management Effectiveness Tracking Tool: Reporting Progress on Protected Area Sites*, 2nd edn. WWF International, Gland, Switzerland.

Yoccoz, N.G., Nichols, J.D., and Boulinier, T. (2001) Monitoring of biological diversity in space and time. *Trends in Ecology and Evolution*, 16, 446–453.

Zimsky, M., Ferraro, P., Mupemo, F, Robinson, J., and Sekhran, N. (2010) *Results of the GEF Biodiversity Portfolio Monitoring and Learning Review Mission, Zambia. Enhancing Outcomes and Impact Through Improved Understanding of Protected Area Management Effectiveness*. GEF. URL: http://www.thegef.org/gef/sites/thegef.org/files/documents/document/Biodiversity_Learning%20_Mission_Report_Zambia%20.pdf. Accessed 19 April 2012.

Scaling Up or Down? Linking Global and National Biodiversity Indicators and Reporting

Philip Bubb

UNEP-WCMC, Cambridge, UK

Introduction

Biodiversity indicators at the national, regional, and global scales are required for the assessment and reporting of progress of strategies and international agreements, such as the Convention on Biological Diversity (CBD) and the Millennium Development Goals (MDGs) (Mace and Baillie, 2007; Mackenzie, 2010, Walpole *et al.*, 2009; see also Chapter 14). The demand and investment for biodiversity indicators at all of these scales has been increasing in recent years, stimulated by the CBD 2010 Biodiversity Target and then the Strategic Plan for Biodiversity 2011–2020.

This chapter examines the motivations and methods for producing global- and national-scale indicators in support of the implementation of the CBD, to identify the constraints and opportunities for strengthening the linkages between these scales. As well as the scientific literature and reports of the CBD we draw from the experience of the Biodiversity Indicators Partnership (BIP) and its members (see www.bipindicators.net). We will focus on the CBD, since it is the only international convention targeted at conserving global biodiversity, as well as its implementation by parties at the national and regional scales. We first review the uses, means of production, and challenges of biodiversity indicators at the global scale, and then at the national and regional scales. The experiences to date and opportunities for increasing cross-scale linkages of indicators as part of implementation of the CBD are then identified, concluding with recommendations for reporting and capacity building.

Biodiversity Monitoring and Conservation: Bridging the Gap between Global Commitment and Local Action, First Edition. Edited by Ben Collen, Nathalie Pettorelli, Jonathan E.M. Baillie and Sarah M. Durant.
© 2013 John Wiley & Sons, Ltd. Published 2013 by John Wiley & Sons, Ltd.

Here we use the Biodiversity Indicators Partnership (BIP) definition of an indicator as 'a measure based on verifiable data that conveys information about more than itself' (Biodiversity Indicators Partnership, 2011). From this definition biodiversity indicators are purpose-dependent and so the interpretation or meaning of the data depends on the issue being examined. In this chapter the term 'biodiversity indicators' is used to cover indicators for all aspects of biodiversity conservation and sustainable use. This includes not only measures of the state of biodiversity, such as threatened species, but also threats to biodiversity such as habitat loss; and conservation actions such as coverage of protected areas; and evolving concepts like ecosystem services.

Global biodiversity indicators use

The definition and reporting of indicators at the global scale by the CBD was first required after the establishment in 2002 of the 2010 Biodiversity Target 'to achieve by 2010 a significant reduction of the current rate of biodiversity loss at the global, regional and national level as a contribution to poverty alleviation and to the benefit of all life on Earth' (Decision VI/26). At its meeting in 2004 the CBD Conference of Parties (COP) adopted a provisional framework of indicators for assessing progress towards, and communicating the 2010 target at the global level (Decision VII/30), and identified 17 headline indicators under seven focal areas (see Table 17.1). In effect most of the headline indicators in this framework were actually categories or topics, such as 'Trends in the extent of selected biomes, ecosystems, and habitats', within which specific indicators were allocated or could be developed. Detailed descriptions of the global indicators developed within this framework, including their metadata, methods, and results, are presented in Butchart *et al.* (2010), and the report of the 2010 Biodiversity Indicators Partnership (2010).

The formulation of the CBD indicator framework was conducted by an expert group and the selection made 'on the basis of data availability and relevance to the objectives of the Convention in general, and in areas that have been identified as central to monitoring progress towards the 2010 biodiversity target' (UNEP/CBD/COP/7/INF/33, 2004). These indicators were not chosen or proposed for national or regional monitoring and reporting on implementation of the Convention. It is important to note that the CBD COP Decision VII/30, which adopted the indicator framework, stated that, 'The global application of those indicators as well as the assessment of the progress towards the 2010 target should not be used to evaluate the level of implementation of the Convention in individual Parties or regions.'

The first use of the CBD global indicator framework for reporting was in the *Global Biodiversity Outlook 2* report (Secretariat of the Convention on Biological Diversity, 2006), where an assessment of trends for the 17 headline indicators was made. This assessment was then updated in *Global Biodiversity Outlook 3* (Secretariat of the

Table 17.1 Summary of whether the Convention on Biological Diversity (CBD) global 2010 Biodiversity Target indicators are applicable at national level, and whether they rely on national-level datasets

CBD global indicator framework, with focal areas, *headline indicators*, and indicators	Applicable at national level?	Global indicator reliance on nationally reported data?
Status and trends of the components of biodiversity		
Trends in the extent of selected biomes, ecosystems, and habitats		
Extent of forests and forest types	Yes	Yes
Extent of marine habitats	Yes	Yes
Trends in abundance and distribution of selected species		
Living Planet Index	Yes	No
Global Wild Bird Index	Yes	Yes
Coverage of protected areas		
Coverage of protected areas	Yes	Yes
Protected Area overlays with biodiversity	Yes	No
Management Effectiveness of Protected Areas	Yes	Yes
Change in status of threatened species		
Red List Index	Yes	No
Trends in genetic diversity		
Ex situ crop collections	Yes	Yes
Genetic diversity of terrestrial domesticated animals	Yes	No
Sustainable use		
Areas under sustainable development		
Area of Forest under Sustainable Management: certification	Yes	Yes
Area of Forest under Sustainable Management: degradation and deforestation	Yes	No
Area of Agricultural Ecosystems under Sustainable Management	Yes	Yes
Proportion of products derived from sustainable sources		
Status of Species in Trade	Yes	Yes
Wild Commodities Index	Yes	No
Ecological Footprint and related concepts		
Ecological Footprint	Yes	Yes
Threats to biodiversity		
Nitrogen deposition		
Nitrogen deposition	Yes	No
Status of invasion and trends in invasive alien species impacts and policy		
Trends in Invasive Alien Species	Yes	Yes

Table 17.1 (*continued*)

CBD global indicator framework, with focal areas, *headline indicators*, and indicators	Applicable at national level?	Global indicator reliance on nationally reported data?
Ecosystem integrity and ecosystem goods and services		
Marine Trophic Index		
Marine Trophic Index	Yes	Yes
Water quality of freshwater ecosystems	Yes	No
Connectivity/fragmentation of ecosystems		
Forest fragmentation	Yes	No
River Fragmentation and Flow Regulation	Yes	Yes
Health and well-being of communities		
Health and well-being of communities directly dependent on local ecosystem goods and services	Yes	No
Biodiversity for food and medicine		
Nutritional status of biodiversity	Yes	No
Biodiversity for food and medicine	Yes	Yes
Status and trends of linguistic diversity and numbers of speakers of indigenous languages		
Status and trends of linguistic diversity and numbers of speakers of indigenous languages		
Status and trends of linguistic diversity and numbers of speakers of indigenous languages	Yes	Yes
Status of resource transfers		
Official Development Assistance provided in support of the Convention on Biological Diversity		
Official Development Assistance provided in support of the Convention on Biological Diversity	Yes	Yes

Convention on Biological Diversity, 2010), which used the evidence from the suite of indicators to conclude that the 2010 Biodiversity Target had not been met at the global level. This conclusion and evidence were then recognized in the justification and adoption of the Strategic Plan for Biodiversity 2011–2020 at CBD COP-10 (Decision X/2), and so had a major international policy impact.

In addition to the CBD the other major use of global biodiversity indicators in international agreements is for reporting of progress on Millennium Development Goal 7 (MDG-7), to ensure environmental sustainability by 2015. This Goal has four targets, including a version of the 2010 Biodiversity Target. Four of the global indicators within the CBD 2010 indicator framework are included as MDG-7 indicators (Extent of forests and forest types, Red List Index, Coverage of Protected Areas, Proportion of fish stocks in safe biological limits). The data for these indicators are reported annually by the Statistics Division of the United Nations Department of Economic and Social Affairs (http://www.un.org/millenniumgoals/reports.shtml).

Global biodiversity indicators production

One of the challenges identified in 2004 for using the CBD global indicators was that the existing indicators were developed and reported by a range of organizations, and there was no single mechanism for coordinating further development or input to the CBD or for synthesis across the indicators as a suite. Another challenge was that the identified global indicators identified were at different stages of development and implementation. In some cases the indicators needed little additional work for their development and use, in other cases significant work was required to develop the indicator methodology and the underlying datasets. For example, the Living Planet Index (LPI) was already established and available as an indicator under the CBD headline indicator of 'Trends in abundance and distribution of selected species', but no global indicator was available for the headline indicator of 'Proportion of products derived from sustainable sources'.

To address these needs the 2010 Biodiversity Indicators Partnerships (2010 BIP) was established with major support from the Global Environment Facility (GEF) as a global initiative, and was recognized by the CBD COP in 2006 in Decision VIII/15. The 2010 BIP brought together over 40 international organizations working on indicator development, to provide the best available information on biodiversity trends to the global community and assess progress towards the 2010 Biodiversity Target. It facilitated the improvement and reporting of global indicators, as well as providing guidance materials, a website, and regional capacity-building workshops for national indicator development in support of the 2010 Biodiversity Target.

The report in 2010 by Biodiversity Indicators Partnership (2010) showed progress from 2006 to 2010 in developing new methodologies and time-series data for 20 specific metrics, covering 13 of the 22 global headline indicators. In 2006 seven metrics were already well developed with established methodologies and globally distributed time-series data. Many of the new metrics were produced by analyses of existing datasets for the purpose of producing global indicators for the topics of the CBD headline indicators. For example, the extent of marine habitats was calculated from compiled data on mangroves, seagrasses, and coral reefs. New analyses were conducted of the overlays of protected areas with already identified areas of importance for biodiversity, such as ecoregions, the Alliance for Zero Extinction Sites, and Important Bird Areas. Indicators of the status of species in trade were produced from subsets of the Red List Index dataset for internationally traded species, and for birds listed on CITES Appendix I and II. For the Wild Commodities Index this was calculated from a subset of the database of the LPI for utilized species (see Chapter 8). In only a few cases, such as the management effectiveness of protected areas indicator, was a new methodology developed and a completely new dataset compiled.

For all of the new metrics or indicators developed for reporting on the 2010 Biodiversity Target, the continued calculation and reporting of the indicators after 2010 relies on the existence of additional funding for this purpose. None of these indicators have been adopted for other purposes or processes. All of the existing global indicators in the CBD framework in 2006 already had institutions and mechanisms responsible for their production and reporting. Examples of these indicators and lead institutions are the extent of forests by the United Nations Food and Agriculture Organization (FAO); the LPI by WWF and the Zoological Society of London (ZSL); coverage of protected areas by the United Nations Environment Programme-World Conservation Monitoring Programme (UNEP-WCMC) and the International Union for the Conservation of Nature (IUCN); the IUCN Red List Index (RLI) by IUCN, BirdLife International, and ZSL; and Ecological Footprint by the Global Footprint Network.

The 2010 BIP demonstrated that new global indicators can be developed to meet the variety of topics covered by the CBD, but the regular and sustainable production of such indicators requires additional funding or institutions to 'champion' these indicators. The need for global-scale reporting every few years to the CBD has not so far been sufficient to maintain the production of indicators for this purpose alone. In contrast, the global indicators and datasets that have been established by UN mandates and have multiple uses for their data, such as coverage of forests and protected areas, are able to obtain sufficient resources for their regular production. The other established global indicators produced by NGOs, such as the LPI, IUCN's RLI, and the Ecological Footprint, are resourced as part of the awareness-raising and campaigning work of these organizations, as well as providing evidence for policy-making.

National biodiversity indicators use

In parallel with the development of global indicators for the CBD there has been increasing attention on the use of indicators as part of national CBD implementation and reporting. For example, the guidelines by the Secretariat of the CBD for the production of Fourth National Reports includes the text, 'Parties are encouraged to use indicators in their national report, including those developed at the national and global levels'. Evidence on the use of national indicators by countries, including their linkages or similarities with global indicators, is supplied by a report commissioned to provide an evidence base for the work of the CBD Ad Hoc Technical Expert Group (AHTEG) on Indicators for the Strategic Plan for Biodiversity 2011–2020 (Bubb *et al.*, 2011). The findings of this report on how and why national-scale biodiversity indicators are used and the challenges for their production are summarized in this chapter.

The evidence for the AHTEG report was compiled from an analysis of the use of indicators in countries' Fourth National Reports to the CBD, an online questionnaire distributed to national agencies responsible for biodiversity information and reporting, and the experiences of UNEP-WCMC in running 2010 BIP regional capacity-building workshops. Over the course of these workshops attendees came from 49 developing and middle income countries in Southeast Asia, the Caribbean, Mesoamerica, and eastern and southern Africa. The resulting guidance materials and workshop reports are available from the Biodiversity Indicators Partnership website (www.bipnational.net).

The online questionnaire asked, 'What was the main reason for the choice of indicators for CBD implementation and reporting?' Of the 134 respondents who had national indicators 56 answered this question, representing 35 countries, and they could select one option only. The results (Figure 17.1) show that 'For CBD goals and targets' was the main reason for the choice of indicators for CBD implementation and reporting for about 30% of respondents. For 50% of respondents specific national objectives and targets or broad topics of national importance have been the main reason, rather than for CBD goals and targets. In the experience of the 2010 BIP capacity-building workshops the existence of CBD goals and targets was not found to be the reason for producing indicators, except for ad hoc production for Fourth National Reports to the CBD. The BIP workshops also found that the CBD 2010 Target global indicator framework was often misunderstood as being a reporting requirement for countries, rather than being part of 'a flexible framework within which national

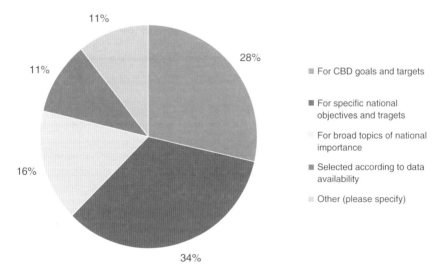

Figure 17.1 Percentages of responses of 56 online survey respondents to the question: 'What was the main reason for the choice of indicators for CBD [Convention on Biological Diversity] implementation and reporting?'

and regional targets may be set, and indicators identified, where so desired by Parties' (CBD COP7 Decision VII/30).

For 10% of respondents the reality of which indicators are actually produced was mostly determined by the existence and accessibility of suitable data. For many developing countries, or countries without a long tradition of environmental research, the availability of existing data is likely to be the principal determinant of which indicators are actually chosen and used. The BIP capacity-building workshops found this to be the case in all countries that did not have an established national institution for the gathering and reporting of biodiversity information. However, it should be recognized that in many cases a dataset can be analysed and interpreted for more than one use if the appropriate scientific and analytical skills exist.

National biodiversity indicators production

By March 2011, Fourth National Reports had been submitted to the Secretariat of the CBD by 159 (83%) of the 193 Parties to the CBD, of which 121 (76%) had reported or referenced at least one indicator for biodiversity in their report, but only 58 (36%) included evidenced indicators (i.e. with data or figures) in their report (Figure 17.2). It is likely, though, that many countries have additional relevant information that was either not readily available for use in the national reports, or could have been obtained from sectors such as forestry and fisheries that may not always be seen as sources of biodiversity-relevant information. Thus, the results of the analysis of the Fourth National Reports should be seen as a minimum view of the current national situation and capacity for indicators for biodiversity.

Of a total of 134 respondents to the online survey, representing 65 countries, 56% said that their country did have indicators relevant to implementation of the CBD, and these respondents represented 35 countries. Twenty percent of respondents said that their country did not have relevant national indicators, and 24% said that they didn't know. It is evident that currently a great many countries do not have established national indicators that are considered to be relevant to the implementation of the CBD and the Strategic Plan for Biodiversity 2011–2020. This does not necessarily mean that they do not have any relevant data to support the implementation of the CBD and the Strategic Plan, but that such data may not be analysed and made available in the form of indicators for particular needs. They may also have indicators on topics such as forestry and fisheries that are very relevant, but have not been used in relation to implementation of the CBD.

In many cases more than one institution may be involved in the production of an indicator or suite of indicators. For example, an academic body or non-governmental organization (NGO) may gather field data and conduct an initial analysis, the national statistics office may validate and approve the analysis, and the environment ministry

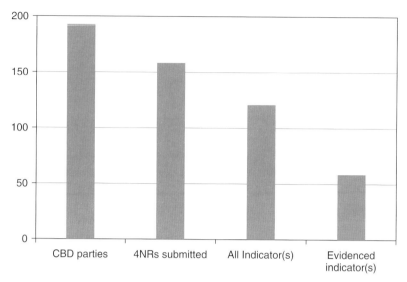

Figure 17.2 **Number of Convention on Biological Diversity (CBD) fourth national reports (4NRs) with indicators.**

adopt the results in its work. The online survey for the AHTEG report asked who is involved in the production of national biodiversity indicators, and 56 respondents with national indicators (representing 35 countries) selected the options that applied (Figure 17.3).

Over 80% of respondents stated that their environment ministry is involved in the production of national indicators for implementation of the CBD. It should be noted that the questionnaire did not detail what was meant by 'involvement', which could range from the selection and commissioning of indicators, the gathering and processing of data, to their communication in reports. Similarly, the questionnaire was about all types of 'biodiversity indicators' that are relevant for implementation of the CBD, and the results reflect the diversity of institutions that may be involved in different indicators.

One or more of government biodiversity offices, national statistics offices, NGOs, and academic bodies were involved in the production of their national indicators according to about 50% of respondents. From the experience of the BIP capacity-building workshops, a key factor in a country's capacity to produce biodiversity indicators over time is whether or not there is a national office or institution with the responsibility for the coordination, analysis, and communication of biodiversity information. Whilst many countries have government agencies responsible for information

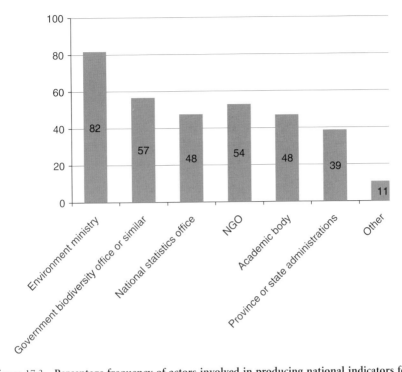

Figure 17.3 **Percentage frequency of actors involved in producing national indicators for implementation of the Convention on Biological Diversity (CBD; 56 respondents).**

and statistics for biodiversity-relevant issues such as forests, fisheries, wildlife, and protected areas, the existence of a government or academic institute for biodiversity information is relatively uncommon outside Europe. Some examples of such institutes are:

China – Nanjing Institute of Environmental Sciences of the Ministry of Environmental Protection
South Africa – South Africa National Biodiversity Institute (SANBI)
Uganda – National Biodiversity Data Bank, Makerere University Institute of Environment and Natural Resources (MUIENR)
Namibia – Namibia Nature Foundation and the Ministry of Environment and Tourism
Mexico – National Commission for the Knowledge and Use of Biodiversity (CONABIO)
Costa Rica – National Biodiversity Institute (INBio)
Brazil – Chico Mendes Institute for Biodiversity Conservation (ICMBio)

National biodiversity indicators data sources

The existence of suitable data is obviously central to how indicators are produced. The online survey for the AHTEG report asked respondents to rank the main data sources for the national indicators, and the results are presented in Figure 17.4. The commonest data source is to adapt data from monitoring and reporting systems that have been designed for other purposes, which reflects both a lack of biodiversity-specific monitoring systems and that biodiversity is a broad concept for which many relevant issues or sectors, such as forestry, can provide data. Similarly, data can be used from existing indicators produced by another sector, which is the fifth commonest use in the questionnaire.

Data from surveys and assessments comprised the second commonest source for indicators. This data type may or may not provide trend data, and may not be entirely suitable for use for a specific indicator. The third commonest data source is academic research, which is unlikely to provide long-term datasets, may be restricted to sub-national-scale analysis, and not be designed to address questions or objectives that national indicators would be designed for. The fourth commonest data source is from monitoring systems developed for the indicators, which may be the ideal if resources are available to maintain the system. Data for national indicators from regional or

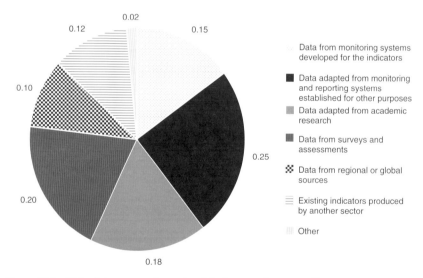

Figure 17.4 **Main data sources for national indicators, from 56 online survey respondents (the proportion of seven options ranked and weighted by order of use).**

global data sources represented the least common source. This probably reflects the absence or inaccessibility of relevant datasets, and that some global datasets are not accessible with their national components disaggregated.

The pattern of these results is broadly similar to the experience from the BIP capacity-building workshops. Amongst the 45 countries involved very few national biodiversity indicators were produced with data from monitoring systems established for the purpose, with coverage of protected areas being the commonest exception. In several African countries regular surveys of wildlife in protected areas, and sometimes in other areas, were undertaken, but the data were rarely analysed and presented in the form of an indicator. In several European countries there are monitoring systems designed to provide data for different taxa.

National biodiversity indicator types

The analysis in the AHTEG report of indicators in Fourth National Reports related to the CBD 2010 global indicator categories found that 'Coverage of protected areas' is by far the commonest indicator that most CBD Parties can produce, as reported by a total of 91 Parties. The second most reported indicator, by 50 countries, is 'Extent of forests and forest types', although many more countries will have such data for reporting to the FAO Forest Resource Assessment. The third most frequent indicator in CBD Fourth National Reports is 'Invasive alien species', found with evidence (data) as an indicator in nine reports, and references without evidence in 32 reports. Aggregation of the indicators in the reports to the level of Focal Area of the CBD 2010 Target Framework showed that over 100 Parties have at least some data on aspects of the status and trends of the components of biodiversity. Some indicators or relevant data on sustainable use, threats to biodiversity, and ecosystem integrity are provided by at least 40 Parties. Only about 15 Parties include some indicators or information on status of traditional knowledge, innovations and practices, or resource transfers, in their Fourth National Reports to the CBD.

Challenges in developing national biodiversity indicators

The lack of suitable data for indicators, including the inaccessibility of existing data, is probably the most widespread problem for the calculation of national-scale indicators for biodiversity, followed by a lack of technical and institutional capacity, and lack of funding. Many developing countries, in particular, report a lack of funding, and without additional international and national funds they will not be able to establish the necessary indicator, monitoring, and reporting systems for their implementation of the Strategic Plan for Biodiversity 2011–2020. However, this situation is linked to the often reported constraint in the AHTEG report online survey of, 'insufficient demand from government for such indicators', with very little awareness or use of biodiversity indicators at all levels of government and society in some countries. Whilst the lack of suitable data for desired indicators is widespread, and requires new investment, it seems that in most countries at least some useful new indicators can be produced from reinterpretation of existing data.

In some countries, including many European countries, there are monitoring schemes to gather data for indicators for biodiversity, but the evidence from Fourth National Reports to the CBD is that most countries do not have such schemes. The problem appears to be that where indicators for biodiversity are not produced it is difficult to demonstrate the value of such indicators for decision-making, and so it can be challenging to obtain funding for the monitoring to supply the data for the indicators.

The lack of progress to date in establishing national biodiversity indicators points to the existence of major obstacles to biodiversity monitoring in many countries. Participants of the 2010 BIP regional workshops gave some insights into the constraints to producing biodiversity indicators. The most frequently cited barriers were lack of capacity, lack of consistent biodiversity trend data, absence of baselines with which to compare current trends and targets against which change is measured, and lack of established monitoring systems. Specific issues that were also reported included:

- 'Marginalization' of environmental ministries
- Limited knowledge of the definition of indicators to measure progress towards the 2010 CBD Target.
- Although there are often data available on various metrics that relate to biodiversity, many of the datasets are 'one-off' studies, often covering only a portion of a country. As a result, it has been a challenge to find ways of integrating different datasets and making them comparable to produce time-series statistics.
- A lack of institutional responsibility and accountability for biodiversity survey and monitoring makes it very difficult for some countries to establish and verify biodiversity trends. Data ownership and management were commonly cited problems.
- Many government institutions do not have data management structures in place so that data and information is often 'person-bound' rather than 'institution-bound'. Sustaining good biodiversity monitoring systems over time is a major challenge in some cases, particularly after donors exit.

Regional indicators for biodiversity

A review in the AHTEG report of regional initiatives to produce indicators for biodiversity found that the Streamlining European 2010 Biodiversity Indicators

(SEBI2010), Circumpolar Biodiversity Monitoring Programme (CBMP), and Nordic Biodiversity Indicators 2010 (NordBio2010) initiatives produce regional-scale biodiversity indicators in response to regional needs and decision processes. The data for these initiatives are not compiled or aggregated from national indicators and so countries in these regions tend to produce their own indicators in response to national management objectives and targets, with limited linkages to the regional process. In contrast the PROMEBIO initiative in Central America is designed to promote standardization and sharing of national biodiversity indicator production to enable a regional perspective and actions.

National use of global indicators for biodiversity

The final selection of indicators for the CBD indicator framework was made principally through a compromise between what was needed and what was possible for the purpose of reporting on the 2010 Biodiversity Target at the global level (UNEP/CBD/COP/7/INF/33, 2004). They were not chosen or proposed for national or regional monitoring and reporting on implementation of the Convention. Despite this proviso, in principle the methods of all of the CBD global headline indicators can be applied at a national scale (see Table 17.1). Some indicators actually rely on an aggregation of data from the national level, such as coverage of protected areas, in which case they are already available at national level.

For some of the global CBD 2010 Biodiversity Target headline indicators there are a few conceptual issues that need to be addressed in order to ensure transfer to national-level metrics. For example, for the Red List Index and River Fragmentation global indicators, the unit of analysis, such as the global population of a species covering many countries, or a multinational river system, may not align easily with national boundaries (see Chapter 2). A national calculation for these indicators would first need to determine the appropriate scale and boundaries for including data, such as nationally endemic species and discrete national populations of species, or river basins and sub-basins. However, these methodological challenges should not be insurmountable. Most of the technical challenges in the application of the global headline indicators to a national or regional scale are likely to result from limitations in the availability of sufficient data, and the capacity to gather data, rather than any particular methodological challenges.

During the 2010 BIP regional capacity-development workshops it was found that almost none of the CBD headline global indicators were being calculated at the national level in any of the 49 participating countries. A very few Parties to the CBD, such as China, the UK, and the European regional process (SEBI, 2010), have sought to develop indicators in accordance with the CBD global 2010 Biodiversity Target framework, as evidenced by their Fourth National Reports to the CBD.

Despite this situation two of the 22 CBD 2010 Target headline indicators are widely used at the national level: the coverage of protected areas; and the extent of forests. Both of these topics are nationally important for countries in their biodiversity and economic strategies. National-level estimation of these indicators is also encouraged for other international reporting obligations, such as the Biodiversity Target under Millennium Development Goal 7 on environmental sustainability.

A further 16 of the 22 CBD 2010 Biodiversity Target headline indicators rely on data produced at the national level (Table 17.1). However, only a few of these global indicators are actually calculated at national level, due to insufficient capacity or resources. A case exists for organizing the necessary support to ensure that well developed global indicators can be implemented at a national level. In some situations there is an increasing drive to help countries generate particular indicators. For example, the Living Planet Index has been particularly active, and global partners have worked with national partners, such as those in Uganda and Canada, to produce national versions of the LPI. Another example is the launch of a national Red Listing website (www.nationalredlist.org) to provide support to develop national- and regional-level species threat assessments, and a large number of countries have developed national red lists for different taxa (109 countries; Zamin *et al.*, 2010). The Wild Bird index (Gregory *et al.*, 2005) has substantial support and guidance from the BirdLife International Partnership and is now being used at the national level in 18 European countries and the United States, and is in development at site and national levels in several African countries. The Ecological Footprint Network, which relies on remote-sensing information, also calculates and makes available for most countries its index and biocapacity values.

National capacity needs for biodiversity indicators

A lack of capacity is a serious obstacle to biodiversity monitoring and indicator production in nearly all but the most developed countries. There is a need for trained personnel to gather, interpret, and communicate data, as well as for systems to produce and hold monitoring data for indicators. An analysis of capacity development needs for developing national biodiversity indicators was conducted in 2010 at the final regional workshops of the BIP Africa biodiversity indicators capacity support. Government and NGO indicator developers reviewed their experience and their conclusions are likely to be relevant for all developing countries and probably for other countries as well. The key capacity findings were:

1. *Capacity development assistance can help to convert existing data into useful indicators but the challenge of a lack of data is universally identified as a major limitation in the production of biodiversity indicators.*

The participating countries found that some relevant national biodiversity indicators can be produced, such as coverage of protected areas and trends in key wildlife species, but there are frequently inadequate or inaccessible data for biodiversity indicators to answer priority national questions for policy and monitoring. The conservation agencies recognized that they cannot achieve their mandates or objectives without relevant and accessible information on biodiversity. This includes being able to communicate the importance of biodiversity in sustaining development and its inclusion in development policies.

2. *Biodiversity indicators need to be developed to address national biodiversity and development priorities.*

   Whilst reporting on progress towards international targets and agreements was agreed to be important, this purpose was viewed as a secondary priority to that of addressing national priorities. Examples of national priorities in African countries included the maintenance of protected area systems, inclusion of biodiversity concerns in land use policies for investment in biofuels, sustainable fisheries management, and land degradation. Long-term investment in the production of biodiversity indicators can only be sustained if the latter are seen to be useful and needed to meet national priorities, which may not always coincide with the priorities of global biodiversity indicator processes.

3. *Countries need to have an effective national institution to coordinate their national biodiversity indicators.*

   In many developing countries the gathering and communication of biodiversity information is on an ad hoc and fragmented basis, such as for periodic reporting requirements. The capacity to have biodiversity indicators and other information available for effective environmental and conservation decision-making requires an institution to be responsible for this. It is not necessary for one institution to conduct all the activities of collection of data, calculation of indicators, and their communication to users, as agreements between government agencies, NGOs, and academic institutions can fulfil many of these roles. However, one national coordinating institution with overall responsibility for biodiversity indicator reporting is essential. In some countries, national statistical offices could have a key leadership role in the institutionalization and mainstreaming of biodiversity indicators, as they can validate and provide credibility to the indicators for non-environmental sectors of government and wider society; as well as often having a familiarity with indicator development and communication that can be shared with the environmental government sector.

4. *Developing countries need financial and technical support to institutionalize and operationalize biodiversity indicators.*

   Without additional financial and technical support it is likely that reasons will remain for a lack of biodiversity indicators in decision-making by government and the rest of society in developing countries. Such support should be provided by way

of training, to enable countries to develop their own indicators, and to engender ownership of national biodiversity indicators.
5. *Consultation and collaboration between government institutions and NGOs within countries and between countries significantly strengthens indicator development and use.*
Capacity development assistance can mean more than technical and financial support, but also can be in terms of establishing indicator networks and appropriate institutional frameworks to facilitate collaboration and mentorship. These can provide opportunities for experienced indicator developers to mentor new arrivals and for the exchange of experiences and support. The organization of regional-scale workshops with multi-stakeholder collaborations within countries have been a very effective means for developing capacity and a stimulation of results (Biodiversity Indicators Partnership, 2011).

Many of these issues highlighted through the regional workshops, have been identified as problems in implementing conservation monitoring elsewhere (see Chapter 14). They are critical not only to ongoing monitoring efforts, but also to engaging national governments in the CBD, and in measuring progress and hence effectiveness of national biodiversity strategies and action plans (NBSAPs).

Scaling up or scaling down? Addressing differing motivations

One reason for the currently weak linkages between global and national biodiversity indicators is that they are often intended for different types of users and purposes. The motivations for global-scale indicators are usually to summarize information and to help understand a subject at a broad level for an international audience. The adaptation of existing global indicators and the development of new ones for reporting on the global CBD targets is one such motivation. Some global biodiversity indicators are also developed as communication tools to help raise awareness of important issues as well as to support decision-making. Global biodiversity indicators can also support strategy development by international organizations, such as the Global Environment Facility, by indicating broad trends, but more detailed information at finer scales is also usually needed.

At the national level the use of biodiversity indicators is focused on national decision-making needs, such as helping the design and monitoring of conservation strategies, and consistency with global indicator frameworks is a secondary consideration if at all. National indicators are also likely to be required for country-specific topics, such as for policies and management plans for sustainable timber production, fisheries, or wildlife tourism. National biodiversity indicators are also sometimes produced to raise awareness and support for sectoral topics of importance to particular interest groups, including NGOs and academia, such as for threatened species or sites, pollution problems, or promoting compliance with international agreements.

From the 2010 BIP workshops it was evident that most developing countries were working to include indicators in their CBD Fourth National Reports, but indicators were often compiled on an ad hoc basis for the reporting exercise, rather than as part of long-term monitoring and decision-support processes. Only a few countries with a dedicated biodiversity information institution, such as China, South Africa, Brazil, and Mexico, had established the regular production and reporting of biodiversity indicators as part of the government decision-making process. There is an opportunity to learn from these countries, to establish what additional mechanisms, beyond a dedicated biodiversity information institution, are in place to facilitate efficient and effective biodiversity indicator reporting.

Harmonizing global-regional-national indicator use will aid assessments of progress towards biodiversity targets, particularly at broader scales. The more consistent that regional and national targets are with the global Aichi targets the easier it will be to promote consistency in indicators across scales.

Global indicators are often driven by international agencies investing in coordinating data collection, or modelling and extrapolating from limited national-level data. Countries and regional organizations will develop and use indicators where they perceive a benefit to themselves. However, meeting reporting requirements for global agreements and processes is unlikely to be a sufficient rationale to develop national indicators and monitoring systems if relevant national datasets and monitoring/reporting mechanisms do not already exist.

The Strategic Plan for Biodiversity 2011–2020 is an overarching framework for action by all partners, from global to local scales. The establishment of national targets in line with the Aichi Biodiversity Targets as part of updated NBSAPs is key to the implementation of the new Strategic Plan. From an indicators perspective, the priority at national scale will be to develop indicators to meet national needs, tailored to nationally adopted targets. However, there is a strong rationale for encouraging harmonized indicator use and reporting across Parties, not least because in a resource-constrained world doing so increases the availability of information for tracking progress towards goals and targets at broader scales.

Acknowledgements

The author is very grateful to Ben Collen and Sarah Durant of the Zoological Society of London, and Matt Walpole, of UNEP-WCMC, for their comments and input to this chapter, and to Anna Chenery and Damon Stanwell-Smith of UNEP-WCMC, for their collaboration in the review of experience on national indicators, monitoring, and reporting for the AHTEG on Indicators for the Strategic Plan for Biodiversity 2011–2020. The report for the AHTEG was commissioned and funded by the UK Department of Food and Rural Affairs (Defra).

References

2010 Biodiversity Indicators Partnership (2010) Biodiversity indicators and the 2010 Biodiversity Target: Outputs, experiences and lessons learnt from the *2010 Biodiversity Indicators Partnership*. Technical Series No. 53. Secretariat of the Convention on Biological Diversity, Montreal, 196 pp.

Biodiversity Indicators Partnership (2011) Guidance for National Biodiversity Indicator Development and Use. *UNEP World Conservation Monitoring Centre*, Cambridge, UK, 40 pp.

Bubb, P., Chenery, A., Herkenrath, P., *et al.* (2011) National indicators, monitoring and reporting for the Strategic Plan for Biodiversity 2011–2020: a review of experience and recommendations in support of the CBD Ad hoc Technical Expert Group (AHTEG) on indicators for the Strategic Plan 2011–2020. UNEP-WCMC with IUCN and ECNC for the UK Department for Environment Food and Rural Affairs (Defra). Available at: http://www.cbd.int/doc/meetings/ind/ahteg-sp-ind-01/information/ahteg-sp-ind-01-inf-02-en.pdf.

Butchart, S.H.M., Walpole, M., Collen, B., *et al.* (2010) Global biodiversity deline continues. *Science*, 328, 1164–1168.

Gregory, R.D., van Strien, A., Vorisek, P., *et al.* (2005) Developing indicators for European birds. *Philosophical Transactions of the Royal Society of London B*, 360, 269–288.

Mace, G.M. and Baillie, J.E.M. (2007) The 2010 biodiversity indicators: Challenges for science and policy. *Conservation Biology*, 21, 1406–1413.

Mackenzie, R. (2010) Monitoring and assessment of biodiversity under the Convention on Biological Diversity and other international agreements. In: Lawence, A. (ed.), *Taking Stock of Nature: Participatory Biodiversity Assessment for Policy*, Planning and Practice. Cambridge University Press, Cambridge, pp. 1–29.

SEBI (2010) Streamlining European 2010 Biodiversity Indicators. Available at: http://ec.europa.eu/environment/nature/knowledge/eu2010_indicators/index_en.htm.

Secretariat of the Convention on Biological Diversity (2006) Global Biodiversity Outlook 2. CBD, Montreal, 81 + vii pp.

Secretariat of the Convention on Biological Diversity (2010) *Global Biodiversity Outlook 3*. CBD, Montreal, 94 pp.

UNEP/CBD/COP/7/INF/33 (2004) Provisional global indicators for assessing progress towards the 2010 Biodiversity Target – note by the Executive Secretary. *Conference of the Parties to the Convention on Biological Diversity*.

Walpole, M., Almond, R., Besançon, C., *et al.* (2009) Tracking progress towards the 2010 biodiversity target and beyond. *Science*, 325, 1503–1504.

Zamin, T., Baillie, J.E.M., Miller, R., Rodriguez, J.P., Ardid, A., and Collen, B. (2010) National Red Listing beyond the 2010 Target. *Conservation Biology*, 24, 1012–1020.

18

Conserving Biodiversity in a Target-Driven World

Simon N. Stuart[1,2,3,4,5] *and Ben Collen*[6]

[1]IUCN Species Survival Commission, Gland, Switzerland
[2]UNEP World Conservation Monitoring Centre, Cambridge, UK
[3]Conservation International, Arlington, VA, USA
[4]Department of Biology and Biochemistry, University of Bath, Bath, UK
[5]Al Ain Zoo, Abu Dhabi, United Arab Emirates
[6]Institute of Zoology, Zoological Society of London, London, UK

Background

We live in a world that sets targets. A sense of dissatisfaction with the way that things are leads us to want to set measurable goals and targets so we can see how we are doing in changing our world. These targets encompass the whole spectrum of life, from education, to health, to economic growth, to biodiversity. In many ways, targets are a sign that the world is becoming more professional and accountable. The world's political leaders have set various national, regional, and global targets in relation to the conservation of biodiversity over the past 10 years, notably the following:

- June 2001 – the EU Summit in Gothenburg in June 2001 where EU Heads of State first adopted the target of 'biodiversity decline should be halted [in the EU] with the aim of reaching this objective by 2010'.
- May 2002 – the Convention on Biological Diversity's (CBD's) sixth Conference of the Parties (COP) included a 2010 target – this time 'to achieve by 2010 a significant reduction of the current rate of biodiversity loss at the global, regional and national level as a contribution to poverty alleviation and to the benefit of all life on earth' – in the Strategic Plan that was adopted.

Biodiversity Monitoring and Conservation: Bridging the Gap between Global Commitment and Local Action, First Edition. Edited by Ben Collen, Nathalie Pettorelli, Jonathan E.M. Baillie and Sarah M. Durant.
© 2013 John Wiley & Sons, Ltd. Published 2013 by John Wiley & Sons, Ltd.

- September 2002 – the World Summit on Sustainable Development held in Johannesburg confirmed the 2010 biodiversity target and called for 'the achievement by 2010 of a significant reduction in the current rate of loss of biological diversity'.
- May 2003 – Environment Ministers and Heads of Delegation from 51 countries adopted the Kiev Resolution on Biodiversity at the fifth Ministerial Conference 'Environment for Europe' and decided to 'reinforce our objective to halt the loss of biological diversity at all levels by the year 2010'.
- September 2007 – the United Nations decided to adopt the 2010 target (in terms of reducing the rate of loss) as a sub-target of Millennium Development Goal (MDG) 7 – Environmental Sustainability.
- October 2010 – the CBD's 10th Conference of the Parties, closed the Nagoya Biodiversity Summit by adopting the 'Aichi Targets', including 20 headline targets, organized under five strategic goals.

In this chapter we look back on the CBD's 2010 Target, and evaluate how the proposed 2020 Aichi Targets will replace it. We reflect on the underlying problem that biodiversity is not accorded by governments the priority that the conservation movement would like, and this results in the targets not being met. We sketch out a few possible ways ahead, but emphasize that this chapter is not a formal scientific paper. Rather it is based on the closing talk at the London Symposium on Biodiversity Monitoring at which S.N.S. attempted to draw together some strands in the discussion of the previous two days, and overlaid it with his own biases and opinions.

The CBD 2010 Target – positive aspects

It is rather common to hear the CBD 2010 Target discussed in negative ways, which we cover in the next section. However, there are a number of positive aspects related to the Target. Despite the fact that it is clear that the Target was not achieved (Butchart *et al.*, 2010; Convention on Biological Diversity, 2010), there is no doubt that it has had a major impact on the biodiversity conservation world. The Target itself has given rise to a cottage industry focusing on interpreting and measuring the target and developing indicators (see Chapters 2–7 and Table 18.1). As a result, a broader and growing, though still limited, body of knowledge on the current status of biodiversity's components is now available. The fact that we are now much more serious about measuring our progress in tracking changing biodiversity, and about determining what impact our conservation efforts are having, has got to be a good thing – not least because it demonstrates that we are focused on achieving real results, however difficult that might be. This has engendered a more realistic mindset on how the conservation movement is performing, and how conservation science can

Table 18.1 The 2010 Target indicators and progress towards them. Adapted from Walpole *et al.* (2009)

Indicator	Development stage		
	Full	Under	Not
Components of biodiversity			
Trends in extent of selected biomes, ecosystems, habitats	x	x	
Trends in abundance of selected species	x	xx	
Coverage of protected areas	x	xx	
Changes in status of threatened species	x		
Trends in genetic diversity		xx	
Sustainable use			
Area under sustainable management		xxx	
Proportion of products from sustainable sources	x	xx	
Ecological footprint and related concepts	x		
Threats to biodiversity			
Nitrogen deposition	x		
Trends in invasive alien species		x	
Ecosystem integrity, goods, and services			
Marine trophic index	x		
Water quality of freshwater ecosystems	x		
Trophic integrity of other ecosystems			x
Connectivity/fragmentation of ecosystems		xx	
Human-induced ecosystem failure			x
Health and well-being of communities		x	
Biodiversity for food and medicine		xx	
Status of knowledge, innovations, and practices			
Linguistic diversity		x	
Indigenous and traditional knowledge			x
Status of access and benefits sharing			
Access and benefits sharing			x
Status of resource transfers			
Official development assistance		x	
Technology transfer			x

Full = fully developed indicator with well-established methods and global time-series data; Under = under development; Not = not being developed. Multiple 'x's mean multiple measures under each headline.

help measure that performance, informing what should be done in order to be more effective in the future.

The wording of the CBD 2010 Target was interesting: 'to achieve by 2010 a significant reduction of the current rate of biodiversity loss at the global, regional and national level as a contribution to poverty alleviation and to the benefit of all life on Earth' (UNEP, 2002). This is a very cleverly worded statement, and we tend to focus only on the first half of it. The first half of the Target focuses on the loss of biodiversity per se, and seeks to recognize that progress has to be achieved at a range of biological and geographical scales. The second half focuses on the rationale for our concerns. It is important to note that human benefits are included in the 2010 Target (*poverty alleviation*), and are also central to the new 2020 Targets (Table 18.2); in both cases, reducing the rate of biodiversity loss is seen as a tool to alleviate poverty, presumably through the provision of healthy ecosystem services, and hence the reason why there were indicators for human well-being and ecosystem services in relation to the 2010 Target, which were maintained and adapted for the 2020 Target. But note that the 2010 Target included the notion of intrinsic value (*to the benefit of all life on Earth*). Put another way, biological diversity does not have to exist just for the benefit of people; it has rights of its own. In summary, the 2010 Target covered conservation, human well-being, ecosystem services, and intrinsic value. This is an admirable balance that reflects the integrative nature of biodiversity.

While the 2020 Aichi Biodiversity Targets have been agreed (Table 18.2), the precise indicators that will measure progress towards those targets have yet to be selected at the time of writing. However, they are likely to be based on indicators similar to the 2010 set, with some strategic adaptation and new development where necessary. For 2010, a broad array of indicators were developed, all of which increase our understanding of what is happening to biodiversity and the ecosystem services that derive from it. There were indicators under seven focal areas, covering: status and trends of biodiversity itself; sustainable use; threats; ecosystem integrity, goods, and services; traditional knowledge; access and benefit sharing; and resource transfer. Although there could certainly have been some improvements in the selection of indicators used, taken collectively they cover a very broad range of biodiversity concerns. This is important, because an oft-repeated criticism of the 2010 indicators, made especially during the negotiations for the 2020 Targets, was their supposed lack of attention to ecosystem services and human benefits.

The 22 indicators of the 2010 Target do not always sit logically under their focal area headings. In fact, just three of these indicators measure wild biodiversity per se (Table 18.1: rows 1, 2, and 4 under 'components of biodiversity'; row 5 covers domestic biodiversity). At least four cover sustainable use (rows 6, 7, 8, and 17), seven cover ecosystem services (all grouped under 'ecosystem integrity, goods and services'), four cover human well-being (rows 16, 18, 19, 20), two cover threats (rows 9 and 10), and three cover responses (3, 21, 22). While it is not the intention to cover the interrelatedness or utility of these measures in this chapter (see Butchart *et al.*, 2010; Sparks *et al.*, 2011; Walpole *et al.*, 2009), there were clearly trade-offs to be made as indicators were developed. A strategy of adopt, adapt, and strategically supplement existing indicators was clearly taken (Balmford *et al.*, 2005; Mace and Baillie, 2007).

Table 18.2 **The 2020 Aichi Biodiversity Targets**

Strategic goal A: *Address the underlying causes of biodiversity loss by mainstreaming biodiversity across government and society*

Target 1	By 2020, at the latest, people are aware of the values of biodiversity and the steps they can take to conserve and use it sustainably
Target 2	By 2020, at the latest, biodiversity values have been integrated into national and local development and poverty reduction strategies and planning processes and are being incorporated into national accounting, as appropriate, and reporting systems
Target 3	By 2020, at the latest, incentives, including subsidies, harmful to biodiversity are eliminated, phased out, or reformed in order to minimize or avoid negative impacts, and positive incentives for the conservation and sustainable use of biodiversity are developed and applied, consistent and in harmony with the Convention and other relevant international obligations, taking into account national socioeconomic conditions
Target 4	By 2020, at the latest, governments, businesses, and stakeholders at all levels have taken steps to achieve or have implemented plans for sustainable production and consumption and have kept the impacts of use of natural resources well within safe ecological limits

Strategic Goal B: *Reduce the direct pressures on biodiversity and promote sustainable use*

Target 5	By 2020, the rate of loss of all natural habitats, including forests, is at least halved and where feasible brought close to zero, and degradation and fragmentation are significantly reduced
Target 6	By 2020 all fish and invertebrate stocks and aquatic plants are managed and harvested sustainably, legally, and applying ecosystem-based approaches, so that overfishing is avoided, recovery plans and measures are in place for all depleted species, fisheries have no significant adverse impacts on threatened species and vulnerable ecosystems, and the impacts of fisheries on stocks, species, and ecosystems are within safe ecological limits
Target 7	By 2020 areas under agriculture, aquaculture, and forestry are managed sustainably, ensuring conservation of biodiversity
Target 8	By 2020, pollution, including from excess nutrients, has been brought to levels that are not detrimental to ecosystem function and biodiversity
Target 9	By 2020, invasive alien species and pathways are identified and prioritized, priority species are controlled or eradicated, and measures are in place to manage pathways to prevent their introduction and establishment
Target 10	By 2015, the multiple anthropogenic pressures on coral reefs, and other vulnerable ecosystems impacted by climate change or ocean acidification are minimized, so as to maintain their integrity and functioning

(continued overleaf)

Table 18.2 (*continued*)

Strategic Goal C: *To improve the status of biodiversity by safeguarding ecosystems, species, and genetic diversity*

Target 11	By 2020, at least 17% of terrestrial and inland water, and 10% of coastal and marine areas, especially areas of particular importance for biodiversity and ecosystem services, are conserved through effectively and equitably managed, ecologically representative, and well-connected systems of protected areas and other effective area-based conservation measures, and integrated into the wider landscapes and seascapes
Target 12	By 2020 the extinction of known threatened species has been prevented and their conservation status, particularly of those most in decline, has been improved and sustained
Target 13	By 2020, the genetic diversity of cultivated plants and farmed and domesticated animals and of wild relatives, including other socioeconomically as well as culturally valuable species, is maintained, and strategies have been developed and implemented for minimizing genetic erosion and safeguarding their genetic diversity

Strategic Goal D: *Enhance the benefits to all from biodiversity and ecosystem services*

Target 14	By 2020, ecosystems that provide essential services, including services related to water, and contribute to health, livelihoods, and well-being, are restored and safeguarded, taking into account the needs of women, indigenous and local communities, and the poor and vulnerable
Target 15	By 2020, ecosystem resilience and the contribution of biodiversity to carbon stocks has been enhanced, through conservation and restoration, including restoration of at least 15% of degraded ecosystems, thereby contributing to climate change mitigation and adaptation and to combating desertification
Target 16	By 2015, the Nagoya Protocol on Access to Genetic Resources and the Fair and Equitable Sharing of Benefits Arising from their Utilization is in force and operational, consistent with national legislation

Strategic Goal E: *Enhance implementation through participatory planning, knowledge management, and capacity building*

Target 17	By 2015 each Party has developed, adopted as a policy instrument, and commenced implementing an effective, participatory, and updated national biodiversity strategy and action plan
Target 18	By 2020, the traditional knowledge, innovations, and practices of indigenous and local communities relevant for the conservation and sustainable use of biodiversity, and their customary use of biological resources, are respected, subject to national legislation and relevant international obligations, and fully integrated and reflected in the implementation of the Convention with the full and effective participation of indigenous and local communities, at all relevant levels

Table 18.2 (*continued*)

Target 19	By 2020, knowledge, the science base, and technologies relating to biodiversity, its values, functioning, status and trends, and the consequences of its loss, are improved, widely shared and transferred, and applied
Target 20	By 2020, at the latest, the mobilization of financial resources for effectively implementing the Strategic Plan for Biodiversity 2011–2020 from all sources, and in accordance with the consolidated and agreed process in the Strategy for Resource Mobilization, should increase substantially from the current levels. This target will be subject to changes contingent to resource needs assessments to be developed and reported by Parties

Additional positive aspects of the CBD 2010 Target included:

1. Compared with 2002 when the Target was agreed, increased awareness among decision-makers of what biodiversity is and the serious situation facing our environment was apparent.
2. Many national governments gave greater focus to biodiversity since 2002 as a result of the 2010 Target, and developed sophisticated sets of indicators to measure impacts and progress.
3. There was significant mobilization of new audiences in support of biodiversity conservation, particularly local governments and business, for example through the Countdown 2010 initiative.
4. A number of new policy frameworks that support biodiversity conservation were adopted, for example the European Commission Biodiversity Strategy and Actions Plans.
5. Although the 2010 Target was clearly not met, some successes were undoubtedly apparent; these need to be understood in detail to assist future progress.

However, despite these very valuable benefits, there can be no denying that the CBD 2010 Target also presented some difficulties, as summarized in the next section.

The CBD 2010 Target – problematic aspects

Mace and Baillie (2007) reviewed the development of indicators for the 2010 Target. They scored the indicators against a number of criteria: the need to be relevant to the purpose; to distinguish between measures of pressure, state, and response; to design and validate the indicators in context; to ensure effective communication with relevant audiences; to turn lists of measures into simple or composite indicators; and to maximize the cost-effectiveness of the indicator process (see also Jones *et al.*, 2011, for this latter aspect). They concluded that urgent steps were needed to complete the indicator set, to reduce and refine the agreed measures. We would add the following aspects as being among the challenges and unintended consequences associated with the 2010 Target:

1. The target established no clear baseline from which to measure progress (a reduction in the rate of loss since when?).
2. By basing the target on 'rate of biodiversity loss' it was possible for a failure to appear as a success. This is not just an academic point. The catastrophic decline of many amphibian species due to the fungal disease chytridiomycosis has led to a severe drop in the IUCN Red List Index for these species (Stuart *et al.*, 2004). However, as the disease works through all the susceptible species (Bielby *et al.*, 2008), the rate of amphibian loss will slow, and technically speaking the CBD 2010 Target will be met. Yet not even the most cynical government would want to claim this as a success. Likewise, the numbers of fish in many marine fisheries have plummeted so severely (Worm *et al.*, 2006) that there are not enough individuals left to maintain the rate of loss, so the rate of decline will slow, thereby achieving the target.
3. The target was hardest to achieve in the most intact and biodiversity-rich areas of the world, in which any loss at all meant failure to meet the target.
4. For indicators based on percentage changes in a baseline value (such as the Living Planet Index; Loh *et al.*, 2005; see also Chapter 4) and the Wild Bird Index (Gregory *et al.*, 2005), a reduction in the rate of loss required the second derivative of the curve to be significantly different from zero. While this is relatively straightforward to calculate (Collen *et al.*, 2009) it is not exactly intuitive, particularly when trying to translate scientific procedure into a policy-friendly format.
5. A reduction in the rate of loss required indicators with at least three data points, rather than just two, and this greatly constrained the choice of available indicators.
6. The target focused on the negative – biodiversity loss – and not on achieving a positive goal. Providing a target with a positive outcome – for example, improved status of species or ecosystems or ecosystem services – might be a better approach, by focusing on success instead of failure.

There were also problems with implementing some of the CBD indicators listed in Table 18.1. For example:

1. Many of the indicators suffered from a lack of available information – even some of the more developed indicators, such as the Living Planet Index (Loh *et al.*, 2005) and the Red List Index (Butchart *et al.*, 2004) could have delivered a lot more if additional funding could have been provided to assemble the underlying data, in particular improving taxonomic and geographical coverage.

2. Some of the indicators identified for the 2010 Target were never developed (at least five remained undeveloped – most notably the indicator for access and benefit sharing). For others, only very limited data were available to enable them to progress, resulting in a high degree of uncertainty over the trends shown (e.g. three of the headline indicators were scored as low certainty; Convention on Biological Diversity, 2010). Of the indicators listed (Table 18.1), nine were fully developed, and 20 needed further development, including most of those relating to the provision of ecosystem services. The poor delivery of the ecosystem service-related indicators has led to the misperception in some communities that the CBD 2010 Biodiversity Target did not cover ecosystem services at all.
3. The 2010 indicator framework did not include metrics of wild genetic diversity, nor measures of eco-regional loss, biological community loss, nor loss of biological phenomena. Furthermore, some key threats such as climate change were very weakly covered by response indicators that were limited in scope.

The most serious issue, however, relating to the CBD 2010 Biodiversity Target relates to the fact that, despite its many virtues and flaws, it was not met (Butchart *et al.*, 2010; Convention on Biological Diversity, 2010). We explore why this was, and how the 2020 Targets were developed in light of this.

The 2010 Target – why was it not met?

There were a number of issues that prevented the 2010 Target from being achieved. From the outset, it was not realistic to expect such radical change within such a tight timeframe (the eight years from 2002 to 2010) (see Chapter 1). Stemming biodiversity loss is not a simple task because there are multiple drivers contributing to the trends that we observe. It is much easier to achieve targets based on simple cause-and-effect relationships, and much easier to understand if they are not met, why that might be. For example, combating ozone depletion required controlling a single factor – the proliferation of CFCs. For biodiversity maintenance, a much broader and more complex set of actions is needed, and some of the actions are politically costly because they impact directly on human lifestyles. For these reasons, the Target set was more ambitious than perhaps was realized at the time.

Furthermore, there has been no clear vision of what actually needs to be done (in terms of legislation, policies, site-level management, conventions, etc.), at either global or national levels, to achieve the Target. Part of the problem is that there is no clear understanding of how our metrics of biodiversity (measured through indicators) relate to policies and legislation, or how they will be affected by changes in policy. No-one ever told the governments what needed to be done; and of course, it turns out that no-one actually knew, except in the most general of terms. Only now are the first datasets becoming available that begin to help answer the following question: what exactly does each country need to do in order to meet the Target? This lack of specific identified actions, further constrained by an impossibly short 8-year timeframe, made reaching the target a distant dream.

Furthermore, the drivers of biodiversity loss (including climate change and inequitable economic growth) are getting worse much faster than the resources (including funding) are growing to support conservation (Millennium Ecosystem Assessment, 2005). The Millennium Ecosystem Assessment, reporting in 2005, showed that these drivers are growing in almost every ecosystem. The limited funding available for conservation gets proportionally smaller every year in relation to the size of the problem. We really are chasing a moving target. However, the fundamental problem is that biodiversity is inadequately valued in our current global economic paradigm. The costs of biodiversity loss are not adequately considered when economic decisions are made and developments are planned (environmental costs are treated as 'externalities'). So long as this remains the case, the drivers of biodiversity loss will continue to get worse.

Some other less convincing reasons were put forward to explain why the Target was not reached. For example, the wording of the Target has been described as unhelpful and negative, and this has apparently alienated some governments. Although improvements could have been made in the wording of the Target, it is too far-fetched to believe that this has hindered success in reaching it, especially when the more fundamental problems listed above are considered.

It is common to hear in policy-oriented discussions that a major problem with the CBD 2010 Target was that it pays insufficient focus to ecosystem services. The assumption seems to be that if there was more emphasis on ecosystem services, governments would pay greater attention to achieving the Target. The basis for this belief is hard to understand, especially as nine of the 22 indicators selected to measure progress (Table 18.1) relate to ecosystem services, with another three each on sustainable use and human well-being. As mentioned above, there is a problem with the performance of the indicators relating to ecosystem services, not least because the services themselves are very difficult to measure, but governments and others have understood that the Target encompasses ecosystem services from the outset. Interestingly, from its very beginnings the CBD has adopted an 'ecosystem approach'.

A final frequently-stated reason for failure to achieve the Target is that the governments were never actually serious about it in the first place. Proponents of this view can cite the decline in funding from public sources for biodiversity conservation since 2002 as evidence for their case. We believe that this view is too cynical. There are enough reasons listed above that explain why the Target was not met. We think that it is more likely that governments underestimated the size of the problem, did not have well-formed policies on what to do in any case, and were distracted by other worries, especially after the global recession hit in 2008–2009.

That said, we cannot let the governments off too lightly. It is a fundamental problem that biodiversity is not a global priority, and is rarely a national one. As mentioned above, there is a vast and growing mismatch between the funding needed to tackle biodiversity decline, and the funding available. Climate change, not biodiversity, has pushed its way to dominate the environmental agenda, arguably at the expense of other conservation priorities, even though habitat loss remains unequivocally the largest determinant of high extinction risk (Hoffmann *et al.*, 2010), and with projections of human population growth over the coming 40 years, is likely to remain so. A good example of the low priority accorded to biodiversity was evident in a seven-page memorandum S.N.S. was shown in 2009 from the US State Department to the US Congress, requesting the latter to prioritize the ratification of a number of international agreements. The Convention on Biological Diversity was not on the list, and this from a government that is generally believed to be relatively favourable to environmental concerns.

How do we raise the priority of biodiversity?

So what should we do? There is unlikely to be a single silver bullet strategy for raising the priority of biodiversity, but a concerted, multifaceted approach could be the answer. The strength of the biodiversity concept is that it is highly integrative, but this is lost when attention becomes focused on just one component of it. So it is not surprising that we probably need to advance an agenda on several fronts simultaneously to ensure that the biodiversity message is heard. Creating a mechanism through which national-level links can build to global-level indicators is a great first step. The following are some ideas to explore.

Biodiversity underpins ecosystem services

We should redouble our efforts to communicate the fact that biodiversity underpins the delivery of ecosystem services for human benefit. There is very limited understanding of how much biodiversity could be lost without seriously jeopardizing ecosystem services, and establishing those boundaries must be a leading research aim; the answer is likely to differ substantially across the different services that we perceive as being important (Mace *et al.*, 2012). However, there is evidence that the loss of whole functional groups of species can negatively affect overall ecosystem resilience (e.g. Folke *et al.*, 2004), that restoration of biodiversity can greatly enhance ecosystem productivity (Worm *et al.*, 2006), and that regions of high priority for biodiversity conservation often also have high ecosystem service value (Turner *et al.*, 2007). Keeping future options open clearly depends on maintaining the full range of biodiversity: 'to keep every cog and wheel', wrote Aldo Leopold, 'is the first precaution of intelligent tinkering' (Leopold, 1953).

Of course, ecosystem services for humans should not be an exclusive argument for conservation, but in recent years there has been a trend in this direction. Sometimes, there seems to be an expectation that if only we could get senior decision-makers convinced of the 'ecosystem services' argument, they will immediately turn the funding taps on, and biodiversity will rise up the political agenda. Experience suggests that such thinking is naive. Arguments based on ecosystem services go back at least as far as the 1972 UN Conference on the Human Environment. Ecosystem services were prominently emphasized in the World Conservation Strategy (1980), the Brundtland Report (1987), and Agenda 21 (1992), and form the basis of the excellent Millennium Ecosystem Assessment report released in 2005 (Millennium Ecosystem Assessment, 2005). There is no evidence that decision-makers have been captivated by this link, even though much of it is scientifically valid. Indeed, the powers-that-be have had over 30 years to respond to the logic of safeguarding ecosystems services, and all the evidence shows that they have not. Our point is not that the 'ecosystem services' argument will not work in the political arena (although one could be cynical about the effectiveness of any argument that relies on long-term benefits beyond those captured within the timespans of political cycles), rather, this argument is not sufficient on its own. It is an essential argument, and we must have it in our toolkit, but we need to integrate it with other arguments, such as those outlined below.

The economic costs of inaction

The true costs of conserving biodiversity effectively are high, much higher than the current conservation investments made by governments, donors, the corporate sector, and civil society combined. However, what are the costs of not conserving biodiversity? Failure to act also has significant costs. There is growing evidence that the cost of inaction in terms of conservation is likely to be much greater than the cost of action. For example, the loss of coral reefs, which is likely to happen in the coming decades if we continue on our current path, will have enormous costs in terms of lost livelihoods from both fisheries and tourism, and in terms of the greatly increased need for coastline protection. One important current study, The Economics of Ecosystems and Biodiversity, has brought out a number of reports that are delivering significant results (TEEB, 2010). For example, they report that conserving forests avoids greenhouse gas emissions worth US$3.7 trillion to the global economy. Halving deforestation rates by 2030 would reduce global greenhouse gas emissions by 1.5–2.7 Gt CO_2 per year, thereby avoiding damages from climate change estimated at more than US$3.7 trillion in net present value (NPV) terms (this figure does not include the many co-benefits of forest ecosystems). There are signs that the global community might at last be willing to pay for at least one ecosystem service, carbon sequestration, as demonstrated by the ongoing climate negotiations around REDD (Reduced Emissions from Deforestation and Degradation). The cynic might argue that rich countries are prepared to pay for forest conservation in the tropics because rising CO_2 levels are threatening lifestyles in the wealthy world. But it is at least a step in the right direction, and driving home the important argument that inaction is very far from a cost-free option.

Public connection with biodiversity

There is a growing disconnect between the view of conservation among the general public, and the 'professional' view in conservation. The launch of the Millennium Ecosystem Assessment (MA) in 2005 was a stark example of this. The headline message of the MA was all about ecosystem services, and the fact that these are seriously at risk almost everywhere. But the media tended to pick up on the trends in species, part of just one of the MA chapters. We have found this to be true time and time again, particularly in relation to the IUCN Red List of Threatened Species. Put out a press release on ecosystem services at risk, and there is often modest media pick-up at best. Put out a story on species and media interest is frequently very broad. To caricature the type of interaction: talk about freshwater ecosystems and you get a yawn; tell them about a frog that lives in the water, and they want to listen. Yet much of the conservation community wants to get away from a focus on species in order to influence decision-making. Indeed, some of the greatest opposition to a species focus in conservation comes from within conservation organizations. Ultimately, governments respond to public opinion; the conservation community as a whole must act to exploit the public's ability to identify with species, and mobilize this around the extinction crisis. While we have to use every tool that is available to us, we must get over our embarrassment with species. It is fine to lobby policy-makers on ecosystems services; but we must also mobilize the public to besiege governments on extinction rates. We don't need to choose between the two; we can do both. Indeed, we must do both.

How will the Aichi 2020 Biodiversity Target help avoid pitfalls post-2010?

The 2020 Targets, structured under five strategic goals, provide a more advanced structure than the 2010 Target (Table 18.2). To what extent does it, and should it, emphasize all of the dimensions of biodiversity, not just some of them? To get a feel for the imbalance in thinking in formal documents on biodiversity, we counted the number of times that the words 'species' and 'ecosystems' appeared in a 21-page note circulated by the Executive Secretary of the Convention on Biological Diversity on 5 June 2009, in the lead-up to the negotiations on the 2020 targets, entitled 'Revision and updating of the strategic plan: synthesis/analysis of views' (UNEP/CBD/SP/PREP/1). The word species appeared twice, both times in footnotes, and both times referring to invasive alien species, not to species more generally. The word ecosystems appeared no less than 62 times. Clearly, there is a serious imbalance here. Incidentally, the word 'biodiversity' appeared 162 times, but the word 'genetic' only once. The negotiators who drafted the CBD text defined biological diversity in a highly integrative way, focusing on genes, species, and ecosystems, and our fear is that the subsequent evolution of the Convention may have lost that radical, integrative spirit. There is something in human nature that makes us more comfortable if we retreat into our narrow, sectoral interest groups. In the post-2010 Target world, we must recover the integrative view of those who drafted the CBD text, and of those who drafted the 2010 Biodiversity Target, and look especially closely about where there could be more profitable links between science and policy (see Chapter 13).

It is also important to ensure that fundamental conservation principles are not conceded as the implementation of the 2020 CBD Strategic Plan proceeds. While we now have the general framework (Table 18.2), the details on metrics, indicators, and the links to the emerging science base are still being determined. One formulation of a post-2010 target that was in wide circulation before Nagoya, where the Aichi Targets were agreed, used the phrase 'to avoid dangerous biodiversity loss' or 'not exceeding dangerous limits' and the notion of tipping points is still routine, if rarely demonstrated. The implication of this was clear. Biodiversity loss is defined in terms of whether or not it is likely to be dangerous to humans, and that the world should be comfortable with any loss of biodiversity that is not dangerous to humans. Leaving aside the fact that it is probably impossible to define whether or not such loss is likely to be dangerous, this formulation throws away the notion of the intrinsic value of biodiversity, which is enshrined in the 2010 CBD Target. We should be concerned about all biodiversity loss, whether or not it is dangerous to humans.

Another common discussion in relation to the 2010 Target was the search for a single composite indicator that does it all. However, the notion of a single perfect indicator misunderstands what biodiversity is all about. Biodiversity is a complex concept and different indicators reveal different facets that are important. For example, the Living Planet Index (see Chapter 4), the Red List Index (see Chapter 2), ecosystem service indicators (e.g. see Chapter 5), impact indicators (e.g. see Chapters 6–9), and others all measure different, complementary, and essential elements of biodiversity. Just as there is a suite of indicators in the economic world (Gross Domestic Product, Dow Jones Index, inflation rate, unemployment rate, etc.), a suite of biodiversity indicators is something that is inevitable from the very nature of biodiversity itself. No one measure of biodiversity can tell us what is happening to all aspects of what is, by definition, a multifaceted term (Purvis and Hector, 2000).

Concerns have been raised about the costs of too many indicators. While we need a degree of realism here, and cost-effectiveness and optimality should certainly be part of any indicator design, there is a greater concern that the conservation world has a habit of thinking too small. Governments need to be realistic about the cost of monitoring biodiversity. Currently, the investment of the world's governments in monitoring biodiversity is trivial, with much of data collection being carried out by volunteers (Pereira *et al*., 2010; see also Chapter 10). An enormous proportion of the costs associated with delivering the 2010 indicators was borne by non-governmental organizations (NGOs). This situation cannot be allowed to continue in the development of 2020 Target indicators, and money must not be the main disincentive in implementing the suite of indicators that the Aichi Target requires. A much larger challenge is likely to be the design of indicators so that they can actually be implemented. In particular, indicators need to be developed for which data can feasibly be generated and collated (see Chapter 14). There needs to be investment in developing new and expanding existing monitoring schemes on the ground to collect new data (e.g. see Chapter 16).

So, taking into account the above concerns, how will the 2020 framework of goals perform, and how do we integrate national-level monitoring to achieve global-level commitments?

Linking targets to actions

The 2010 indicators were not organized according to any conceptual framework. Many have called for a new approach, specifically incorporating indicators into the Drivers, Pressure, State, Impact, Response (DPSIR) framework (e.g. Sparks *et al*., 2011), which to some extent has been done (though the targets are not comprehensive). Some appropriate criteria for useful and effective indicators are that they need to be:

- measurable;
- sensitive over time;
- scaleable between global and national levels;
- as far as possible, based on data that already exist, or for which there are realistic and feasible programmes to bring the data together (a number of the existing indicators have fallen at this hurdle).

While it would be pertinent to avoid indicator proliferation under the new target (Jones *et al*., 2011), care must be taken that the opportunity to put meaningful baselines in place should not be missed. The new mission under the Aichi Target focuses not on results to be achieved, but on actions to be undertaken, though the 20 individual targets associated with the five strategic goals identified, are largely results-oriented. In order to link targets to actions, profitable links between science and policy should be exploited. However, there is clearly a trade-off between comprehensiveness of coverage and practicality of implementing indicator monitoring programmes (see Chapters 14 and 15).

A lack of capacity exists for countries to meet reporting obligations for international biodiversity agreements, particularly in less developed countries. Their job is the largest, with countries least able to afford conservation actions located in the world's richest areas for biodiversity (Zamin *et al.*, 2010). Consequently not only should access to funding to implement active conservation, particularly for such countries, be a main focus of the CBD and its funding mechanisms, but also significant funding should be directed to aid such countries in the implementation of monitoring networks, and development of national-level indicators. In recognizing that setting such targets is a political process that reflects national perceptions of both the relative importance of different aspects of biodiversity and their relative vulnerability, we must ensure that open-access common pool resources, such as the high seas, which are especially vulnerable, are also accounted for.

While the 2020 targets provide an important mechanism by which to assess global biodiversity change, ensuring that appropriate monitoring systems are in place and translating monitoring results into effective conservation on the ground remains a major global challenge for a number of reasons, including financial and technical capacity constraints and policy and legal barriers. Developing an understanding of how countries can most efficiently assess national biodiversity change, and helping bridge the gap between international commitments and local action remains a key challenge in the conservation of biodiversity.

Conclusion

The adoption of the 2010 Target by governments and agencies was visionary and courageous. As a consequence, much was learned about biodiversity indicators in the period between the setting of the target in 2002, and the reporting carried out in 2010. Underlying datasets have improved (and continue to do so), and the general understanding of global and regional patterns and trends in biodiversity is hugely greater than even five years ago. But as we move into measuring our progress towards the 2020 Aichi Targets, investments in monitoring still need to grow, to fill both gaps in knowledge and in national-level capacity. Most importantly, with the new set of Targets being ambitious, but achievable, greatly enhanced commitment to biodiversity conservation is required. We have put forward some suggestions of how we might move biodiversity up the political agenda. In particular, this requires us to exploit all aspects of the multifaceted concept of biodiversity. The tendency to emphasize one aspect of biodiversity at the expense of others is self-defeating and we need to become comfortable with an integrated, cross-sectoral approach.

Acknowledgements

B.C. is supported by the Rufford Foundation.

References

Balmford, A., Bennun, L.A., ten Brink, B., *et al.* (2005) The Convention on Biological Diversity's 2010 target. *Science*, 307, 212–213.
Bielby, J., Cooper, N., Cunningham, A.A., Garner, T.W.J., and Purvis, A. (2008) Predicting declines in the world's frogs. *Conservation Letters*, 1, 82–90.
Butchart, S.H.M., Stattersfield, A.J., Bennun, L.A., *et al.* (2004) Measuring global trends in the status of biodiversity: Red List Indices for birds. *PLoS Biology*, 2, 2294–2304.
Butchart, S.H.M., Walpole, M., Collen, B., *et al.* (2010) Global biodiversity decline continues. *Science*, 328, 1164–1168.
Collen, B., Loh, J., Whitmee, S., McRae, L., Amin, R., and Baillie, J.E.M. (2009) Monitoring change in vertebrate abundance: the Living Planet Index. *Conservation Biology*, 23, 317–327.
Convention on Biological Diversity (2010) Global Biodiversity Outlook 3. UNEP.
Folke, C., Carpenter, S., Walker, B., *et al.* (2004) Regime shifts, resilience, and biodiversity in ecosystem management. *Annual Review of Ecology, Evolution and Systematics*, 35, 557–581.
Gregory, R.D., van Strien, A., Vorisek, P., *et al.* (2005) Developing indicators for European birds. *Philosophical Transactions of the Royal Society of London B*, 360, 269–288.
Hoffmann, M., Hilton-Taylor, C., Angulo, A., *et al.* (2010) The impact and shortfall of conservation on the status of the world's vertebrates. *Science*, 330, 1503–1509.
Jones, J.P.G., Collen, B., Atkinson, G., *et al.* (2011) The why, what and how of global biodiversity indicators beyond the 2010 target. *Conservation Biology*, 25, 450–457.
Leopold, A. (1953) *Round River*. Oxford University Press.
Loh, J., Green, R.E., Ricketts, T., *et al.* (2005) The Living Planet Index: using species population time series to track trends in biodiversity. *Philosophical Transactions of the Royal Society of London B*, 360, 289–295.
Mace, G.M. and Baillie, J.E.M. (2007) The 2010 Biodiversity Indicators: challenges for science and policy. *Conservation Biology*, 21, 1406–1413.
Mace, G.M., Norris, K., and Fitter, A.H. (2012) Biodiversity and ecosystem services: a multi-layered relationship. *Trends in Ecology and Evolution*, 27, 19–26.
Millennium Ecosystem Assessment (2005). *Ecosystems and Human Well-being: Biodiversity Synthesis*. World Resources Institute, Washington, DC.
Pereira, H.M., Proença, V., Belnapp, J., *et al.* (2010) Global biodiversity monitoring: filling the gap where it counts the most. *Frontiers in Ecology and the Environment*, 8, 459–460.

Purvis, A. and Hector, A. (2000) Getting the measure of biodiversity. *Nature*, 405, 212–219.
Sparks, T.H., Butchart, S.H.M., Balmford, A., *et al.* (2011) Linked indicator sets for addressing biodiversity loss. *Oryx*, 45, 411–419.
Stuart, S.N., Chanson, J.S., Cox, N.A., *et al.* (2004) Status and trends of amphibian declines and extinctions worldwide. *Science*, 306, 1783–1786.
TEEB (2010) The Economics of Ecosystems and Biodiversity: *Mainstreaming the Economics of Nature: a Synthesis of the Approach, Conclusions and Recommendations of TEEB.*
Turner, W.R., Brandon, K., Brooks, T.M., Costanza, R., Da Fonseca, G.A.B., and Portela, R. (2007) Global conservation of biodiversity and ecosystem services. *BioScience*, 57, 868–873.
UNEP (2002) Report on the sixth meeting of the Conference of the Parties to the Convention on Biological Diversity (UNEP/CBD/COP/20/Part 2) Strategic Plan Decision VI/26. Available at: https://www.cbd.int/decision/cop/?id=7200.
Walpole, M., Almond, R., Besançon, C., *et al.* (2009) Tracking progress towards the 2010 biodiversity target and beyond. *Science*, 325, 1503–1504.
Worm, B., Barbier, E.B., Beaumont, N., *et al.* (2006) Impacts of biodiversity loss on ocean ecosystem services. *Science*, 314, 787–790.
Zamin, T., Baillie, J.E.M., Miller, R., Rodriguez, J.P., Ardid, A., and Collen, B. (2010) National Red Listing beyond the 2010 Target. *Conservation Biology*, 24, 1012–1020.

Index

Note: page numbers in *italics* refer to figures, those in **bold** refer to tables and boxes.

Aichi targets *see* Convention on Biological Diversity, 2020 targets
alien species *see* invasive species
apples 205–6
Atoll Ecosystem Conservation Project **375–6**, 387

Bangladesh **367–8**
bats 215
   call characteristics 215, **232–3**
   comparison with other monitoring methods 224–5
   ecological role 215–16
   European biodiversity *227*
   population counting 216–17
   species identification 223–4
   sustainable monitoring 220
   United Kingdom 225–6, 322
   see also ultrasound monitoring
Batsound software 231
bees 5
Belgian Forum on Invasive Species 146–7
Bhutan 320, 366, **369–70**, 387
biocapacity 191–2
biodiversity indicators 7–9, 9–10
   global 403–6
      Aichi targets 428
      national use 415–16
      for production 406–7
      purposes 418–19
   invasive species 413
   national 418–19
      capacity issues 416–18
      data sources 413
      production 409–11
      types 413
      use of global indicators 416
   see also Living Planet Index; Red List Index; Wildlife Picture Index
Biodiversity Indicators Partnership 403, 406
Biodiversity Intactness Index 303
biodiversity monitoring *see* monitoring programmes
Biodiversity Observation Network 302
Biodiversity Program Study 2004 393
biodiversity sampling 49–50
biodiversity targets 421–2
biological invasions *see* invasive species
BirdLife International 132, 147
birds
   acoustic monitoring 214–15
   bioclimatic studies 125
   biodiversity 47
   body size distributions **48**
   climate change indicators 130–1

Biodiversity Monitoring and Conservation: Bridging the Gap between Global Commitment and Local Action, First Edition. Edited by Ben Collen, Nathalie Pettorelli, Jonathan E.M. Baillie and Sarah M. Durant.
© 2013 John Wiley & Sons, Ltd. Published 2013 by John Wiley & Sons, Ltd.

birds (*Cont'd*)
  efficacy of Wildlife Protection Index 64
  Global Red List Index **24**
  human utilization 163, 165–7, 167–9, 168
  invasive species effects 149
  Living Planet Index 85
  participatory monitoring 343–4
  population monitoring, climate change and 125
  power monitoring of analysis 338–9
  Red List indices 163, 167–8, *170*
  remote sensing 105, **110**
  satellite imaging **110**
  Venezuela **28**
blue tits 105
bobcat 161
Bolivia 193–4
Brazil 411
British Trust for Ornithology 343–4
buffalo 108
Bukit Barisan National Park 52–3, 57–8
bush meat hunting 179
butterflies 131

camera traps 52–3, 63
carbon dioxide 97
Cervus elaphus 108
cheetahs 53, **374–5**, 387
China 25–6, 411
chlamydia 336
Circumpolar Biodiversity Monitoring Programme (CBMP) 84, 415
climate change 8, 120–1, 304
  criteria-based models 126–7, 128–9
  e 128
  envelope models 125–6, 128
  impact on species *121*
  indicators 121–2
  meta-analyses 128
  model validation 132–3
  species impact, population change comparison 123–5
  susceptibility frameworks 127, 129–30
  temperature 97–8
  theoretical frameworks 127–30

Climate Change Indicator for Birds in Europe 130–1
communication 328
Community Temperature Index 131, *131*
community-based conservation (CBC) 266
  displacement 273–4
  local knowledge 272–3
  monitoring methods 276–8
    interviews 276–7
    secondary data 278
  socio-economic impact monitoring motivators 268–70
  scales of impact 271–2
community-based natural resource management (CBNRM) 321–2
conifers, Global Red List Index **24**
conservation action 1–2, 35–6
  see also community-based conservation
Conservation Assessment and Management Plans (CAMP), South Africa 32, 33–4
Conservation Breeding Specialist Group 32
Conservation International 214
conservation management objectives 252–3
ConserveOnline 391
consumption measures
  ecological footprint calculation 191–4
  equivalence 200–2, 204–5
  human footprint map 194–7
  surrogacy 205
  trade and 197–8
Convention on Biological Diversity (CBD) 2, 21, 25, 150, 152, 214, 314, 316, 330
  2010 targets 2, 45–6, 76–7, 84, 161, 214, 291–2, 295–6, **404–5**, 423
    positive aspects 422–7
    problematic aspects 427–9
    reasons for failure 429–31
  2020 targets 3, 7–8, **293–5**, 296–305, *301*, 419, **425–7**
    baselines 303–4
    state changes 304

target adjustment 298
target setting 297–8
context for targets 299–300
habitat loss 99
implementation of indicators 428–9
invasive species 150, 152
national biomonitoring indicators
  production 409–11
  use 407–9
red, green and blue targets 300
target iteration 298–9
targets within a continuing
  process 300–1
Convention Concerning the Protection of
  the World Cultural and National
  Heritage 314–16, **315**
Convention on International Trade in
  Endangered Species (CITES) 37,
  182, **315**, 330
Convention on the Law of the Sea 149
Convention on Migratory Species
  (CMS) 24, 37, **315**
Convention on Wetlands of International
  Importance **315–16**, 330
COP10 conference 292, 299
corals **24**
Costa Rica 411
Countdown 2010 292
crabs **24**
crayfish **24**, 340
cycads **24**

DAISIE 142–5, 154–5
data collection 46, 322–3
  quality verification 342
  storage 341–2
deforestation 432
development 265–6, 273–4
Development Assistance
  Frameworks 356–7
disease 304
disease risk 304
DPSIR model 295, 435
dragonflies **24**
Durrell Wildlife Conservation Trust
  341–2

ecological footprint calculation 190,
  191–4, **193**, *203*
  human footprint and 200–2
  limitations 205–6
economic effects 432–3
ecosystem services 431–2
ecosystem shifts 304
El Nacional 30
electromagnetic radiation 100
elephants 57–8, 108, **110**
Endangered Wildlife Trust (EWT) 32
Enhanced Vegetation Index (EVI) 109–10
envelope models 125–6, 128
ethical accountability 269–70
Euderma maculatum 215
EUROBATS 216
Europe 5, *203*
  invasive species trends 142–5, *144*, *145*
Evolutionarily Distinct Globally
  Endangered (EDGE) species 58
extinction 1, 52–3
  climate change and 120, 122
  Living Planet Index 87
  Proportion of Area Occupied 253–4
  rates *see* species extinction rates 1
  Red Lists and 21, 37
  risk, Red List Index 22–3

Financial Sustainability Scorecard 360
financing 393–5
fish
  Global Red List Index **24**
  human use 176
  human utilization 160
  sustainable use 175–7, 178–9
  Venezuela **28**
FishBase 179
flycatchers 47
Food and Agriculture Organization
  (FAO) 8, 159, 160
Forest Stewardship Council 174–5, **184**
forests 160
  biodiversity indicators 413
  deforestation 432
  fragmentation 248
  sustainable use 174–6

France 225
frequency division 221
freshwater environments, invasive species *148*
functional diversity 303

General Circulation models 125
generalized additive models (GAM) 55–7, 74–5, 87
  confidence limits 56
GeoEye-1 101–2
Geographical Information System 33–4, 222–3
Geological Survey (US) 102
Global Biodiversity Outlook 299, 403–5
Global Earth Observation System of Systems 302
Global Environment Facility (GEF) 12, 350, 351–5
  Enabling Activities 353–4
  GEF-4 353, 360, **361**
  GEF-5 354, 360, **362–3**
  increasing contribution to broader biodiversity monitoring 388–92
    cost-effectiveness 389–90, 393
    project mentoring 392
    review and codification of tools 390–2
  monitoring and evaluation 358–65
  projects *see* United Nations Development Projects
  Scientific and Technical Advisory Panel 390
  technical support 393–5
  see also United Nations Development Fund
global hectares 192, 198–9
Global Observation Research in Alpine Environments (GLORIA) 132
Global Red Lists *see* Red Lists, global
Global Wild bird Index **404**
governance 329
great tits 105
Green Balkans 229

greenhouse gases 97, 121
grey-crowned crane 85
grivet monkey 108
Group on Earth Observations 302

habitat loss 8, 60–1, 248
habitat monitoring 99
heterodyne detectors 221
house sparrow 162
Human Appropriation of Net Primary Production 303
human footprint map 190, 194–7
  ecological footprint calculation and 200–2
  Europe *203*
  limitations 205–6
  United States *204*
Human Influence Index 196–7, *196*
human population 97, 189–90
  animal use 159–60
  consumption indicators *see* consumption indicators
  displacement by conservation programmes 273–4
  individual action 330
  use of forests 160
  use of wild animals 161–2
  utilized species *see* utilized species
  well-being 4–5
  see also community-based conservation
Hungary 199, 205–6
hunting 180–1, 284

Ikonos 101
India 327, 342, **372–3**
Indicator Bats Program 226–38
  acoustic recording equipment 230, 235, 240
  data analysis 231
  data storage 232–4
  future prospects 239
  geo-reference data 230–1, 236–7
  monitoring methodology 226–8
  project outcomes 234–8
  sustainable monitoring 228–9

ultrasound detection 229–30
   weaknesses 239–41
individual action 330
Indonesia 57–8
Intergovernmental Panel on Climate Change 299
international agreements 149–50, 176–7, 314–16, **315**
   cooperation 325
   see also Convention on Biological Diversity
International Panel on Climate Change (IPCC) 125–6
International Plant Protection Convention 149
International Union for the Conservation of Nature 160
International Union for the Conservation of Nature (IUCN) 19–20
interviews 277
invasions *see* biological invasions
invasive animal species, definition 152
invasive species 138–9, 304
   data availability 140–1, *141*, 153
   indicators
      DAISIE 142–5
      development challenges 151–4
      invasive species per country 139–47
      SEBI2010 141–2
   international agreements 149–50
   recommendations 154–5
   terminology 153, 154
   trend monitoring 141–2, *144*, *145*, 149–51
   Red List species 147–9
invertebrates
   invasive 140
   Venezuela **28**
Iran **374–5**, 387–8
Ireland, bat monitoring 225

Japan-Australia Migratory Birds Treaty **367–8**
Johannesburg Plan 292

Kenya 63, 266
key informants 280–1
Kyoto Protocol 111

Laikipia 63
Landsat 102–3, 107
Laos (Lao PDR) 63
Lear's macaw 167
leopard 161
less developed countries (LDC) 272, 277
LIDAR 105–6, 111
LINKPA project 366, **369–70**, 387–8
Living Planet Index 71–3, 122, **184**, 225, 321, **404**, 416, 428
   data sources 74, 89
      integrity 80–1
   geographical variation 89–90
   global results 77–81
   implementation
      regions, nations and thematic groups 81–5, **82**
      Uganda 85–7
   interpretation 77
   methods 73–6
   strengths 87–8, **88**
   utilized species 169–73
   weaknesses 88–90, **88**
Long Range Transboundary Air Pollution Convention (LRTAP) 296–7
Lyme disease 304

Madagascar 53, 340
Makerere University Institute of Environment and Natural Resources 84, 85
Maldives **375–6**, 387
mammals 57–8, 108
   biodiversity 47
   body size distribution **49**
   Global Red List Index **24**
   human utilization 169
   invasive 143–5, *145*
   Red List indices *170*
   remote imaging 108, **110**
   South African Red Data Book 31–6
      conservation status 34–6, **34**

mammals (Cont'd)
  Venezuela 28
  Wildlife Picture Index 53–5
  see also bats
Management Effectiveness Tracking Tool (METT) 360
marine biosphere reserve 372
marine environments, invasive species 148
Marine and Stewardship Council 181
Material Energy Flow Accounting 191
Maximum Entropy Modelling 236–7, *238*
Mediterranean habitats 82, *83*, 84, 325
  invasive species *146*
meta-analyses 128
methane 97
migratory species, Living Planet Index 84
Millennium Development Goals (MDG) 21, 292, 355, 405–6
Millennium Ecosystem Assessment 161–2, 299, 302–3, 327, 430, 433
Milne Bay Community-Based Marine and Coastal Conservation Project 382–3
MiniDisc format 230, 235
money 201
Mongolia 37
monitoring programmes 316–18
  differences in capacity 21–2
  global *see* global biodiversity monitoring
  goals 318–21
  national *see* national biodiversity monitoring
  plan design 321–2
  power analysis *see* power analysis
  programme stages **319–20**
  satellite 99–100
  strategic development 317–18
Montreal protocol on ozone-depleting substances 297
Mpala Research Centre 58
multivariate analysis 223

Namibia 411
NASA 102
National Audit Office 336
National Bat Monitoring Programme 224–5
national biodiversity monitoring
  communication and dissemination **320**, 328
  data analysis **319**, 323–4
  data collection **319**, 322–3
  data storage **319**, 323
  goals 318–21
  governance and **329**
  human/financial capacity **319**, 324–5
  individual action 330
  institutional frameworks **319**, 325–6
  plan design **319**, 321–2
  production for CBD 409–11
  standardization **319**, 324
  trust and credibility **320**
  use in CBD 407–9
National Biodiversity Network 225
National Biodiversity Strategies and Action Plan (NBSAP) 318, 418
National Environment Research Council (UK) 342
national legislation
  invasive species 150–1
  monitoring programmes and 316–17
  socio-economic impact assessment of projects 268–9
Nepal 376–9
NOAA/AVHRR satellites 107
noctule bats 218, 236
Nordic Biodiversity Indicators 415
Normalized Difference Vegetation Index, animal data and **110**
Normalized Difference Vegetation Index (NDVI) 106–10, *109*

occupancy methods 251–2
  Proportion of Area Occupied 253–4
  species richness 260
Official Development Assistance (ODA) 352
Ojani, Juhani 26
ozone-depleting substances 297

Pakistan **379–81**
Papua New Guinea **382–3**
participatory monitoring 284, 340–1, *341*
  birds 343–4
Participatory Rural Appraisal 279–81
PBSR framework 162
PERSEVET 229
pipistrelle bats 217–18, 236, *237*
plants
  bioclimatic studies 125–6
  invasive 140
  see also forests
policy commitments 2
population change 85–6
  birds 130–1
  causal identification 79
  climate change comparisons 123–5
  invasive species 140–7, 144–5, *145*, 151–4
  Living Planet index 79, 88–9
  utilized species 179–80
population counts, *see also* Proportion of Area Occupied
power analysis 217–19, 228, 335–6
  examples 338–9
  factors affecting power 336–7
  resource limitations 339–41
    data storage 341–2
  sample size 336
power relations 327
PRESENCE software 55, 57, 60
Pressure-State-Response-Benefit framework 162
priority raising 431–3
Programme on Reducing Emissions from Deforestation and Forest Degradation 99
project mentoring 392
PROMEBIO 415
Proportion of Area Occupied (PAO) 250, 253–4
  assumptions 256–60
  causes of bias 257–8
  imperfect detection 253, 254–5

methods 254–5
repeat surveys 255–6
Protected Areas 265, **354**, 357
Protocol on Environmental Protection to the Antarctic 149–50
public awareness 433

QuickBird 101

RADAR 104–5, 111
Ramsar convention **315–16**, 330
REBIOMA project 341
Red Book of Venezuelan Fauna 26–30
red deer 108, **110**
Red List 20, 122, 129, 162, **184**, **404**, 415, 428
  amphibians *170*
  applications 37–8
  birds *168*
  definition 19–20
  global results 23–5
  guidelines 20–1
  impacts of invasive species 147–9
  implementation 25–6
  Index calculation 22–3
  national indices 25–6
    Venezuela 27
  South Africa **35**
  strengths 36–8
  utilised species 165–7, 183
    food and medicine 169
  Venezuela 26–30
  weaknesses 38–9
Reducing Emissions from Deforestation and Degradation plus Poverty Alleviation (REDD+) 269
regional biodiversity indicators 414–15
reindeer 108
remote sensing 100–1
  data storage 323
  Normalized Difference Vegetation Index 106–10
reptiles
  Global Red List Index 24
  Venezuela **28**

resource exploitation 8–9
response indicators, utilized species 181–3
Romania 228–9
Romanian Bat Protection Association 226

sample size 336
Sariska Tiger Reserve 342
satellites **104**
　advantages for habitat monitoring 99–100
　aerial imaging 101–2
　Landsat 102–3
　limitations 111
　principle of operation 100–1
　RADAR/LIDAR imaging 104–6
Sea Around Us project 179
SEBI2010 141–2
Slender Loris Conservation Programme 248, 261–2
smartphones 240
socio-economic impact monitoring 268–70
　local knowledge 272–3
　participatory monitoring 284
　Participatory Rural Appraisals 279–81
　quantitative survey 281–2
　variables, context 270–1
Soil-Adjusted Vegetation Index (SAVI) 109
South Africa 411
　Red Data Book 31–3
　　assessment process 33–4
　　conservation actions 35–6
　　status of mammals 34–5
South African Mammal CAMP 33
species (mis)identification 257, 339–40
Species Temperature Index 131
SPOT system 101–2
Sri Lanka 261, **384–6**
　Slender Loris Conservation Programme 248–51, 261–2
state change indicators 304
Steppe Forward Programme 58–9
Strahl, Stuart D. 26

Strategic Plan for Biodiversity 419
straw-coloured fruit bat 85
Sumatran short-eared rabbit 52–3
susceptibility frameworks 127, 129–30
Sustainable Livelihoods Assets (SLA) framework 274–6, *275*
sustainable use
　fish 175–7
　indicators 178–80
　timber 174–5
Sweden, Red List 25
synthetic aperture RADAR (SAR) 105
synthetic-aperture RADAR (SAR) 105

Tanzania 271–2, 284, 322, 326
Tanzanian Wildlife Research Institute 326
taxon datasheets 33
TEAM programme 58
territory ranges 123–4
tiger tail seahorse 161
tigers 57, 64, 342
timber products 160
　sustainable use 174–5
time expansion 221–2, 229
topi 108, **110**
Tour du Valat 84
tourism 432
trade 160, 277
　consumption measures and 197–8, 201–2, 205–6
　wildlife 160–1, 180–1
　see also socio-economic impact monitoring
trade chain 164
trees 24
trend analysis 336–7
　biological invasions 141–2
　Living Planet index 77
　Wildlife Picture Model 55–7
trophy hunting 180
Tropical Ecology Assessment and Monitoring Network (TEAM) 132, 214, 214–15

tropical habitats 64
tropical regions
  biodiversity 45
  species sampling 49–50
  see also Wildlife Protection Index
trust 327–8
TRY 303
type I and II errors 337

Uganda 85–7, 411
UK Bats and Roadside Mammals Survey 227–8
UK Tracking Mammal partnership 344
ultrasound monitoring 214–16
  analysis and species identification 223–4
  comparison with other monitoring 224–5
  data storage and management 224
  detection technologies 220–1, 229–30
  frequency ranges of bat calls **232–3**
  methodology 216–19
  power analysis 217–19, 228
  species identification 240–1
  sustainability 220
  see also Indicator Bats Program
United Kingdom
  bat monitoring 225–6, 322
  invasive species 146–7, *147*
United Nations Development Programme (UNDP) 350–1, 355–8, 395
  direct monitoring 365, 387–8
  monitoring and evaluation 358–65
  projects
    Bangladesh **367–8**
    Bhutan 366, **369–70**, 387–8
    India **372–3**
    Iran **374–5**, 387–8
    Maldives **375–6**, 387
    Nepal **376–8**
    Pakistan **379–81**
    Papua New Guinea **382–3**
    Sri Lanka **384–6**
  see also Global Environment Facility

United States 108–9
  consumption footprints 1, *202*, *203*, *204*
University of New Mexico 223
utilized species 8–9, 159–60, **166**, 320–1, 431–2
  birds 165–7
  economic indicators of wild species 180–1
  exploitation indicators 182–3, **184**
  food and medicine 169, 178–9
  Living Planet Index 169–73
  population indicators 164
    species-level 165–7
  response indicators 181–3
  sustainable use 173–7
    fisheries 175–7
    future directions 177–8
    sustainability indicators 178–80
    timber 174–5
  utilization as driver of change in Red List status 167–9

vegetation sensing 106–10
Venezuela
  currency *31*
  Red Data Book, 1995 and 1999 editions 26–7
vertebrates, abundance measures 72–3

Wales 342
Wild Commodities Index 406–7
wildebeest 108
Wildlife Management Areas 271–2
Wildlife Picture Index 46–7, 50–3
  implementation 57–9
  implementation cost 61–3
  measurement 53–5
  occupancy vs. abundance assessment 65
  simulations 59–61
  sources of error 52–3
  species choice 51
  trend analysis 55–7

Wildlife Picture Index (*Cont'd*)
  weaknesses 65–6
  weaknesses and strengths 63–6
World Bird Database 167
World Conservation Monitoring
      Centre 174–5

yellow-crested cockatoo 161, 167
Yield Factor 198

Zambia 389
Zoological Society of London 58, 226, 407
Zoom H2 recorder 230, 235